CHARACTERIZATIONS
OF
INFORMATION
MEASURES

CHARACTERIZATIONS OF INFORMATION MEASURES

Bruce Ebanks

Department of Mathematics, Marshall University,
Huntington, WV 25755, USA

Prasanna Sahoo

Department of Mathematics, University of Louisville,
KY 40292, USA

Wolfgang Sander

Institute for Analysis, Technical University, Pockelsstrasse
14, D-38106, Braunschweig, Germany

World Scientific
Singapore • New Jersey • London • Hong Kong

Published by

World Scientific Publishing Co. Pte. Ltd.
P O Box 128, Farrer Road, Singapore 912805
USA office: Suite 1B, 1060 Main Street, River Edge, NJ 07661
UK office: 57 Shelton Street, Covent Garden, London WC2H 9HE

British Library Cataloguing-in-Publication Data
A catalogue record for this book is available from the British Library.

CHARACTERIZATIONS OF INFORMATION MEASURES

ISBN 981-02-3006-0

Printed in Singapore.

Dedicated by

Bruce Ebanks
to his wife Annalisa
and
his teacher Che Tat Ng,

and

Prasanna Sahoo
in memory of his father Hari
and
to his teacher Pallaniappan Kannappan,

and

Wolfgang Sander
to his wife Helda
and
daughters Michelle and Christin

PREFACE

This book is a monograph on the characterizations of various measures of information. It has arisen from over fifteen years' research on the subject, much of which has been done by the authors. In this book we treat various properties of measures of information and the extent to which these properties characterize known measures of information. To the best of our knowledge this is the first book investigating this topic on open domains in depth; an earlier book by Aczél and Daróczy (1975) treated the subject on closed domains.

The book is written for students, educators, and practitioners of information theory, statistics, and functional equations. The goal of the book is twofold. The first goal is to provide the interested reader with up-to-date information about the characterization of information measures on open domains. The second is to present a comprehensive illustration of the use of functional equations to model and solve problems in information theory. Not many prerequisites are needed to understand this book. Some basic calculus, measure, probability and set theory facts, which can be found in almost all textbooks on these subjects, are sufficient.

We rely heavily upon the theory of functional equations, but we have presented in this book essentially everything that is needed in that regard. That is, the book includes all results about functional equations that we use in the characterizations of information measures. Indeed this book can be treated as a book on functional equations with applications to information theory. Additional information can be found in several books on functional equations, such as Aczél (1966), Kuczma (1985) or Aczél and Dhombres (1989). Another interesting feature is a detailed coverage of sum form information measures in a very general setting.

The book contains nine chapters. Chapter 1 includes some basic results from the theory of functional equations. In Chapter 2, we treat the branching property of information measures and discuss the characteristic functional equation for these measures. This chapter concludes with the symmetric solution of the cocycle equation. Chapter 3 presents some results regarding recursivity properties of information measures. Here, we treat recursivity in a very general form, the recursivity of multiplicative type. Chapter 4 deals with the fundamental equation of information and presents the forms of recursive measures which satisfy some regularity property such as measurability.

The last five chapters cover the characterizations of sum form information measures which arise from various types of additivities. Chapter 5 presents, in addition to the classical additivity property of Shannon, two other more general types of additivity, known as additivity of type λ and additivity of type (M_1, M_2). In this chapter the reader will also find a justification for the study of information measures with these additivity properties. Chapter 6 covers some basic sum form functional equations. These equations are instrumental in characterizations of all additive sum form information measures of type λ as well as all additive sum form information measures of type (M_1, M_2). All additive sum form information measures with measurable generating function are explicitly determined in Chapter 7. Such measures are shown to be a linear combination of Kerridge's inaccuracy, entropies of degree one and zero, and a constant. In Chapters 8 and 9, the forms of all additive sum form information measures of type λ and of all weighted additive sum form information measures of type (M_1, M_2) are determined, respectively.

We do not claim that everything written in this book originated with us. In fact, we freely use the works of many researchers, and we have attempted to acknowledge all contributions. For some results we present new proofs which do not appear elsewhere. In other cases, we did not succeed in improving on the original explanations. Nonetheless we feel that it is worthwhile to have all these results gathered in one place and linked together in a coherent fashion. We hope the readers will agree.

Donald Knuth's TeXprogram was used to create the manuscript. We wish to thank Betty Hampton for her valuable assistance in typing a large part of the manuscript.

Bruce Ebanks, *Huntington*
Prasanna Sahoo, *Louisville*
Wolfgang Sander, *Braunschweig*

CONTENTS

CHAPTER 1

INTRODUCTION, PRELIMINARIES AND NOTATION

"... the hard core of information theory is, essentially, a branch of mathematics, a strictly deductive system. A thorough understanding of the mathematical foundation and its communication application is surely a prerequisite to other applications." – Shannon (1956)

1.1. Introduction and notation

It is certainly true that the concept of information is one of the dominant ideas of the second half of the twentieth century. People from all walks of life, from economic advisors, office managers and librarians, to researchers in linguistics, computer scientists and communication engineers, are concerned with information processing. Many of the inventions of the current era deal with the storage, transmission, transformation and retrieval of information. Information is manifested in various forms – oral, written, visual, electronic, mechanical, electromagnetic, etc. From a mathematical point of view, the essence of information is its quantity, and the basic problem is how to measure information quantity. Following the lead of Claude Shannon, the father of information theory, our aim in this book is to solidify the mathematical foundation of the theory of information measurement. We attempt to do this by introducing desirable properties for an information measure, then using those properties to determine explicit forms for information measures. Such information measures characterized by their properties should be the most useful ones for practical applications.

We shall concern ourselves with probabilistic measures of information for discrete probability distributions, with their properties, and conversely with combinations of such properties determining useful measures of information. A rather complete treatment of this theory as of 1975 can be found in the book of Aczél and Daróczy. The purpose of the current book is to bring the reader up to date by gathering and organizing the developments of the past two decades. As the reader will see, the picture is essentially finished now (although we shall mention a few as yet unsolved problems).

Let us now introduce some notation and some well-known measures of

1

information. Throughout the book, we let k be an arbitrary but fixed positive integer. Let $\mathbb{R}$ denote the set of real numbers, $\mathbb{Q}$ the set of rational numbers, $\mathbb{N}$ the set of natural numbers $\{1, 2, 3, ...\}$, and let

$$J =]0, 1[^k, \qquad \overline{J} = [0, 1]^k$$

be the open, respectively closed, k-dimensional unit intervals. Let

$$\mathcal{P} =]0, \infty[^k$$

denote the positive cone of $\mathbb{R}^k$. We sometimes use $\mathcal{P}_\ell$ for the positive cone of $\mathbb{R}^\ell$, $\ell \neq k$. For instance $\mathcal{P}_1 =]0, \infty[$. Furthermore, for each $n = 2, 3, ...$, we define

$$D_n = \left\{ P = (p_1, p_2, ..., p_n) \mid p_1, p_2, ..., p_n, \sum_{i=1}^{n} p_i \in J \right\},$$

$$\overline{D}_n = \left\{ P = (p_1, p_2, ..., p_n) \mid p_1, p_2, ..., p_n, \sum_{i=1}^{n} p_i \in \overline{J} \right\},$$

$$\Gamma_n^o = \left\{ P = (p_1, p_2, ..., p_n) \mid p_1, p_2, ..., p_n \in J; \sum_{i=1}^{n} p_i = 1 \right\},$$

$$\Gamma_n = \left\{ P = (p_1, p_2, ..., p_n) \mid p_1, p_2, ..., p_n \in \overline{J}; \sum_{i=1}^{n} p_i = 1 \right\}.$$

Here $\mathbf{1}$ represents the k-vector $(1, 1, ..., 1)$. So D_n is an open set in $\mathbb{R}^{kn}$, $\overline{D}_n$ is the closure of D_n, Γ_n^o is an open $k(n-1)$-dimensional simplex, and Γ_n is the closure of Γ_n^o in $\mathbb{R}^{kn}$. All operations on vectors are to be done componentwise. Moreover, we agree on writing

$$x^\alpha = (\xi_1, \ldots, \xi_k)^{(\alpha_1, \ldots, \alpha_k)} := \xi_1^{\alpha_1} \xi_2^{\alpha_2} \cdots \xi_k^{\alpha_k}$$

for all $x, \alpha \in \mathbb{R}^k$. Note that $x^\alpha \in \mathbb{R}$ whereas $x^n \in \mathbb{R}^k$ for $n \in \mathbb{N}$. For instance

$$x^2 = (\xi_1, \ldots, \xi_k)(\xi_1, \ldots, \xi_k) = (\xi_1^2, \ldots, \xi_k^2).$$

Finally we follow the convention

$$x \odot y = (\xi_1, \ldots \xi_k) \odot (\eta_1, \ldots, \eta_k) = \xi_1 \eta_1 + \ldots + \xi_k \eta_k$$

so that $x \odot y$ is the usual inner product in $\mathbf{R}^k$. By log we always understand the logarithm to the base 2, that is $\log = \log_2$. Furthermore, if $x = (\xi_1, \ldots, \xi_k)$ then we define $\log x$ by $\log x := (\log \xi_1, \ldots, \log \xi_k)$. (The symbols $a := b$ and $b =: a$ means that a is defined by b.)

A real-valued sequence of functions I_n on Γ_n^o or Γ_n is called an *information measure* on the open or closed domain, respectively. The usual information-theoretical interpretation is that $I_n(p_1, p_2, ..., p_n)$ is a measure of uncertainty as to the outcome of an experiment having n possible outcomes with probabilities $p_1, p_2, ..., p_n$. Equivalently, $I_n(p_1, p_2, ..., p_n)$ is the amount of information received from the knowledge of which of the possible outcomes occurred. Some simple illustrations of applications of information measures to logical games and coding are given in the book of Aczél and Daróczy (1975). For example, they explain how the Shannon entropy (cf.(1.1.1) below) is used to calculate how many yes-or-no questions are needed to identify a particular object among a certain finite set of objects.

We list now some well-known information measures. Measures depending on one probability distribution (i.e. $k = 1$) are generally called *entropies*. Probably the most well-known of all information measures is the *Shannon entropy* $\{H_n\}$, defined by

$$H_n(P) := -\sum_{i=1}^{n} p_i \log p_i \qquad (k = 1) \qquad (1.1.1)$$

for $P = (p_1, p_2, ..., p_n)$ in Γ_n^o or Γ_n. We have a small problem in Γ_n when some of the p_i's are zero. But this can be overcome by taking the continuous extension of the function $x \mapsto x \log x$ to the boundary, namely by defining

$$0 \log 0 := 0. \qquad (1.1.2)$$

Of course we have no such problem on the open domain Γ_n^o.

Other entropies which are well-known (and which have applications, for example, in image processing (see Sahoo, Soltani, Wong and Chen (1988) and Sahoo, Wilkins and Yeager (1997)), pattern recognition and statistics) are the *entropies* $\{H_n^\alpha\}$ *of degree* α, defined for any real $\alpha \neq 1$ by

$$H_n^\alpha(P) := (2^{1-\alpha} - 1)^{-1} \left(\sum_{i=1}^{n} p_i^\alpha - 1 \right) \qquad (1.1.3)$$

for $P = (p_1, p_2, ..., p_n)$ in Γ_n^o or Γ_n. Again, there is a problem in Γ_n when some p_i's are zero, so one must adopt the convention (if $\alpha \leq 0$)

$$0^\alpha := 0. \tag{1.1.4}$$

An application of the Bernoulli-l'Hospital theorem shows that

$$\lim_{\alpha \to 1} H_n^\alpha(P) = H_n(P),$$

so it is reasonable and natural to define H_n^1 by

$$H_n^1(P) := H_n(P)$$

for P in Γ_n^o or Γ_n (adopting the conventions (1.1.2) and (1.1.4) in the case of Γ_n). So the entropies of degree α form a one-parameter family containing the Shannon entropy as a special case.

A two-parameter family generalizing Shannon's entropy is formed by the entropies $\{H_n^{(\alpha,\beta)}\}$ of *degree* (α, β), defined by

$$H_n^{(\alpha,\beta)}(P) := \begin{cases} -2^{\alpha-1} \sum\limits_{i=1}^n p_i^\alpha \log_2 p_i, & \text{if } \beta = \alpha \\[2mm] \left(2^{1-\alpha} - 2^{1-\beta}\right)^{-1} \sum\limits_{i=1}^n \left(p_i^\alpha - p_i^\beta\right), & \text{if } \beta \neq \alpha . \end{cases}$$

Now we turn to higher dimensions $(k > 1)$. In the case $k = 2$, some measures of information which have proved to be useful are *Kerridge's inaccuracy*

$$K_n(P) := -\sum_{i=1}^n \pi_{1i} \log_2 \pi_{2i}, \tag{1.1.5}$$

Kullback's error (or directed divergence)

$$E_n(P) := \sum_{i=1}^n \pi_{1i} \log_2 \frac{\pi_{1i}}{\pi_{2i}}, \tag{1.1.6}$$

and the *error of degree* α $(\neq 1, 0)$

$$E_n^\alpha(P) := \left(2^{\alpha-1} - 1\right) \left(\sum_{i=1}^n \pi_{1i}^\alpha \pi_{2i}^{1-\alpha} - 1\right), \tag{1.1.7}$$

for $P = \begin{pmatrix} \pi_{11} & \cdots & \pi_{1n} \\ \pi_{21} & \cdots & \pi_{2n} \end{pmatrix}$ in Γ_n^o or Γ_n. An information measure depending upon two probability distributions is generally called a deviation, error, or inaccuracy, because of its interpretation. For example (1.1.6) can be interpreted as the error due to using $(\pi_{21}, \ldots, \pi_{2n})$ as an estimation of the true probability distribution $(\pi_{11}, \ldots, \pi_{1n})$. Notice that $E_n(P) \geq 0$ (which is known as the Shannon inequality) and that $E_n(P) = 0$ if and only if the two probability distributions coincide. Note also that (1.1.6) has an interpretation as an information gain.

Also, for $k = 3$ we mention *Theil's information improvement* or *directed divergence* (of Shannon type)

$$G_n(P) := \sum_{i=1}^{n} \pi_{1i} \log_2 \frac{\pi_{2i}}{\pi_{3i}}, \qquad (1.1.8)$$

and the analogous *directed divergence of degree* α $(\neq 1)$

$$G_n^\alpha(P) := \left(2^{\alpha-1} - 1\right)^{-1} \left(\sum_{i=1}^{n} \pi_{1i} \, \pi_{2i}^{\alpha-1} \, \pi_{3i}^{1-\alpha} - 1 \right), \qquad (1.1.9)$$

for $P = \begin{pmatrix} \pi_{11} & \cdots & \pi_{1n} \\ \pi_{21} & \cdots & \pi_{2n} \\ \pi_{31} & \cdots & \pi_{3n} \end{pmatrix}$ in Γ_n^o or Γ_n. For instance, (1.1.8) is the directed divergence between two estimations, $P_2 = (\pi_{21}, \ldots, \pi_{2n})$ and $P_3 = (\pi_{31}, \ldots, \pi_{3n})$, of a probability distribution $P_1 = (\pi_{11}, \ldots, \pi_{1n})$. Moreover, (1.1.8) can be interpreted as an information improvement where P_3 is the revised prediction of the probability distribution of n events whose originally predicted probability distribution was P_2, and where P_1 is the true distribution.

Thus an information measure $\{I_n\}$ depending upon k probability distributions $P_1, \ldots, P_k$ has one natural interpretation as a measure of information improvement due to $k - 2$ revised predictions $P_3, \ldots, P_k$, where P_2 is the original prediction and where the true distribution is P_1. Such situations of repeated revisions of data collections occur rather often in weather forecasting, economics and experiments. Let us remark that an element P of the domain of I_n is a $k \times n$-matrix

$$P = \begin{pmatrix} \pi_{11} \cdots & \pi_{1i} \cdots & \pi_{1n} \\ \pi_{j1} \cdots & \pi_{ji} \cdots & \pi_{jn} \\ \pi_{k1} \cdots & \pi_{ki} \cdots & \pi_{kn} \end{pmatrix} \qquad (1.1.10)$$

where $\pi_{ji} \in \,]0, 1[, 1 \leq j \leq k, 1 \leq i \leq n$. We denote by p_i the column vectors

$$p_i = (\pi_{1i}, \pi_{2i}, \ldots, \pi_{ki})^T \in J, \quad 1 \leq i \leq n \qquad (1.1.11)$$

(T denotes the transpose) and by P_j the row vectors

$$P_j = (\pi_{j1}, \pi_{j2}, \ldots, \pi_{jn}) \in \Gamma_n^o, \quad 1 \leq j \leq k. \qquad (1.1.12)$$

We will not distinguish row and column displays when there is no danger of confusion. For example, we agree on writing

$$P = (p_1, \ldots, p_n) = (P_1, \ldots, P_k)$$

(although $P = (P_1, \ldots P_k)^T$).

Investigating information measures depending upon at least two probability distributions we run into real difficulties on the closed domain Γ_n when we have zero probabilities. For instance, in the case of Kullback's error (1.1.6), we must restrict our domain by requiring that $\pi_{1i} = 0$ whenever $\pi_{2i} = 0$, and furthermore adopt the awkward convention

$$0 \log_2 \frac{0}{0} := 0. \qquad (1.1.13)$$

Because of the awkwardness of the description of these closed domains (particularly in higher dimensions) and such conventions as (1.1.2), (1.1.4) and (1.1.13), it was observed that it would be desirable to investigate measures of information on the open domain Γ_n^o.

On the other hand, the proofs of characterization theorems become more difficult on the open domain. Many of the techniques used to prove such theorems on the closed domain could not be adapted to the open domain. What we present in this book is a summary of developments (including some new results) on the open domain, due mainly to the work of Aczél, Daróczy, Kannappan, Losonczi, Maksa, Ng, and the authors.

1.2. Functional equation preliminaries

In this section, we present some of the basic results from the theory of functional equations which we shall use from time to time throughout the book. Many of these results can be found elsewhere (for example, Aczél and Daróczy (1975), Kuczma (1985)), but we think it will be convenient for the reader to have the pertinent facts collected in one place. Also, it furnishes a certain completeness to the book. The reader may wish to skip this section on first reading and refer back to it when necessary.

Most of the results presented here are well-known on "whole" algebraic structures (i.e. semigroups, groups, rings, fields), but we shall be dealing with equations on "restricted" domains. As is also well-known, such restrictions can, at the very least, make proofs more difficult. Moreover, many problems on restricted domains have solutions which do not appear on the larger domains. On the other hand, sometimes we find that all solutions on the restricted domain can be extended to solutions on a larger domain.

We start with the following simple result.

Lemma 1.2.1. *Let $\epsilon > 0$, $I =]0, \epsilon[^k$, $\Delta = \{(x,y) \,|\, x, y, x+y \in I\}$. The general solution $g : I \to \mathbb{R}$ of*

$$g(x) = g(y), \qquad (x,y) \in \Delta, \tag{1.2.1}$$

is given by $g(x) = c$ (for all $x \in I$) for some constant $c \in \mathbb{R}$. In particular, this holds if $I = J$ and $\Delta = D_2$.

Proof: Let $u, v \in I$. We have to show that $g(u) = g(v)$. For this purpose, choose $z \in I$ such that both (u, z) and (v, z) belong to Δ. (This is possible because each component of u and v is less than ϵ.) Then two applications of (1.2.1) yield

$$g(u) = g(z) = g(v),$$

so g is a constant on I. Clearly, any constant function satisfies (1.2.1), hence the proof is finished.

Starting with chapter 3, we shall often be working with *multiplicative* functions. These are functions $M : J \to \mathbb{R}$ (or $M : \mathcal{P} \to \mathbb{R}$) satisfying (here $xy = (\xi_1, ..., \xi_k)(\eta_1, ..., \eta_k) := (\xi_1\eta_1, ..., \xi_k\eta_k)$)

$$M(xy) = M(x)M(y) \tag{1.2.2}$$

for all $x, y \in J$ (respectively, for all $x, y \in \mathcal{P}$). Sometimes such a function will also be additive on a restricted domain. We say that $A : J \to \mathbb{R}$ (or $A : \mathcal{P} \to \mathbb{R}$) is *additive on D_2* (respectively, *additive on $\mathcal{P}$*) if

$$A(x+y) = A(x) + A(y) \tag{1.2.3}$$

for all pairs $(x, y) \in D_2$ (respectively, for all $x, y \in \mathcal{P}$). Any map $A : \mathbb{R}^k \to \mathbb{R}$ which satisfies (1.2.3) for every $x, y \in \mathbb{R}^k$ will be called additive on $\mathbb{R}^k$, or simply *additive*. Similar terminology applies to functions defined on domains of other dimensions.

Now we list some facts about multiplicative functions and additive functions.

Lemma 1.2.2. *Let $M : J \to \mathbb{R}$ (or $M : \mathcal{P} \to \mathbb{R}$) be (1.2.2) multiplicative. Then either $M = 0$ or M is nowhere zero.*

Moreover, if $M \neq 0$, then we have $M(x) > 0$ for all $x \in J$ (respectively, $x \in \mathcal{P}$).

Proof: Let us begin by proving the first statement on $\mathcal{P}$. Suppose $M(x_o) = 0$ for some $x_o \in \mathcal{P}$. Then (1.2.2) gives

$$M(x_o y) = 0 \qquad\qquad (1.2.4)$$

for all $y \in \mathcal{P}$. But every $z \in \mathcal{P}$ can be written as $x_o y$ for $y = x_o^{-1} z \in \mathcal{P}$. Thus we have $M = 0$. So if $M \neq 0$, then we cannot have $M(x_o) = 0$ for any $x_o \in \mathcal{P}$. That is, M is nowhere zero.

Now on J, if $M(x_o) = 0$ for some $x_o \in J$, then (1.2.4) gives only

$$M(p) = 0, \qquad p < x_o, \qquad\qquad (1.2.5)$$

where the inequality here means that each component of p is less than the corresponding component of x_o. Thus $M(p) = 0$ if p is sufficiently close to **0**. Now, from (1.2.2) we also get $M(x^2) = M(x)^2$, and by induction

$$M(x^n) = M(x)^n \qquad (n = 1, 2, 3, ...). \qquad\qquad (1.2.6)$$

So, given any $x \in J$, we first choose a positive integer n so that $x^n < x_o$. Then (1.2.5) and (1.2.6) give $M(x)^n = M(x^n) = 0$, which means that $M(x) = 0$. This concludes the proof of the first part of the lemma.

For the second part, simply note that (1.2.6) for $n = 2$ implies

$$M(x^2) \geq 0,$$

for all $x \in J$ (respectively, $x \in \mathcal{P}$). Now the proof is completed by the observation that every element in J (respectively $\mathcal{P}$) can be written as x^2 for some $x \in J$ (respectively $\mathcal{P}$).

Lemma 1.2.3. *Every multiplicative M on J can be extended to a multiplicative $\overline{M}$ on $\mathcal{P}$. Moreover, if M is also additive on D_2, then $\overline{M}$ will be additive on $\mathcal{P}$.*

Proof: First, suppose M is multiplicative on J. If $M = 0$, then we take $\overline{M} = 0$. Otherwise, by Lemma 1.2.2, M is nowhere zero. Given any $t \in \mathcal{P}$, we can write $t = \frac{p}{q}$ for $p, q \in J$. We define $\overline{M}$ on $\mathcal{P}$ by

$$\overline{M}(t) = \overline{M}\left(\frac{p}{q}\right) := \frac{M(p)}{M(q)}. \qquad (1.2.7)$$

First, we have to show that $\overline{M}$ is well-defined. Suppose $t = \frac{p}{q} = \frac{r}{s}$ for $p, q, r, s \in J$. Then $ps = rq$ implies $M(p)M(s) = M(ps) = M(rq) = M(r)M(q)$, so

$$\frac{p}{q} = \frac{r}{s} \Rightarrow \frac{M(p)}{M(q)} = \frac{M(r)}{M(s)}, \qquad p, q, r, s \in J.$$

This shows that (1.2.7) defines $\overline{M}$ unambiguously on $\mathcal{P}$.

Next, $\overline{M}$ is an extension of M, since for any $t \in J$ we have

$$\overline{M}(t) = \frac{M\left(\frac{t}{2}\right)}{M\left(\frac{1}{2}\right)} = M(t),$$

where $\frac{1}{2}$ here means the vector $\left(\frac{1}{2}, ..., \frac{1}{2}\right) \in J$.

Furthermore, suppose $s, t \in \mathcal{P}$, and let $s = \frac{p}{q}$, $t = \frac{u}{v}$ with $p, q, u, v \in J$. Then $st = \frac{pu}{qv}$ with $pu, qv \in J$, hence

$$\overline{M}(st) = \frac{M(pu)}{M(qv)} = \frac{M(p)M(u)}{M(q)M(v)} = \overline{M}(s)\,\overline{M}(t).$$

This shows that $\overline{M}$ is multiplicative on $\mathcal{P}$.

For the second part of the lemma, we assume that M also satisfies (1.2.3), that is

$$M(p + q) = M(p) + M(q), \qquad (p, q) \in D_2.$$

Given any $t, s \in \mathcal{P}$, we choose first $r \in J$ such that $r(s + t) \in J$. Then, setting $p = rs$, $q = rt$, we have by (1.2.7) and (1.2.3)

$$\overline{M}(s + t) = \frac{M(p + q)}{M(r)} = \frac{M(p) + M(q)}{M(r)} = \overline{M}(s) + \overline{M}(t),$$

which shows that $\overline{M}$ is also additive on $\mathcal{P}$. This completes the proof.

In fact, all additive functions (not only those which are multiplicative) can be extended, but without multiplicativity the extension must be done in a different way. We see this in the next two lemmas.

Lemma 1.2.4. *Let $\epsilon > 0$, $I =]0, \epsilon[^k$, and $\Delta = \{(x, y) \mid x, y, x + y \in I\}$. A map $A : I \to \mathbb{R}$ which is additive on Δ can be extended (uniquely) to an additive $\overline{A}$ on $\mathcal{P}$. In particular, this is the case if $I = J$ and $\Delta = D_2$.*

Proof: First, we observe that the (1.2.3) additivity implies that $A(2x) = 2A(x)$, as long as $2x \in I$. By induction, we get $A(nx) = nA(x)$ whenever $nx \in I$. That is

$$A\left(\frac{1}{n}y\right) = \frac{1}{n}A(y) \qquad n \in \mathbb{N}, \, y \in I. \tag{1.2.8}$$

Now, given $x \in \mathcal{P}$, we define $\overline{A}$ by

$$\overline{A}(x) := n \, A\left(\frac{1}{n}x\right), \tag{1.2.9}$$

where n is any natural number for which $\frac{1}{n}x \in I$. If $\frac{1}{n}x, \frac{1}{m}x \in I$ for two natural numbers n, m, then by (1.2.8) we have

$$\frac{1}{m}A\left(\frac{1}{n}x\right) = A\left(\frac{1}{m}\frac{1}{n}x\right) = \frac{1}{n}A\left(\frac{1}{m}x\right).$$

Thus $nA\left(\frac{1}{n}x\right) = mA\left(\frac{1}{m}x\right)$, which shows that $\overline{A}$ is well-defined by (1.2.9).

Finally, let $x, y \in \mathcal{P}$. We choose $n \in \mathbb{N}$ such that $\frac{1}{n}(x + y) \in I$ (so $\left(\frac{1}{n}x, \frac{1}{n}y\right) \in \Delta$). Then

$$\overline{A}(x + y) = nA\left(\frac{1}{n}(x + y)\right) = n\left[A\left(\frac{1}{n}x\right) + A\left(\frac{1}{n}y\right)\right] = \overline{A}(x) + \overline{A}(y),$$

showing that $\overline{A}$ is additive on $\mathcal{P}$. This completes the proof.

Next we are going to prove that an additive function A on $\mathcal{P}$ can be extended to an additive function $\overline{A}$ on $\mathbb{R}^k$. Since we need this result later in a more general setting, we prove the following result (cf. Aczél, Baker, Djokovic, Kannappan and Rado (1971)).

Lemma 1.2.5. *If $(G, +)$ is an abelian group which is generated by an abelian semigroup $(S, +)$, then every homomorphism $\varphi : S \to \mathbb{R}$ can be uniquely* extended to a homomorphism $\overline{\varphi} : G \to \mathbb{R}$. In particular, every additive $A : \mathcal{P} \to \mathbb{R}$ extends uniquely to an additive $\overline{A}$ on $\mathbb{R}^k$. Furthermore, if $A(x) = A(\xi_1, \ldots, \xi_k)$ is additive on $\mathcal{P}$ (or $\mathbb{R}^k$ or D_2), then

$$A(x) = \sum_{i=1}^{k} a_i(\xi_i) \tag{1.2.10}$$

for all $x \in \mathcal{P}$ (respectively $\mathbb{R}^k$, respectively J), where each a_i $(i = 1, \ldots, k)$ is additive on $\mathbb{R}$.

Proof: The assumption that S generates G means that $G = S - S$. (Note in particular that $\mathbb{R}^k = \mathcal{P} - \mathcal{P}$.) Now, for $x \in G$ with the representation $x = s - t$ for some $s, t \in S$ we define

$$\overline{\varphi}(x) := \varphi(s) - \varphi(t). \tag{1.2.11}$$

We must show that $\overline{\varphi}$ is well-defined. Suppose $x = s - t = s' - t'$ $(s, t, s', t' \in S)$ then $s + t' = s' + t \in S$ and we obtain

$$\varphi(s) + \varphi(t') = \varphi(s + t') = \varphi(s' + t) = \varphi(s') + \varphi(t),$$

hence $\varphi(s) - \varphi(t) = \varphi(s') - \varphi(t')$, and $\overline{\varphi}(x)$ is unambiguously defined by (1.2.11).

Next, we show that $\overline{\varphi}$ is a homomorphism. Let $x, y \in G$ and choose $s, t, u, v \in S$ so that $x = s - t$, $y = u - v$. Thus $x + y = (s + u) - (t + v) \in S - S$. Therefore we have

$$\overline{\varphi}(x + y) = \varphi(s + u) - \varphi(t + v) = \varphi(s) + \varphi(u) - (\varphi(t) + \varphi(v))$$
$$= (\varphi(s) - \varphi(t)) + (\varphi(u) - \varphi(v)) = \overline{\varphi}(x) + \overline{\varphi}(y).$$

Moreover, $\overline{\varphi}$ is an extension of φ. For arbitrary $x \in S$ choose $s, t \in S$ such that $x = s - t$, or $s = x + t$. Thus we get $\varphi(s) = \varphi(x + t) = \varphi(x) + \varphi(t)$, that is,

$$\overline{\varphi}(x) = \varphi(s) - \varphi(t) = \varphi(x).$$

Now let φ' be another extension of the homomorphism φ. If $x = s - t \in G$ $(s, t \in S)$ then the equation

$$\varphi'(x) = \varphi'(s - t) = \varphi'(s) - \varphi'(t) = \varphi(s) - \varphi(t) = \overline{\varphi}(x)$$

shows that $\overline{\varphi}$ is the unique homomorphic extension of φ. This proves the first two parts of Lemma 1.2.5

For the third part of the lemma, if A is additive on D_2 or $\mathcal{P}$, we first extend it to an additive map, say $\overline{A}$, on $\mathbb{R}^k$ by Lemma 1.2.4 and the first part of the current lemma. Now, for $\overline{A}$ additive on $\mathbb{R}^k$, we deduce from (1.2.3) by induction that $\overline{A}(x_1 + x_2 + \cdots + x_n) = \overline{A}(x_1) + \cdots + \overline{A}(x_n)$, so in particular

$$
\begin{aligned}
\overline{A}(x) &= \overline{A}\left(\xi_1, ..., \xi_k\right) \\
&= \overline{A}\left(\xi_1, 0, ..., 0\right) + \overline{A}\left(0, \xi_2, 0..., 0\right) + \cdots + \overline{A}\left(0, ..., 0, \xi_k\right) \\
&= a_1(\xi_1) + a_2(\xi_2) + \cdots + a_k(\xi_k),
\end{aligned}
$$

where $a_i(\xi) := \overline{A}\left(0, ..., 0, \xi, 0, ..., 0\right)$ for all $\xi \in \mathbb{R}$, where ξ appears in coordinate position i ($i = 1, ..., k$). The additivity of a_i follows from its definition and the additivity of $\overline{A}$, and this completes the proof.

Now we present some "regularity" results about additive functions. We begin by giving the form of continuous additive functions. Then we consider additive functions with weaker regularity properties.

Lemma 1.2.6. *Any continuous additive function $A : \mathbb{R}^k \to \mathbb{R}$ has the form*

$$
A(x) = \sum_{i=1}^{k} c_i \xi_i = c \odot x, \quad x \in \mathbb{R}^k
$$

for some constant vector $(c_1, \ldots, c_k) \in \mathbb{R}^k$.

Proof: Because of Lemma 1.2.5 it is sufficient to prove that any continuous additive function $a : \mathbb{R} \to \mathbb{R}$ has the form

$$
a(x) = cx, \quad x \in \mathbb{R}
$$

for some constant $c \in \mathbb{R}$. We first prove the $\mathbb{Q}$-homogeneity of a, that is

$$
a(rt) = ra(t), \quad t \in \mathbb{R}, \quad r \in \mathbb{Q}. \tag{1.2.12}
$$

We have already observed (in the proof of Lemma 1.2.4) that additivity implies

$$
a(nx) = n\, a(x), \quad n \in \mathbb{N},\ x \in \mathbb{R},
$$

which means that a is $\mathbb{N}$-homogeneous. Let $r = \frac{m}{n}$ be an arbitrary positive rational number, let $t \in \mathbb{R}$ and $x = rt = \frac{mt}{n}$. Then $nx = mt$, so

$$
na(x) = a(nx) = a(mt) = ma(t);
$$

that is

$$a(x) = \frac{m}{n}a(t),$$

or

$$a(rt) = ra(t), \qquad t \in \mathbb{R},\ r \in \mathbb{Q},\ r > 0. \tag{1.2.13}$$

But (1.2.13) yields, for $t = 0$, also $a(0) = 0$. Also, the additivity equation $a(x + y) = a(x) + a(y)$ yields, for $y = -x$,

$$a(-x) = -a(x),$$

so now (1.2.13) extends to (1.2.12). Finally, with $t = 1$ and $c = a(1)$, we have

$$a(r) = cr, \qquad r \in \mathbb{Q}. \tag{1.2.14}$$

Since $\mathbb{Q}$ is dense in $\mathbb{R}$ and a is continuous we get the desired result $a(x) = a(1)x$ for all $x \in \mathbb{R}$.

Remark 1.2.7. In Lemma 1.2.6 it is sufficient to suppose that A is continuous at $0 \in \mathbb{R}^k$ (or at any particular point $x_0 \in \mathbb{R}^k$): If $x \in \mathbb{R}^k$ is arbitrary and if $\lim_{n\to\infty} x_n = x$ then we get

$$\begin{aligned}
\lim_{n\to\infty} A(x_n) &= \lim_{n\to\infty} A(x + x_n - x) \\
&= A(x) + \lim_{n\to\infty} A(x_n - x) \\
&= A(x) + A(0) = A(x),
\end{aligned}$$

since $A(0) = 0$. Thus, A is continuous at x.

Let us now present a further improvement of Lemma 1.2.6.

Lemma 1.2.8. *If an additive function $A : \mathbb{R}^k \to \mathbb{R}$ is bounded from below on an open interval, then A is continuous.*

Proof: By hypothesis there are $x_0 \in \mathbb{R}^k$ and $B \in \mathbb{R}$ such that $A(x) > B$ for all $U_\gamma(x_0) := \{y \in \mathbb{R}^k : |y - x_0| < \gamma\}$, if γ is sufficiently small. It follows that $|A(x)|$ is bounded for $x \in U_\gamma(0)$ by $K := A(x_0) - B$. Indeed, if $x \in U_\gamma(0)$ then $x_0 \pm x \in U_\gamma(x_0)$ and we get

$$\left.\begin{aligned}
\pm A(x) = A(\pm x) &= A(x_0 \pm x - x_0) \\
&= A(x_0 \pm x) - A(x_0) \\
&\geq B - A(x_0),
\end{aligned}\right\} \tag{1.2.15}$$

which means that $K = A(x_0) - B \geq \pm A(x)$. To show that A is continuous at 0, let $\epsilon > 0$ and choose $n \in \mathbb{N}$ such that $\frac{K}{n} < \epsilon$. Then we get for all $|x| < \delta := \frac{\gamma}{n}$

$$|A(x) - A(0)| = |A(\frac{nx}{n}) - 0| = \frac{|A(nx)|}{n} \leq \frac{K}{n} < \epsilon. \qquad (1.2.16)$$

Now the result follows from Remark 1.2.7.

Remark 1.2.9. Corresponding to Lemma 1.2.8, we have also the following statement: If an additive $A : \mathbb{R}^k \to \mathbb{R}$ is bounded from above on any interval, then A is continuous. For the proof, we have only to apply Lemma 1.2.8 to the function $-A$, which is also additive.

Lemma 1.2.10. *If $M : J \to \mathbb{R}$ is both multiplicative on J and additive on D_2, then M is either identically zero or a projection*

$$M(x) = M(\xi_1, ..., \xi_k) = \xi_j, \qquad x \in J, \qquad (1.2.17)$$

onto one of its components $(j \in \{1, ..., k\})$.

Proof: First, considering M as an additive map on D_2, by Lemma 1.2.5 we have (cf. (1.2.10))

$$M(\xi_1, ..., \xi_k) = \sum_{i=1}^{k} a_i(\xi_i), \qquad x \in J, \qquad (1.2.18)$$

where each $a_i : \mathbb{R} \to \mathbb{R}$ is additive.

Next, we show that each a_i is bounded from below on $\left]-\frac{1}{2}, \frac{1}{2}\right[$. Let $j \in \{1, ..., k\}$. Then by (1.2.18) we have

$$a_j(\xi) = a_j\left(\frac{1}{2} + \xi\right) - a_j\left(\frac{1}{2}\right)$$
$$= M\left(\frac{1}{2}, ..., \frac{1}{2}, \frac{1}{2} + \xi, \frac{1}{2}, ..., \frac{1}{2}\right) - M\left(\frac{1}{2}, ..., \frac{1}{2}\right)$$

valid for all $\xi \in \left]-\frac{1}{2}, \frac{1}{2}\right[$. But M is also multiplicative on J, which means by Lemma 1.2.2 that $M(x) \geq 0$ for all $x \in J$. Thus

$$a_j(\xi) \geq -M\left(\frac{1}{2}, ..., \frac{1}{2}\right), \qquad \xi \in \left]-\frac{1}{2}, \frac{1}{2}\right[.$$

That is, a_j is bounded below on an open interval. Hence Lemma 1.2.8 tells us that a_j is linear. But j was arbitrary in $\{1, 2, ..., k\}$, so we have

$$a_j(\xi) = c_j\, \xi, \qquad j \in \{1, ..., k\},\ \xi \in \mathbb{R},$$

for some constants $c_1, ..., c_k \in \mathbb{R}$. Inserting this into (1.2.18), we get

$$M(\xi_1, ..., \xi_k) = \sum_{i=1}^{k} c_i\xi_i, \qquad x \in J. \tag{1.2.19}$$

Finally, we fully use the multiplicativity of M. With $y = (\eta_1, ..., \eta_k)$, by (1.2.19) the equation $M(xy) = M(x)M(y)$ on J means

$$\sum_{i=1}^{k} c_i\xi_i\eta_i = \left(\sum_{i=1}^{k} c_i\xi_i\right)\left(\sum_{j=1}^{k} c_j\eta_j\right),$$

or

$$\sum_{i=1}^{k} \left(c_i - c_i^2\right)\xi_i\eta_i - \sum_{i=1}^{k} \sum_{\substack{j=1 \\ (j \neq i)}}^{k} c_i c_j \xi_i \eta_j = 0,$$

which holds for all $\xi_i, \eta_i \in\,]0, 1[\ (i = 1, ..., k)$. Since we are free to substitute values of ξ_i, η_i from $]0, 1[$ independently, this means that we must have

$$c_i = c_i^2 \qquad \text{and} \qquad c_i\, c_j = 0 \quad (i \neq j), \tag{1.2.20}$$

for all $i, j \in \{1, ..., k\}$. An obvious solution is $c_i = 0$ for all $i = 1, ..., k$; in (1.2.19) this yields $M \equiv 0$. Now suppose that $c_j \neq 0$ for some $j \in \{1, ..., k\}$. Then the first equation of (1.2.20) gives $c_j = 1$, and the second gives $c_i = 0$ for $i \neq j$. In this case, (1.2.19) reduces to (1.2.17). Since there are no other cases, this concludes the proof.

A representation analogous to (1.2.10) can be established for multiplicative functions on k-dimensional domains.

Lemma 1.2.11. *Let $M : J \to \mathbb{R}$ (or $M : \mathcal{P} \to \mathbb{R}$) be multiplicative. Then*

$$M(x) = M(\xi_1, ..., \xi_k) = \prod_{i=1}^{k} m_i(\xi_i) \tag{1.2.21}$$

for all $x \in J$ (respectively, $x \in \mathcal{P}$), where each $m_i : \mathcal{P}_1 \to \mathbb{R}$ is multiplicative.

Proof: First, if necessary, we extend M to a multiplicative function on $\mathcal{P}$, as in Lemma 1.2.3. Let us again call this function M. From (1.2.2), it follows by induction that $M(x_1 x_2 \cdots x_n) = M(x_1) M(x_2) \cdots M(x_n)$, so that we can write

$$
\begin{aligned}
M\left(\xi_1, ..., \xi_k\right) \\
&= M\left(\left(\xi_1, 1, ..., 1\right)\left(1, \xi_2, 1, ..., 1\right) \cdot ... \cdot \left(1, ..., 1, \xi_k\right)\right) \\
&= M(\xi_1, 1, ..., 1)\, M(1, \xi_2, 1, ..., 1) \cdot ... \cdot M(1, ..., 1, \xi_k), \quad x \in \mathcal{P}.
\end{aligned}
$$

Define now $m_i : \mathcal{P}_1 \to \mathbb{R}$ $(i = 1, ..., k)$ by

$$
m_i(\xi) = M(1, ..., 1, \xi, 1, ..., 1), \quad \xi \in \mathcal{P}_1,
$$

where ξ stands in the ith coordinate position. Then we have on the one hand (1.2.21), and on the other hand

$$
\begin{aligned}
m_i(\xi\eta) &= M(1, ..., \xi\eta, ..., 1) \\
&= M((1, ..., \xi, ..., 1)(1, ..., \eta, ..., 1)) \\
&= M(1, ..., \xi, ..., 1)\, M(1, ..., \eta, ..., 1) \\
&= m_i(\xi)\, m_i(\eta), \quad \xi, \eta > 0.
\end{aligned}
$$

Thus each m_i $(i = 1, 2, ..., k)$ is multiplicative on $\mathcal{P}_1$, and the proof is complete.

Now we turn to another class of functions which will also be very important in developments to come. A map $L : J \to \mathbb{R}$ (or $L : \mathcal{P} \to \mathbb{R}$) is called *logarithmic* on J (respectively, *logarithmic on $\mathcal{P}$*) if it satisfies

$$
L(xy) = L(x) + L(y) \tag{1.2.22}
$$

for all $x, y \in J$ (respectively, for all $x, y \in \mathcal{P}$). Again, we use similar terminology on domains of other dimensions. For these functions, we have results analogous to Lemmas 1.2.3, 1.2.5, and 1.2.11 for multiplicative functions and additive functions.

Lemma 1.2.12. *Every logarithmic $L : J \to \mathbb{R}$ extends to a logarithmic $\overline{L} : \mathcal{P} \to \mathbb{R}$.*

Moreover, if L is logarithmic on $\mathcal{P}$ (or on J), then

$$
L(x) = L\left(\xi_1, ..., \xi_k\right) = \sum_{i=1}^{k} \ell_i\left(\xi_i\right), \tag{1.2.23}
$$

for all $x \in \mathcal{P}$ (respectively, $x \in J$), where each $\ell_i : \mathcal{P}_1 \to \mathbb{R}$ is logarithmic.

Proof: If L is logarithmic on J, then we extend to $\overline{L}$ as follows. Given any $x \in \mathcal{P}$, choose $t \in J$ so that also $tx \in J$, and define

$$\overline{L}(x) := L(tx) - L(t). \qquad (1.2.24)$$

To show that $\overline{L}$ is well-defined, suppose $t, s \in J$ with $tx, sx \in J$. Then

$$L(tx) + L(s) = L(tsx) = L(sx) + L(t),$$

which shows that $L(tx) - L(t) = L(sx) - L(s)$, thus giving a unique definition of $\overline{L}(x)$ through (1.2.24).

To see that $\overline{L}$ is logarithmic on $\mathcal{P}$, given any $x, y \in \mathcal{P}$, we choose $s \in J$ such that $sx, sy \in J$. Then $s^2xy \in J$ and we have

$$\begin{aligned}
\overline{L}(xy) &= L(s^2xy) - L(s^2) \\
&= L(sx) + L(sy) - 2L(s) \\
&= [L(sx) - L(s)] + [L(sy) - L(s)] \\
&= \overline{L}(x) + \overline{L}(y),
\end{aligned}$$

as desired.

Now, suppose L is logarithmic on $\mathcal{P}$. From (1.2.22), it follows by induction that $L(x_1x_2\cdots x_n) = L(x_1) + \cdots + L(x_n)$. So we compute

$$\begin{aligned}
L\left(\xi_1, ..., \xi_k\right) &= L\left((\xi_1, 1, ..., 1)\,(1, \xi_2, 1..., 1)\cdots(1, ..., 1, \xi_k)\right) \\
&= L(\xi_1, 1, ..., 1) + L(1, \xi_2, 1, ..., 1) + \cdots + L(1, ..., 1, \xi_k),
\end{aligned}$$

which is (1.2.23) with $\ell_i : \mathcal{P}_1 \to \mathbb{R}$ defined by

$$\ell_i(\xi) := L(1, ..., 1, \xi, 1, ..., 1), \qquad \xi > 0,$$

where ξ stands in the ith coordinate position. Moreover, this definition shows that ℓ_i is logarithmic on $\mathcal{P}_1$, since

$$\begin{aligned}
\ell_i(\xi\eta) &= L(1, ..., 1, \xi\eta, 1, ..., 1) \\
&= L((1, ..., 1, \xi, 1, ..., 1) \cdot (1, ..., 1, \eta, 1, ..., 1)) \\
&= L(1, ..., 1, \xi, 1, ..., 1) + L(1, ..., 1, \eta, 1, ..., 1) \\
&= \ell_i(\xi) + \ell_i(\eta).
\end{aligned}$$

This establishes the lemma.

Remark 1.2.13. Note that Lemma 1.2.12 could be obtained also as a consequence of Lemma 1.2.5. Starting with L logarithmic on J (or $\mathcal{P}$), define A on $\mathcal{P}$ (respectively, on $\mathbb{R}^k$) by

$$A(t) := L\left(e^{-t}\right), \qquad t \in \mathcal{P} \text{ (respectively } t \in \mathbb{R}^k\text{)}.$$

[Here $e^{-(\tau_1,\dots,\tau_k)}$ means $(e^{-\tau_1}, \dots, e^{-\tau_k})$, since we are performing the exponentiation componentwise.] Then A is additive on its domain. From (1.2.10), we get (1.2.23) by means of $L(x) = A(-\log x)$.

By a similar method, we could deduce Lemma 1.2.11 from Lemma 1.2.5 in the case $M \neq 0$.

The equations (1.2.2), (1.2.3), and (1.2.22) are known as *Cauchy equations*. (There is a fourth one, $e(x + y) = e(x)e(y)$, which we shall not encounter in this book.) Similar equations with several unknown functions are known generally as *Pexider equations*. We shall need the solutions of certain equations of this type, starting with the following.

Lemma 1.2.14. *Maps* $F_i : J \to \mathbb{R}$ $(i = 1, 2, 3)$ *satisfy*

$$F_1(x) + F_2(y) = F_3(x + y), \qquad (x, y) \in D_2, \qquad (1.2.25)$$

if and only if there exists an additive map $A : \mathbb{R}^k \to \mathbb{R}$ *and constants* $c_1, c_2 \in \mathbb{R}$ *for which*

$$\left. \begin{array}{l} F_1(x) = A(x) + c_1 \\ F_2(x) = A(x) + c_2 \\ F_3(x) = A(x) + c_1 + c_2 \end{array} \right\}, \qquad x \in J. \qquad (1.2.26)$$

Proof: As the right hand side of (1.2.25) is symmetric in x and y, the same is true of the left. That is,

$$F_1(x) + F_2(y) = F_1(y) + F_2(x),$$

or

$$F_2(y) - F_1(y) = F_2(x) - F_1(x), \qquad (x, y) \in D_2.$$

By Lemma 1.2.1, this means that $F_2 - F_1$ is a constant, say

$$F_2(x) = F_1(x) + c, \qquad x \in J. \qquad (1.2.27)$$

Now let $(x, y) \in D_2$ and choose $z \in J$ so that $(x, y, z) \in D_3$. Applying (1.2.25) twice, together with (1.2.27), we find that

$$F_1(x) + F_1(y + z) + c = F_3(x + y + z) = F_1(x + y) + F_1(z) + c.$$

That is,

$$F_1(y + z) - F_1(z) = F_1(x + y) - F_1(x), \qquad (x, y, z) \in D_3.$$

Since the left-hand side is independent of x and the right-hand side is independent of z, both sides are a function of y alone. Thus we have

$$F_1(x + y) - F_1(x) = A(y), \qquad (x, y) \in D_2. \tag{1.2.28}$$

We claim that A is additive on D_2.

Indeed, (1.2.28) implies

$$F_1(x) + A(y) = F_1(x + y) = F_1(y + x) = F_1(y) + A(x),$$

whence by Lemma 1.2.1 again,

$$F_1(x) - A(x) = F_1(y) - A(y) =: c_1, \qquad (x, y) \in D_2. \tag{1.2.29}$$

With this, (1.2.28) yields

$$[A(x + y) + c_1] - [A(x) + c_1] = A(y), \qquad (x, y) \in D_2,$$

which means exactly that A is additive on D_2. (We have seen already in Lemmas 1.2.4 and 1.2.5 that we can extend A to an additive function on $\mathbb{R}^k$.)

Finally, (1.2.29) and (1.2.27) give the first two lines of (1.2.26) with $c_2 := c + c_1$. That together with (1.2.25) then yields

$$F_3(x + y) = [A(x) + c_1] + [A(y) + c_2] = A(x + y) + c_1 + c_2$$

for all $(x, y) \in D_2$. To prove the third formula in (1.2.26) let $x \in J$. Choose $y \in J$ such that $x - y \in J$. Then $(y, x - y) \in D_2$. Thus

$$\begin{aligned}
F_3(x) &= F_3(y + x - y) \\
&= F_1(y) + F_2(x - y) \\
&= A(y) + c_1 + A(x - y) + c_2 \\
&= A(y) + c_1 + A(x) - A(y) + c_2 \\
&= A(x) + c_1 + c_2,
\end{aligned}$$

which completes (1.2.26).

The converse is immediately verifiable.

Remark 1.2.15. Lemma 1.2.14 also holds for $F_i : \mathcal{P} \to \mathbb{R}$ (or $F_i : \mathbb{R}^k \to \mathbb{R}$) satisfying (1.2.25) on $\mathcal{P} \times \mathcal{P}$ (respectively, on $\mathbb{R}^k \times \mathbb{R}^k$), and yields (1.2.26) for all $x \in \mathcal{P}$ (respectively, $x \in \mathbb{R}^k$). One can use the same proof, taking all x, y, z to be arbitrary elements of $\mathcal{P}$ (respectively, $\mathbb{R}^k$).

The Pexider equation

$$H_1(x) + H_2(y) = H_3(xy) \qquad (1.2.30)$$

could be solved by the same method used in Lemma 1.2.14, but we give a different proof, based on Remark 1.2.13.

Lemma 1.2.16. *Maps $H_i : J \to \mathbb{R}$ ($i = 1, 2, 3$) (or $H_i : \mathcal{P} \to \mathbb{R}$) satisfy (1.2.30) for all $x, y \in J$ (respectively, $x, y \in \mathcal{P}$), if and only if there exists a logarithmic map $L : \mathcal{P} \to \mathbb{R}$ and constants $c_1, c_2 \in \mathbb{R}$ such that*

$$\left.\begin{array}{l} H_1(x) = L(x) + c_1 \\ H_2(x) = L(x) + c_2 \\ H_3(x) = L(x) + c_1 + c_2 \end{array}\right\}, \qquad x \in J \qquad (1.2.31)$$

(respectively, $x \in \mathcal{P}$).

Proof: If H_i ($i = 1, 2, 3$) satisfy (1.2.30) on J (or $\mathcal{P}$), we introduce $F_i : \mathcal{P} \to \mathbb{R}$ (respectively, $F_i : \mathbb{R}^k \to \mathbb{R}$) by

$$F_i(x) := H_i\left(e^{-x}\right), \qquad (i = 1, 2, 3), \quad x \in \mathcal{P} \qquad (1.2.32)$$

(respectively, $x \in \mathbb{R}^k$). Then F_i ($i = 1, 2, 3$) satisfy (1.2.25) for all $x, y \in \mathcal{P}$ (respectively, $\mathbb{R}^k$). By Remark 1.2.15, we have therefore (1.2.26) for all $x \in \mathcal{P}$ (respectively, $\mathbb{R}^k$). Finally, (1.2.32) transforms (1.2.26) into (1.2.31), where $L : \mathcal{P} \to \mathbb{R}$ is defined by

$$L(x) := A(-\log x), \quad x \in \mathcal{P}. \qquad (1.2.33)$$

Since A is additive, L is logarithmic.

Conversely, it is easy to check that, with L logarithmic (that is L satisfying (1.2.22)), $H_i (i = 1, 2, 3)$ given by (1.2.31) actually satisfy (1.2.30). This concludes the proof.

1.3. A general regularity theorem

In the preceding section we have presented mainly one "regularity" theorem about additive functions: One-sided boundedness on an open interval implies the continuity of additive functions.

This type of result is known for rather large classes of functional equations. The topic of this section is to prove a rather general regularity theorem of the type "λ^r-measurability in $\mathbb{R}^r$ implies continuity" where λ^r denotes the Lebesgue measure in $\mathbb{R}^r$ so that especially we get that a λ^r-measurable additive function is continuous and thus linear.

Let us now motivate the class of functional equations which we will consider. With the substitution $x = t + y$ the Cauchy equation

$$f(t + y) = f(t) + f(y), \quad x, t \in \mathbb{R}^r \tag{1.3.1}$$

goes over into

$$f(x) = f(x - y) + f(y)$$

or

$$f(x) = f(g(x, y)) + f(y), \quad x, y \in \mathbb{R}^r \tag{1.3.2}$$

where $g(x, y) = x - y$. That is, (1.3.1) can be transformed into (1.3.2) since $t + y = x$ can be solved globally for t as a function of x and y. Thus we arrive - in generalization of this consideration - at functional equations of the type

$$f(x) = \sum_{i=1}^{n} f(g_i(x, y)).$$

To cover still more general classes of functional equations (for example, Pexider equations or the equation $d(xy) = xd(y) + yd(x)$ for $x, y \in \mathbb{R}^r$) we finally arrive at functional equations of the form

$$f(x) = \sum_{i=1}^{n} h_i\left(x, y, f_i(g_i(x, y))\right).$$

Our aim is to prove that f is continuous whenever f_i are measurable, h_i are smooth enough and g_i are locally solvable ($i = 1, \ldots, n$). Moreover, the domains of f_i, g_i, h_i should be so general that they cover the most typical situations.

We will see that we need such a general regularity theorem several times in this book. The proof is based on ideas of Jarai (1979, 1982, 1986) and uses only basic results from Lebesgue measure theory, such as Lusin's Theorem and the transformation formula for Lebesgue integrals.

In the next two results we use the following notation. If g is defined on $D \subset X \times Y$ and has values in Z then we define

$$D_x := \{y \in Y \mid (x, y) \in D\}$$

and $g_x : D_x \to Z$ is defined by $g_x(y) := g(x, y)$, $y \in D_x$.

Lemma 1.3.1. *Let T be an open subset of $\mathbb{R}^s$, let D be an open subset of $T \times \mathbb{R}^r$ and let $(x_0, y_0) \in D$. Suppose that the function $g : D \to \mathbb{R}^r$ is continuous and has a continuous partial derivative with respect to y. If the Jacobian $J\,g_{x_0}(y_0) \neq 0$ then there exist open neighborhoods V and W of x_0 and y_0, respectively, such that*
(i) for each $a > 0$ there exists a real number $b > 0$ such that $\lambda^r(g_x(B)) \geq b$ whenever $x \in V$, $B \subset W$ and $\lambda^r(B) \geq a$; and
(ii) if A is λ^r-measurable in $\mathbb{R}^r$ then $g_x^{-1}(A) \cap W$ is λ^r-measurable in $\mathbb{R}^r$ for every $x \in V$.

Proof: For all $(x, y) \in D$ let

$$L(x) = \frac{\partial}{\partial y}\, g(x, y_0).$$

From the proof of the inverse function theorem (cf. Rudin (1964), Theorem 9.24) we get that, if W is a neighborhood of y_0 in $\mathbb{R}^r$, $(x, y) \in D$ and

$$\left\| \frac{\partial}{\partial y}\, g(x, y) - L(x) \right\| < (2\|L(x)^{-1}\|)^{-1}$$

for each $y \in W$, then g_x maps W homeomorphically onto the open subset $g_x(W)$ of $\mathbb{R}^r$. (Here $\|A\|$ denotes the operator norm $\sup_{\|y\| \leq 1} \|A(y)\|$ of the operator A in $L(\mathbb{R}^r, \mathbb{R}^r)$.)

Now choose c and d so that

$$0 < c < (2\|L(x_0)^{-1}\|)^{-1}, \quad \text{and} \quad 0 < d < \left| \det \frac{\partial}{\partial y}\, g(x_0, y_0) \right|.$$

Because of the continuity of $\frac{\partial}{\partial y} g(x,y) - L(x)$ and $det \frac{\partial}{\partial y} g(x,y)$ there exist open neighborhoods V and W of x_0 and y_0, respectively, so that $\overline{V} \times \overline{W}$ is a compact subset of D and

$$\left\| \frac{\partial}{\partial y} g(x,y) - L(x) \right\| < c < (2\|L(x)^{-1}\|)^{-1}, \quad \left| det \frac{\partial}{\partial y} g(x,y) \right| \geq d$$

for all $x \in V$, $y \in W$ ($\overline{V}$ and $\overline{W}$ denote the closures of V and W). To prove (i) let $a > 0$, $B \subset W$ and suppose $\lambda^r(B) \geq a$. We will prove $\lambda^r(g_x(B)) \geq b := ad$ whenever $x \in V$. Supposing to the contrary that $\lambda^r(g_x(B)) < b$ for some $x \in V$, we choose (using the regularity of λ^r) an open set U such that $g_x(B) \subset U \subset g_x(W)$ and $\lambda^r(U) < b$. Now consider the open set $E :=$ $g_x^{-1}(U) \cap W$. Because of $B = B \cap W = (g_x^{-1}g_x(B)) \cap W \subset g_x^{-1}(U) \cap W = E$ and $g_x(E) \subset U$ we get the contradiction (using the transformation formula for Lebesgue integrals)

$$b > \lambda^r(U) \geq \lambda^r(g_x(E)) = \int_E \left| det \frac{\partial g}{\partial y}(x,y) \right| d\lambda^r(y)$$
$$\geq d \cdot \lambda^r(E) \geq d \cdot \lambda^r(B) \geq d \cdot a = b.$$

Thus (i) is proven.

For the proof of (ii) let $A \subset \mathbb{R}^r$ be λ^r-measurable. Because λ^r is regular we can choose a Borel set B in $\mathbb{R}^r$ with $A \subset B$ and $\lambda^r(B \backslash A) = 0$ and get (using $A = B \backslash (B \backslash A)$)

$$g_x^{-1}(A) \cap W = (g_x^{-1}(B) \cap W) \backslash (g_x^{-1}(B \backslash A) \cap W).$$

From this equation (ii) follows since $g_x^{-1}(B) \cap W$ is a Borel set and

$$\lambda^r(g_x^{-1}(B \backslash A) \cap W) = 0$$

by an application of (i).

Theorem 1.3.2. *Let T be an open subset of $\mathbb{R}^s$, and let D be an open subset of $T \times \mathbb{R}^r$ and $Y_i \subset \mathbb{R}^r$. Let $f : T \to \mathbb{R}$, $f_i : Y_i \to \mathbb{R}^r$, $g_i : D \to Y_i \subset \mathbb{R}^r$ and $h_i : D \times \mathbb{R}^r \to \mathbb{R}$ for $i = 1, \ldots, n$.*

Suppose that the following conditions hold:
(i) f satisfies the functional equation

$$f(x) = \sum_{i=1}^n h_i(x, y, f_i(g_i(x,y))), \quad (x,y) \in D, \tag{1.3.3}$$

(ii) g_i, h_i are continuous on D and $D \times \mathbb{R}^r$, respectively, and f_i is λ^r-measurable on Y_i, $i = 1, \ldots, n$;

(iii) for each fixed $x \in T$ the mappings $y \rightarrow g_i(x,y)$ are differentiable with derivative $\frac{\partial}{\partial y} g_i(x,y)$ and with Jacobian $J_2 g_i(x,y)$; moreover the mapping $(x,y) \rightarrow \frac{\partial}{\partial y} g_i(x,y)$ is continuous on D, and for every $x \in T$ there exists a $(x,y) \in D$ such that $J_2 g_i(x,y) \neq 0$ for $i = 1, \ldots, n$.

Then f is continuous on T.

Proof: Let $x_0 \in T$ be fixed. We shall show that f is continuous at x_0. By (iii) there is $(x_0, y_0) \in D$ satisfying $J_2 g_i(x_0, y_0) \neq 0$ for $i = 1, \ldots, n$. Since $\mathbb{R}^s$ and $\mathbb{R}^r$ are locally compact, Lemma 1.3.1 shows that there exist compact neighborhoods $V = \{x \in J : \|x - x_0\| \leq r_1\}$ and $W = \{y \in J : \|y - y_0\| \leq r_2\}$ (for some $r_1, r_2 > 0$) satisfying $K := V \times W \subset D$ and conditions (i) and (ii) of the lemma. (Here $\| \ \|$ denotes the usual Euclidean norm.) Thus we get

$$\lambda^r (K_x \cap K_{x_0}) = \lambda^r(W) =: c$$

for every $x \in V$. Now let $a = \frac{c}{2n}$. By Lemma 1.3.1 there exists $b > 0$ for which $\lambda^r \left(g_{i,x}(B) \right) \geq b$ whenever $x \in V$, $B \subset K_x = W$ and $\lambda^r(B) \geq a$ $(1 \leq i \leq n)$. We define the compact subsets $X_i = g_i(K) \subset Y_i$, $i = 1, \ldots, n$. Using Lusin's Theorem there exist compact subsets C_i of X_i for which $\lambda^r(X_i \backslash C_i) < b$ and $f_i|_{C_i}$ is continuous $(1 \leq i \leq n)$. But then

$$\lambda^r \left(g_{i,x}^{-1}(X_i \backslash C_i) \right) < a, \quad x \in V, \quad i = 1, \ldots, n$$

(for otherwise we obtain from Lemma 1.3.1 the contradiction

$$\lambda^r (g_{i,x}(g_{i,x}^{-1}(X_i \backslash C_i))) = \lambda^r(X_i \backslash C_i) \geq b).$$

Next, we prove that for every $x \in V$

$$\left(\bigcap_{i=1}^{n} g_{i,x}^{-1}(C_i) \right) \bigcap \left(\bigcap_{i=1}^{n} g_{i,x_0}^{-1}(C_i) \right) \neq \emptyset. \qquad (1.3.4)$$

Assuming the contrary and observing that the sets $g_{i,x}^{-1}(X_i \backslash C_i)$ and $g_{i,x}^{-1}(C_i)$ are disjoint with union $K_x = W$ we arrive at

$$c = \lambda^r(W)$$

$$= \lambda^r(K_x \cup K_{x_0}) = \lambda^r \left[\left(\bigcup_{i=1}^{n} g_{i,x}^{-1}(X_i \backslash C_i) \right) \cup \left(\bigcup_{i=1}^{n} g_{i,x_0}^{-1}(X_i \backslash C_i) \right) \right]$$

$$\leq \sum_{i=1}^{n} \lambda^r \left(g_{i,x}^{-1}(X_i \backslash C_i) \right) + \sum_{i=1}^{n} \lambda^r g_{i,x_0}^{-1}(X_i \backslash C_i)$$

$$< 2na = c,$$

which is impossible. But (1.3.4) means that for every $x \in V$ there exists $y \in W$ for which

$$(x, y), (x_0, y) \in K, \quad g_i(x, y), g_i(x_0, y) \in C_i, \quad i = 1, \ldots, n. \qquad (1.3.5)$$

After these preparations we are ready to prove that f is continuous at x_0. We shall show that for every $\epsilon > 0$ there corresponds $\delta > 0$ for which $|f(x) - f(x_0)| < \epsilon$ whenever $\|x - x_0\| < \delta$. Let $\epsilon > 0$ and let $M_i := V \times W \times f_i(C_i)$, $1 \le i \le n$. Since M_i is compact, h_i is uniformly continuous on M_i ($1 \le i \le n$), and the mapping

$$h(x, y, t_1, \ldots, t_n) := \sum_{i=1}^{n} h_i(x, y, t_i), \ (x, y) \in D, t_i \in \mathbb{R}^r, \qquad (1.3.6a)$$

is uniformly continuous on $M := V \times W \times f_1(C_1) \times \ldots \times f_n(C_n)$. Hence there exists $\delta_1 > 0$ such that for all $x \in V$, $z_i, z_i' \in f_i(C_i)$ with $\|x - x_0\| < \delta_1$ and $\|z_i - z_i'\| < \delta_1$ ($1 \le i \le n$), we get

$$|h(x, y, z_1, \ldots, z_n) - h(x_0, y, z_1', \ldots, z_n')| < \epsilon. \qquad (1.3.6)$$

Since f_i and g_i are uniformly continuous on C_i and $V \times W$, respectively, there exist $\delta_2 > 0$ and $\delta_3 > 0$ such that $x_i, x_i' \in C_i$ with $\|x_i - x_i'\| < \delta_2$ imply

$$\|f_i(x_i) - f_i(x_i')\| < \delta_1 \qquad (1.3.7)$$

and

$$\|g_i(x, y) - g_i(x_0, y)\| < \delta_2 \qquad (1.3.8)$$

whenever $\|x - x_0\| < \delta_3$ and $(x, y), (x_0, y) \in K = V \times W$.

Now let $\delta = min(\delta_1, \delta_3)$, let $\|x - x_0\| < \delta$ and choose $y \in W$ so that (1.3.5) is satisfied. Then by (1.3.7) and (1.3.8) we have

$$\|f_i(g_i(x, y)) - f_i(g_i(x_0, y))\| < \delta_1, \quad 1 \le i \le n.$$

Since $\|x - x_0\| < \delta \le \delta_1$ we get from (1.3.6) and the functional equation (1.3.3) that $|f(x) - f(x_0)| < \epsilon$. This completes the proof.

Remark 1.3.3. The proof of the preceding Theorem shows that Theorem 1.3.2 remains valid if we replace functional equation (1.3.3) by

$$f(x) = h\Big(x, y, f_1(g_1(x, y)), f_2(g_2(x, y)), \ldots, f_n(g_n(x, y))\Big) \qquad (1.3.9)$$

for all $(x, y) \in D$, where now $h : D \times \mathbb{R}^{rn} \to \mathbb{R}$ is continuous on $D \times \mathbb{R}^{rn}$. Indeed, defining h by (1.3.6a), equation (1.3.9) goes over into (1.3.3). But in the proof of Theorem 1.3.2 we have in fact treated the more general case of equation (1.3.9).

Remark 1.3.4. For more general results than Theorem 1.3.2 we refer to Jarai (1979, 1982, 1986). For our purposes in this book, Theorem 1.3.2 is strong enough and can be applied readily to our equations. It is also interesting that in Theorem 1.3.2 the condition "f_i is λ^r-measurable on $\mathbb{R}^{r}$", can be replaced by the topological condition "f_i has the Baire property in $\mathbb{R}^{r}$" (see Bourbaki (1966)).

Finally, let us mention that regularity theorems for generalized Pexider equations were studied in Sander (1978) and Grosse-Erdmann (1989) with completely different methods.

Example 1.3.5. Simple applications of Theorem 1.3.2 yield the measurable solutions of the Pexider equations (1.2.25) and (1.2.30), and also equation (1.2.2).

(a) If F_1, F_2, F_3 satisfy (1.2.25) and if any F_i is λ^k-measurable, then we get

$$F_i(x) = a \odot x + c_i = \sum_{j=1}^{k} a_j \xi_j + c_i \qquad (1.3.10)$$

where $a = (a_1, \ldots a_k) \in \mathbb{R}^k$ and the c_i are constants such that $c_3 = c_1 + c_2$ ($i = 1, 2, 3$). To prove this statement, we first see from (1.2.26) that measurability of a single F_i implies measurability of the other two. Now we apply Theorem 1.3.2 with $T = Y_1 = Y_2 = J$, $D = D_2$, $n = 2$, $s = r = k$, $g_1(x, y) = y$, $g_2(x, y) = x + y$, $h_1(x, y, z) = -z$, $h_2(x, y, z) = z$, $f = F_1$, $f_1 = F_2$, $f_2 = F_3$ and get that F_1 hence also F_2 and F_3 (again via (1.2.26)) are continuous. From Lemmas 1.2.6 and 1.2.14 we get (1.3.10).

(b) If H_1, H_2, H_3 satisfy (1.2.30) and if any H_i is λ^k-measurable, then we obtain

$$H_i(x) = a \odot \log x + c_i = \sum_{j=1}^{k} a_j \log \xi_j + c_i \qquad (1.3.11)$$

where $a = (a_1, \ldots, a_k) \in P$, the c_i are constants satisfying $c_3 = c_1 + c_2$, and $\log x = (\log \xi_1, \ldots, \log \xi_n)$. This can be proved as in (a) above (using (1.2.31), (1.2.33) and (1.2.23)).

(c) Similarly, we find that the λ^k-measurable solution $M : J \to \mathbb{R}$ of (1.2.2) is given by

$$M(x) = x^\alpha = \xi_1^{\alpha_1} \cdots \xi_k^{\alpha_k} \qquad (1.3.12)$$

for $x = (\xi_1, \ldots, \xi_k) \in \mathbb{R}^k$, where $\alpha = (\alpha_1, \ldots, \alpha_k)$ is a constant vector in $\mathbb{R}^k$. We define $A : \mathcal{P}_1 \to \mathbb{R}$ by

$$A(s) := \log M \left(e^{-s} \right), \qquad s \in \mathcal{P}_1.$$

Then A is additive. As in case (a) above, we deduce that

$$A(s) = a \odot s = \sum_{i=1}^k a_i s_i.$$

Thus M has the form (1.3.12), where $\alpha_i = -a_i$ $(1 \leq i \leq k)$.

Remark 1.3.6. Finally we remark that the results in Example 1.3.5 are not the most general ones. For instance, if $k = 1$ and A satisfies the Cauchy equation then one-sided boundedness on a set of positive measure implies that A is continuous. This is a famous result due to Ostrowski (1929). (With the aid of a theorem of Steinhaus, (see for instance Sander (1977) or Jarai (1995)), the proof of this result can be given in few lines.) Nonetheless, for the sake of simplicity, we shall use measurability (in the sense of Lebesgue) as our standard "weakest" regularity condition throughout this book.

CHAPTER 2

THE BRANCHING PROPERTY

2.1. Introduction

One of the most fundamental of all axioms which can be postulated for
a measure of information is the branching property. Intuitively, it expresses
the following idea. Suppose that we want to know the outcome of a given
experiment A which may result in one of the events E_1, E_2, ..., E_n, with
corresponding probabilities p_1, p_2, ..., p_n. Alternatively, we could perform
experiment B with events $E_1 \cup E_2$, E_3, ..., E_n and corresponding probabilities
$p_1 + p_2$, ..., p_n. This may entail some loss of information, for if we observe
$E_1 \cup E_2$ as the outcome of B then we don't have as much knowledge as we
would have obtained from experiment A - namely, whether it was E_1 or E_2
which occurred. The basic assumption underlying the branching property is
that the amount of information which is lost by observing the outcome of B
instead of A depends only on the probabilities p_1 and p_2 of the respective
events E_1 and E_2. Let us now formulate this idea precisely.

Let

$$D_n = \{P = (p_1, p_2, ..., p_n) \mid p_1, p_2, ..., p_n \in J; \sum_{i=1}^{n} p_i \in J\},$$

and let

$$\Gamma_n^o = \{P = (p_1, p_2, ..., p_n) \mid p_1, p_2, ..., p_n \in J; \sum_{i=1}^{n} p_i = 1\}.$$

We shall use the notation $\{I_n\}$ to represent a measure of information, that
is, a sequence of real-valued functions I_n, for $n = 2, 3, ...$. Unless specified
otherwise, it is to be understood that I_n is defined on the *open* domain Γ_n^o.

Definition 2.1.1. A measure $\{I_n\}$ of information is said to have the *branching property* (or to be a *branching* information measure) if there exists a
function $G : D_2 \to \mathbb{R}$ such that

$$I_n(p_1, p_2, ..., p_n) = I_{n-1}(p_1 + p_2, ..., p_n) + G(p_1, p_2) \qquad (2.1.1)$$

for all $P \in \Gamma_n^o$ and for all $n \geq 3$. Such a function G is called a *generating function* for $\{I_n\}$.

It is easy to see that, in order to determine the form of $I_n(P)$ for branching $\{I_n\}$, it suffices to determine the forms of the functions I_2 and G. Indeed, the following simple result due to Aczél and Daróczy (1975) formalizes this fact.

Lemma 2.1.2. *A measure $\{I_n\}$ of information has the branching property (2.1.1) if, and only if,*

$$I_n(p_1, p_2, ..., p_n) = I_2(1 - p_n, \, p_n) + \sum_{j=2}^{n-1} G(p_1 + \cdots + p_{j-1}, p_j), \quad (2.1.2)$$

for all $P \in \Gamma_n^o$ and for all $n \geq 3$.

Proof: First, suppose that $\{I_n\}$ has the branching property. Equation (2.1.1) is exactly (2.1.2) for $n = 3$. If $n \geq 4$, then by repeated application of (2.1.1) we have

$$
\begin{aligned}
I_n(p_1, p_2, ..., p_n) &= I_{n-1}(p_1 + p_2, ..., p_n) + G(p_1, p_2) \\
&= I_{n-2}(p_1 + p_2 + p_3, ..., p_n) + G(p_1 + p_2, p_3) + G(p_1, p_2) \\
&= I_{n-2}(p_1 + p_2 + p_3, ..., p_n) + \sum_{j=2}^{3} G(p_1 + ... + p_{j-1}, p_j) \\
&= .... \\
&= I_2(p_1 + p_2 + ... + p_{n-1}, p_n) + \sum_{j=2}^{n-1} G(p_1 + ... + p_{j-1}, p_j),
\end{aligned}
$$

which is (2.1.2). The converse is trivial.

Furthermore, as we shall presently show, the addition of some symmetry properties enables us to determine the forms of I_2 and G, and therefore I_n for all n, explicitly.

Remark 2.1.3. If $\{I_n\}$ is both branching and 4-symmetric (i.e. I_4 is a symmetric function of its arguments) on the closed domain

$$\Gamma_n = \{P = (p_1, p_2, ..., p_n) \mid p_1, p_2, ..., p_n \in \bar{J}; \sum_{i=1}^{n} p_i = 1\},$$

then it was shown by Ng (1974) that

$$I_n(P) = \sum_{i=1}^{n} F(p_i)$$

for some function $F : \bar{J} \to \mathbb{R}$. Measures of information having this special form are said to have the *sum form* (or *sum property*). It means that $I_n(P)$ is a sum of terms, each of which depends only upon a single event. Since nearly all measures of information used in applications (including all those introduced in Section 1.1) have this form, it seems to be a very fundamental property.

2.2. The characteristic functional equation for branching measures of information

Another basic axiom for information measures is the property of symmetry. Intuitively, it seems quite reasonable to suppose that the information content of an experiment should not depend on the labeling of its outcomes. That is the basis for the symmetry property, which we now formulate.

An information measure $\{I_n\}$ is said to be *m-symmetric* (for some positive integer m) if the map I_m is a symmetric function, that is,

$$I_m(p_1, ..., p_m) = I_m(p_{\pi(1)}, ..., p_{\pi(m)}), \qquad P \in \Gamma_m^o, \qquad (2.2.1)$$

for all permutations π on $\{1, ..., m\}$.

Some interesting mathematical problems are created by considering weaker forms of symmetry. These are discussed in the last section of the chapter.

Lemma 2.2.1. *Suppose $\{I_n\}$ has the branching property (2.1.1), and $\{I_n\}$ is 4-symmetric. Then the generating function G satisfies the characteristic functional equation for branching measures of information*

$$G(x, y) + G(x + y, z) = G(x, z) + G(x + z, y), \qquad (x, y, z) \in D_3. \quad (2.2.2)$$

and G is a symmetric function

$$G(x, y) = G(y, x), \qquad (x, y) \in D_2. \qquad (2.2.3)$$

Proof: Suppose $\{I_n\}$ is branching and 4-symmetric. Using (2.1.2) twice and (2.2.1), we find that

$$
\begin{aligned}
G(p_1 + p_2, p_3) &+ G(p_1, p_2) \\
&= [I_4(p_1, p_2, p_3, p_4) - I_3(p_1 + p_2, p_3, p_4)] \\
&\quad + [I_3(p_1 + p_2, p_3, p_4) - I_2(p_1 + p_2 + p_3, p_4)] \\
&= I_4(p_1, p_2, p_3, p_4) - I_2(1 - p_4, p_4) \\
&= I_4(p_1, p_3, p_2, p_4) - I_2(1 - p_4, p_4) \\
&= G(p_1 + p_3, p_2) + G(p_1, p_3).
\end{aligned}
$$

This is equation (2.2.2). In a similar way, we also get from (2.1.1)

$$
\begin{aligned}
G(p_1, p_2) &= I_4(p_1, p_2, p_3, p_4) - I_3(p_1 + p_2, p_3, p_4) \\
&= I_4(p_2, p_1, p_3, p_4) - I_3(p_2 + p_1, p_3, p_4) \\
&= G(p_2, p_1)
\end{aligned}
$$

which is (2.2.3).

Remark 2.2.2. Note that for (2.2.2) we needed only the partial symmetry

$$
I_4(p_1, p_2, p_3, p_4) = I_4(p_1, p_3, p_2, p_4), \quad P \in \Gamma_4^\circ, \qquad (2.2.4)
$$

and for (2.2.3) we needed only

$$
I_4(p_1, p_2, p_3, p_4) = I_4(p_2, p_1, p_3, p_4), \quad P \in \Gamma_4^\circ.
$$

We shall come back to this point in section 2.4.

Remark 2.2.3. Observe also that with the help of symmetry (2.2.3), the characteristic branching equation (2.2.2) can be transformed in the following way

$$
\begin{aligned}
G(x, y) + G(x + y, z) &= G(y, x) + G(y + x, z) \\
&= G(y, z) + G(y + z, x) \\
&= G(x, y + z) + G(y, z).
\end{aligned}
$$

That is, the system (2.2.2) - (2.2.3) is equivalent to the system consisting of (2.2.3) and

$$
G(x, y) + G(x + y, z) = G(x, y + z) + G(y, z), \quad (x, y, z) \in D_3. \qquad (2.2.5)
$$

This last equation has the same form as the *cocycle equation* in homological algebra (cf. Kurosh (1956), pp. 48-50, for example). But here it is on a restricted domain and therefore more difficult to solve than in the purely algebraic context.

We proceed now to solve the system of functional equations (2.2.2) - (2.2.3). Our proof is essentially that of Ng (1974), who solved this system on the closed domains $\bar{D}_3, \bar{D}_2$ for $G : \bar{D}_2 \to \mathbb{R}$. Since his proof needed some slight modifications for our purpose (because our vectors do not have any zero components), we include the modified proof here.

In the next section, we shall use this theorem to describe the forms of all symmetric branching measures of information.

Theorem 2.2.4. *A map $G : D_2 \to \mathbb{R}$ satisfies (2.2.3) and (2.2.5) if and only if G is representable in the form*

$$G(x, y) = f(x) + f(y) - f(x + y), \qquad (x, y) \in D_2, \qquad (2.2.6)$$

through some function $f : J \to \mathbb{R}$.

Proof: If G is of the form (2.2.6), then obviously G satisfies (2.2.3) and (2.2.5). We prove the converse, first giving an outline of the steps involved.

Outline of the proof of the converse: (Let $\mathbf{0} = (0, 0, ..., 0)$.)
Step 1. Let $J^* = J \cup \{\mathbf{0}\}$, and define E_n (for $n = 2, 3, ...$) by

$$E_n := \left\{ (x_1, x_2, ..., x_n) \mid x_1, x_2, ..., x_n, \sum_{i=1}^{n} x_i \in J^* \right\}.$$

G can be extended to a map $F : E_2 \to \mathbb{R}$ which satisfies (2.2.5) on E_3 and (2.2.3) on E_2.
Step 2. The operation $\oplus$ defined on $(J^* \times \mathbb{R})^2$ by

$$(x, u) \oplus (y, v) := (x + y, u + v - F(x, y)) \qquad (2.2.7)$$

is commutative, associative, cancellative, divisible, and distributive over the rationals.
Step 3. The class $\mathcal{O} = \{(S, f)\}$ of pairs (S, f) satisfying
    O1. $\{\mathbf{0}\} \subset S \subset J^*$;
    O2. If $s \in S$, then $\frac{1}{n} s \in S$ for all $n \in \mathbb{N} := \{1, 2, 3, ...\}$;
    O3. If $s_1, s_2 \in S$ and $s_1 + s_2 \in J^*$, then $s_1 + s_2 \in S$;

O4. $f : S \to \mathbb{R}$ satisfies

$$(s_1 + s_2, \ f(s_1 + s_2)) = (s_1, f(s_1)) \oplus (s_2, f(s_2)) \qquad (2.2.8)$$

for all $s_1, s_2 \in S$ with $s_1 + s_2 \in S$;
is nonempty and can be partially ordered by

$$(S, f) \prec (T, g) \iff S \subset T \quad \text{and} \quad g \mid_S = f.$$

By Zorn's Lemma there is a maximal element (S_o, f_o) of $\mathcal{O}$.
Step 4. In fact, $S_o = J^*$ and

$$(x + y, \ f_o(x + y)) = (x, f_o(x)) \oplus (y, f_o(y)), \qquad (x, y) \in E_2.$$

By (2.2.7) this is equivalent to

$$(x + y, \ f_o(x + y)) = (x + y, f_o(x) + f_o(y) - F(x, y))$$

which implies that F has the representation

$$F(x, y) = f_o(x) + f_o(y) - f_o(x + y).$$

Verification of step 1: We extend G to F on E_2 via $F(x, y) = G(x, y)$ for all $(x, y) \in D_2$, and

$$F(x, 0) := F(0, x) := F(0, 0) := 0, \qquad x \in J.$$

Clearly, this extended F satisfies (2.2.5) on E_3 and (2.2.3) on E_2.

Verification of step 2: The verifications of

$$(x, u) \oplus (y, v) = (y, v) \oplus (x, u) \qquad \text{(commutativity)},$$

$$[(x, u) \oplus (y, v)] \oplus (z, w) = (x, u) \oplus [(y, v) \oplus (z, w)] \qquad \text{(associativity)},$$

$$(x, u) \oplus (y, v) = (x, u) \oplus (z, w) \quad \text{implies} \quad (y, v) = (z, w) \ \text{(cancellativity)},$$

for all $u, v, w \in \mathbb{R}$, $x, y, z, x + y + z \in J^*$, are straightforward calculations using (2.2.7), (2.2.5) and (2.2.3). We prove for instance associativity:

$$\begin{aligned}
[(x, u) \oplus (y, v)] \oplus (z, w) &= (x + y, u + v - F(x, y)) \oplus (z, w) \\
&= (x + y + z, u + v + w - F(x, y) - F(x + y, z)) \\
&= (x + y + z, u + v + w - F(y, z) - F(x, y + z)) \\
&= (x, u) \oplus (y + z, v + w - F(y, z)) \\
&= (x, u) \oplus [(y, v) \oplus (z, w)].
\end{aligned}$$

To show that $\oplus$ is distributive over the rationals, we first define positive integer multiples

$$n\,(x,u) := (x,u) \oplus (x,u) \oplus \cdots \oplus (x,u) \qquad (n-\text{fold})$$

for each $n \in \mathbb{N}$ such that $nx \in J^*$. We also define $0\,(x,u) := (\mathbf{0},0)$ for all $x \in J^*$, $u \in \mathbb{R}$. Then it is easy to see that $(m+n)(x,u) = m\,(x,u) \oplus n\,(x,u)$, and $(mn)(x,u) = m[n(x,u)]$. Furthermore $m[(x,u) \oplus (y,v)] = m(x,u) \oplus m(y,v)$ follows also from the commutativity and associativity.

Next, we note that $\frac{1}{n}(x,u)$ is definable for all $x \in J^*$, $n \in \mathbb{N}$, as the unique point (y,v) such that $n(y,v) = (x,u)$. Observe that

$$n(y,v) = \left[ny,\ nv - \sum_{i=1}^{n-1} F(y,iy) \right],$$

and so by a direct solution of $n(y,v) = (x,u)$, we get

$$\frac{1}{n}(x,u) = \left[\frac{1}{n}x,\ \frac{1}{n}u + \frac{1}{n}\sum_{i=1}^{n-1} F\left(\frac{1}{n}x, \frac{i}{n}x\right) \right].$$

As a consequence, we have

$$\frac{1}{k}\left[\frac{1}{n}(x,u) \right] = \frac{1}{kn}(x,u) \quad \forall x \in J^*, \quad n,k \in \mathbb{N}.$$

For each nonnegative rational r, write $r = \frac{m}{n}$ with $m \in \mathbb{N} \cup \{0\}$, $n \in \mathbb{N}$, and for each (x,u) in $J^* \times \mathbb{R}$ with $rx \in J^*$, we define

$$r(x,u) := m\left[\frac{1}{n}(x,u) \right].$$

This is well-defined, since for any $k \in \mathbb{N}$ we have

$$mk\left[\frac{1}{nk}(x,u) \right] = m\left[k\left(\frac{1}{k}\left[\frac{1}{n}(x,u) \right] \right) \right] = m\left[\frac{1}{n}(x,u) \right],$$

and therefore $r(x,u) = p\left[\frac{1}{q}(x,u) \right]$ where $r = \frac{p}{q}, p \in \mathbb{N} \cup \{0\}, q \in \mathbb{N}$, and g.c.d.$(p,q) = 1$.

Now we can verify that

(i) $(r_1+r_2)(x,u) = r_1(x,u) \oplus r_2(x,u)$ for all $x \in J^*$ and nonnegative rational r_1, r_2 with $r_1 x + r_2 x \in J^*$;

(ii) $(r_1 r_2)(x,u) = r_1[r_2(x,u)]$ for all $x \in J^*$ and nonnegative $r_1, r_2 \in \mathbb{Q}$ with $r_2 x, r_1 r_2 x \in J^*$; and

(iii) $r[(x,u) \oplus (y,v)] = r(x,u) \oplus r(y,v)$ for all $x, y \in J^*$, nonnegative $r \in \mathbb{Q}$ with $x+y, r(x+y) \in J^*$.

For (i), let $r_i = \frac{m_i}{n}$ with $m_i \in \mathbb{N} \cup \{0\}$, $(i = 1,2)$, $n \in \mathbb{N}$, $x \in J^*$ with $r_1 x + r_2 x \in J^*$. Then

$$(r_1 + r_2)(x,u) = \left(\frac{m_1 + m_2}{n}\right)(x,u)$$

$$= (m_1 + m_2)\left[\frac{1}{n}(x,u)\right]$$

$$= m_1\left[\frac{1}{n}(x,u)\right] \oplus m_2\left[\frac{1}{n}(x,u)\right]$$

$$= r_1(x,u) \oplus r_2(x,u).$$

For (ii), supposing $r_2 x, r_1 r_2 x \in J$, we have

$$(r_1 r_2)(x,u) = \frac{m_1 m_2}{n^2}(x,u)$$

$$= m_1 m_2 \left[\frac{1}{nm_2}(nm_2)\left[\frac{1}{n^2}(x,u)\right]\right]$$

$$= r_1(r_2(x,u)).$$

Finally, for (iii),

$$r(x,u) \oplus r(y,v) = m\left[\frac{1}{n}(x,u)\right] \oplus m\left[\frac{1}{n}(y,v)\right]$$

$$= m\left[\frac{1}{n}(x,u) \oplus \frac{1}{n}(y,v)\right]$$

while

$$r[(x,u) \oplus (y,v)] = m\left[\frac{1}{n}((x,u) \oplus (y,v))\right],$$

so it suffices to show that $\frac{1}{n}(x,u) \oplus \frac{1}{n}(y,v) = \frac{1}{n}[(x,u) \oplus (y,v)]$. This is done by the following calculation.

$$\frac{1}{n}(x,u) \oplus \frac{1}{n}(y,v) = \frac{1}{n}\left(n\left[\frac{1}{n}(x,u) \oplus \frac{1}{n}(y,v)\right]\right)$$

$$= \frac{1}{n}\left[\frac{n}{n}(x,u) \oplus \frac{n}{n}(y,v)\right]$$

$$= \frac{1}{n}[(x,u) \oplus (y,v)].$$

Verification of step 3: Class $\mathcal{O}$ is nonempty since it contains ($\{0\}$, zero map). If $\mathcal{C} = \{(S_i, f_i) \mid i \in I\}$ is a nonempty chain in $\mathcal{O}$, then obviously (S, f) defined by $S = \cup\{S_i \mid i \in I\}$ and $f = \cup\{f_i \mid i \in I\}$ (that is, $f(s) = f_i(s)$ if $s \in S_i$) is again in $\mathcal{O}$ and is an upper bound for $\mathcal{C}$ (note that f is a function since $\mathcal{C}$ is a chain; see Hewitt and Stromberg, (1965), p.10). By Zorn's Lemma there exists (S_o, f_o) which is maximal in $\mathcal{O}$.

Verification of step 4: To prove that $S_o = J^*$, let us suppose that S_o is a proper subset of J^*. We construct $(S_o^*, f_o^*) \in \mathcal{O}$ which is a proper extension of (S_o, f_o). This contradicts the maximality of (S_o, f_o).

Case 1. Suppose there is a point $x_o \in J^* \setminus S_o$ for which $\frac{1}{n}x_o \notin S_o - S_o$ for all $n \in \mathbb{N}$. Then $s_1 + r_1 x_o = s_2 + r_2 x_o$ (for $s_i \in S_o$, nonnegative $r_i \in \mathbb{Q}$, $s_i + r_i x_o \in J^*$) if and only if $s_1 = s_2$ and $r_1 = r_2$. For if not, say $s_1 > s_2$, then with $r_i = \frac{m_i}{n}$ ($m_i \in \mathbb{N} \cup \{0\}$, $n \in \mathbb{N}$) we have

$$s_1 = s_2 + (r_2 - r_1)x_o = s_2 + (m_2 - m_1)\frac{x_o}{n},$$

with $m_2 - m_1 > 0$, so

$$\frac{s_1}{m_2 - m_1} - \frac{s_2}{m_2 - m_1} = \frac{x_o}{n}.$$

By **O2**, this means $\frac{x_o}{n} \in S_o - S_o$, contrary to our assumption.

Hence on the set

$$S_o^* := \{s + r x_o \in J^* \mid s \in S_o, r \in \mathbb{Q}, r \geq 0\},$$

we may define a map f_o^* by

$$[s + r x_o, f_o^*(s + r x_o)] := [s, f_o(s)] \oplus r[x_o, f_o^*(x_o)],$$

where $f_o^*(x_o)$ is an arbitrary real number. We now show that (S_o^*, f_o^*) is in $\mathcal{O}$. The properties **O1, O2, O3** are easily verified. Let us prove that **O4** is valid. For $s_i + r_i x_o \in S_o^*$ ($i = 1, 2$), we have

$$
\begin{aligned}
&[(s_1 + r_1 x_o) + (s_2 + r_2 x_o), f_o^*((s_1 + r_1 x_o) + (s_2 + r_2 x_o))] \\
&= [(s_1 + s_2) + (r_1 + r_2)x_o, f_o^*((s_1 + s_2) + (r_1 + r_2)x_o)] \\
&= [s_1 + s_2, f_o(s_1 + s_2)] \oplus (r_1 + r_2)[x_o, f_o^*(x_o)] \\
&= [s_1, f_o(s_1)] \oplus r_1[x_o, f_o^*(x_o)] \oplus [s_2, f_o(s_2)] \oplus r_2[x_o, f_o^*(x_o)] \\
&= [s_1 + r_1 x_o, f_o^*(s_1 + r_1 x_o)] \oplus [s_2 + r_2 x_o, f_o^*(s_2 + r_2 x_o)].
\end{aligned}
$$

Thus the pair (S_o^*, f_o^*) is in $\mathcal{O}$ and is a proper extension of (S_o, f_o), contradicting the maximality of (S_o, f_o). So this case cannot occur, and for all $x \in J^* \setminus S_o$ there is an $n \in \mathbb{N}$ such that $\frac{1}{n}x \in S_o - S_o$.

Case 2. Now we may suppose that for each $x \in J^*$ there exist $s_1, s_2 \in S_o$ and $n \in \mathbb{N}$ for which $s_1 = s_2 + \frac{1}{n}x$. This is true if $x \in J^* \setminus S_o$ since Case (1) does not occur, and if $x \in S_o$ we choose $n = 1, s_1 = x$ and $s_2 = \mathbf{0} \in S_o$. (Note that $x = \mathbf{0} + x$.)

In this case we show that it is possible to define a map $f^* : J^* \to \mathbb{R}$ by

$$\left.\begin{array}{l} f^*(x) : = \text{the unique point } a \in \mathbb{R} \text{ such that} \\[2mm] [s_1, f_o(s_1)] = [s_2, f_o(s_2)] \oplus \dfrac{1}{n}(x, a). \end{array}\right\} \tag{2.2.9}$$

First, we have to demonstrate the existence of such a number a. If $x \in S_0$, then we have chosen $n = 1, s_1 = x$ and $s_2 = \mathbf{0}$, so $[s_1, f_0(s_1)] = [x, f_0(x)]$ and $[s_2, f_0(s_2)] \oplus \frac{1}{n}(x, a) = (\mathbf{0}, 0) \oplus (x, a) = (x, a)$. In this case, the (unique) solution $a = f_0(x)$ exists. This also shows that f^* is an extension of f_0, and so (J^*, f^*) is an extension of (S_0, f_0). On the other hand, if $x \in J^* \setminus S_0$, then $s_1 = s_2 + \frac{1}{n}x$ implies that

$$[s_2, f_0(s_2)] \oplus \frac{1}{n}(x, a) = [s_2, f_0(s_2)] \oplus \left[\frac{1}{n}x, \frac{1}{n}a + \frac{1}{n}\sum_{i=1}^{n-1} F\left(\frac{1}{n}x, \frac{i}{n}x\right)\right]$$

$$= \left[s_2 + \frac{1}{n}x, f_0(s_2) + \frac{1}{n}a + \frac{1}{n}\sum_{i=1}^{n-1} F\left(\frac{1}{n}x, \frac{i}{n}x\right) - F\left(s_2, \frac{1}{n}x\right)\right]$$

So $[s_1, f_0(s_1)] = [s_2, f_0(s_2)] \oplus \frac{1}{n}(x, a)$ has the (again unique) solution $a = n[f_0(s_1) - f_0(s_2) + F(s_2, \frac{1}{n}x)] - \sum F(\frac{1}{n}x, \frac{i}{n}x)$. Uniqueness also follows from cancellativity. Finally, we have to show that f^* is well-defined by (2.2.9). Indeed, suppose that for a given $x \in J^*$ we can write

$$s_1 = s_2 + \frac{1}{n}x, \qquad\qquad t_1 = t_2 + \frac{1}{m}x \tag{2.2.10}$$

with $s_i, t_i \in S_o$ and $n, m \in \mathbb{N}$. Let a, b be the unique points such that

$$\left.\begin{array}{l} [s_1, f_o(s_1)] = [s_2, f_o(s_2)] \oplus \dfrac{1}{n}(x, a) \\[3mm] [t_1, f_o(t_1)] = [t_2, f_o(t_2)] \oplus \dfrac{1}{m}(x, b). \end{array}\right\} \tag{2.2.11}$$

We must show that $a = b$. To this end, consider (cf. (2.2.10)) that $\frac{1}{2m} s_1 - \frac{1}{2m} s_2 = \frac{1}{2mn} x = \frac{1}{2n} t_1 - \frac{1}{2n} t_2$. Thus

$$\frac{1}{2m} s_2 + \frac{1}{2n} t_1 = \frac{1}{2m} s_1 + \frac{1}{2n} t_2 \in J^*,$$

while each term is again in S_o. Hence by **O3** and **O4** on S_o, we have

$$\left[\frac{1}{2m} s_2, f_o \left(\frac{1}{2m} s_2 \right) \right] \oplus \left[\frac{1}{2n} t_1, f_o \left(\frac{1}{2n} t_1 \right) \right]$$

$$= \left[\frac{1}{2m} s_1, f_o \left(\frac{1}{2m} s_1 \right) \right] \oplus \left[\frac{1}{2n} t_2, f_o \left(\frac{1}{2n} t_2 \right) \right]. \tag{2.2.12}$$

On the other hand, multiplying the first equation of (2.2.11) by $\frac{1}{2m}$ yields

$$\frac{1}{2m} [s_1, f_o(s_1)] = \frac{1}{2m} [s_2, f_o(s_2)] \oplus \frac{1}{2mn} (x, a).$$

Using **O4** (which implies that $[rs, f_o(rs)] = r[s, f_o(s)]$ for $s \in S_o$, $r \in \mathbb{Q}$, $r \geq 0$, with $rs \in S_o$), we get

$$\left[\frac{1}{2m} s_1, f_o \left(\frac{1}{2m} s_1 \right) \right] = \left[\frac{1}{2m} s_2, f_o \left(\frac{1}{2m} s_2 \right) \right] \oplus \frac{1}{2mn} (x, a).$$

Similarly, from the second equation of (2.2.11) we get

$$\left[\frac{1}{2n} t_1, f_o \left(\frac{1}{2n} t_1 \right) \right] = \left[\frac{1}{2n} t_2, f_o \left(\frac{1}{2n} t_2 \right) \right] \oplus \frac{1}{2mn} (x, b).$$

Since $\frac{1}{2m} s_1 + \frac{1}{2n} t_2 + \frac{1}{2mn} x = \frac{1}{2m} s_2 + \frac{1}{2n} t_1 + \frac{1}{2mn} x \in J^*$, we can combine the last two equations to obtain

$$\left[\frac{1}{2m} s_1, f_o \left(\frac{1}{2m} s_1 \right) \right] \oplus \left[\frac{1}{2n} t_2, f_o \left(\frac{1}{2n} t_2 \right) \right] \oplus \frac{1}{2mn} (x, b)$$

$$= \left[\frac{1}{2n} t_1, f_o \left(\frac{1}{2n} t_1 \right) \right] \oplus \left[\frac{1}{2m} s_2, f_o \left(\frac{1}{2m} s_2 \right) \right] \oplus \frac{1}{2mn} (x, a).$$

Comparing this with (2.2.12) and using the cancellativity of $\oplus$, we have

$$\frac{1}{2mn} (x, b) = \frac{1}{2mn} (x, a),$$

hence $a = b$.

To show that (J^*, f^*) is in $\mathcal{O}$, we must prove that $[x + y, f^*(x + y)] = [x, f^*(x)] \oplus [y, f^*(y)]$ holds for $x, y, x + y \in J^*$. To do this, let

$$s_1 = s_2 + \frac{1}{2n}x, \qquad\qquad t_1 = t_2 + \frac{1}{2m}y$$

with $s_i, t_i \in S_o$ and $m, n \in \mathbb{N}$. From these we get

$$\frac{1}{2m}s_1 = \frac{1}{2m}s_2 + \frac{1}{2mn}x, \qquad \frac{1}{2n}t_1 = \frac{1}{2n}t_2 + \frac{1}{2mn}y,$$

$$\frac{1}{2m}s_1 + \frac{1}{2n}t_1 = \left(\frac{1}{2m}s_2 + \frac{1}{2n}t_2\right) + \frac{1}{2mn}(x + y).$$

Applying the definition (2.2.9) of f^* to each, we obtain

$$\left[\frac{1}{2m}s_1, f_o\left(\frac{1}{2m}s_1\right)\right] = \left[\frac{1}{2m}s_2, f_o\left(\frac{1}{2m}s_2\right)\right] \oplus \frac{1}{2mn}[x, f^*(x)], \quad (2.2.13)$$

$$\left[\frac{1}{2n}t_1, f_o\left(\frac{1}{2n}t_1\right)\right] = \left[\frac{1}{2n}t_2, f_o\left(\frac{1}{2n}t_2\right)\right] \oplus \frac{1}{2mn}[y, f^*(y)], \quad (2.2.14)$$

$$\left.\begin{array}{l}\left[\frac{1}{2m}s_1 + \frac{1}{2n}t_1, f_o\left(\frac{1}{2m}s_1 + \frac{1}{2n}t_1\right)\right] = \\[2mm] \left[\frac{1}{2m}s_2 + \frac{1}{2n}t_2, f_o\left(\frac{1}{2m}s_2 + \frac{1}{2n}t_2\right)\right] \oplus \dfrac{[x + y, f^*(x + y)]}{2mn},\end{array}\right\} \quad (2.2.15)$$

respectively. Adding (2.2.13) to (2.2.14) and again using **O4** on (S_o, f_o), we get

$$\left[\frac{1}{2m}s_1 + \frac{1}{2n}t_1, f_o\left(\frac{1}{2m}s_1 + \frac{1}{2n}t_1\right)\right] =$$
$$\left[\frac{1}{2m}s_2 + \frac{1}{2n}t_2, f_o\left(\frac{1}{2m}s_2 + \frac{1}{2n}t_2\right)\right] \oplus \frac{1}{2mn}[x, f^*(x)] \oplus \frac{1}{2mn}[y, f^*(y)].$$

Comparing this to (2.2.15) and again using the cancellativity of $\oplus$, we have

$$\frac{1}{2mn}[x + y, f^*(x + y)] = \frac{1}{2mn}[x, f^*(x)] \oplus \frac{1}{2mn}[y, f^*(y)],$$

which produces the desired result.

The fact that (J^*, f^*) is in $\mathcal{O}$ and is a proper extension of (S_o, f_o) again contradicts the maximality of (S_o, f_o). Hence case (2) can not occur.

This finishes the verification of step 4, showing that $S_o = J^*$. Now G has the form (2.2.6) where f is the restriction of f_o to J. This completes the proof of the theorem.

Remark 2.2.5. As Ng (1974) points out in his original paper, the interval $J =]0, 1[^k \subset \mathbb{R}^k$ can be replaced by some other sets and the theorem remains true, even with the same proof. For instance J can be replaced by $[0, 1]^k$, by the positive cone $\mathcal{P}$ of $\mathbb{R}^k$, or by $\mathbb{R}^k$ itself. Also, $\mathbb{R}^k$ can be replaced by more general $\mathbb{Q}$-linear spaces, and so can $\mathbb{R}$ as the codomain.

We wish to point out that in $\mathbb{R}^k$ or any $\mathbb{Q}$-linear space, the set J can be replaced by any set which is "starlike" with respect to the origin. That means that for any $x \in J$, the open segment $\{\lambda x \mid 0 < \lambda < 1\}$ is also in J. In fact, all that is required is that all points λx for $\lambda \in \mathbb{Q}$ and $0 < \lambda < 1$ be in J. If 0 is included, then the same proof works, only Step 1 of the proof is not necessary.

Remark 2.2.6. The quantity $G(x, y) = f(x) + f(y) - f(x + y)$ is called a Cauchy difference, because the equation $f(x) + f(y) = f(x + y)$ is called the (fundamental) Cauchy functional equation. Any solution of the Cauchy equation is called an additive function, and so G may be considered as a measure of how much f deviates from being additive. There is an extensive literature on the cocycle equation and Cauchy differences, some of which will be mentioned in section 2.5.

2.3. Symmetric branching information measures

In this section, we present the explicit forms of all measures of information, on the open domain, which are branching and 4-symmetric. As we observed earlier, nearly all measures of information which have found useful applications possess these two fundamental properties. Therefore, the theorem we present is really the foundation for further developments in this book. It will turn out, just as on the closed domain (cf. Remark 2.1.3), that all measures with these two basic properties have the sum form.

Theorem 2.3.1. *A measure* $\{I_n\}$ *of information has the branching property* (2.1.1) *and is 4-symmetric, if and only if there is a map* $g : J \to \mathbb{R}$ *for which* I_n *has the sum form*

$$I_n(p_1, \ldots, p_n) = \sum_{i=1}^{n} g(p_i), \quad P \in \Gamma_n^o, \tag{2.3.1}$$

for all $n = 2, 3, \ldots$.

Proof: Suppose $\{I_n\}$ is branching and 4-symmetric. By Lemma 2.2.1, the generating function G satisfies the characteristic branching equation (2.2.2) and is symmetric (2.2.3). By Remark 2.2.3, the system (2.2.2) - (2.2.3) is equivalent to the system consisting of (2.2.3) and the cocycle equation (2.2.5). The general solution of this system is given by Theorem 2.2.4, and we have

$$G(x, y) = f(x) + f(y) - f(x + y), \qquad (x, y) \in D_2. \qquad (2.2.6)$$

Hence, by Lemma 2.1.2, we have

$$I_n(p_1, p_2, \ldots, p_n)$$

$$= I_2(1 - p_n, p_n) + \sum_{j=2}^{n-1} [f(p_1 + \ldots + p_{j-1}) + f(p_j) - f(p_1 + \ldots + p_j)]$$

$$= I_2(1 - p_n, p_n) + \sum_{j=1}^{n-1} f(p_j) - f(p_1 + \ldots + p_{n-1})$$

$$= I_2(1 - p_n, p_n) + \sum_{j=1}^{n-1} f(p_j) - f(1 - p_n),$$

for all $P \in \Gamma_n^\circ$ $(n = 3, 4, \ldots)$. Defining a new function $h : J \to \mathbb{R}$ by $h(p) := I_2(1 - p, p) - f(1 - p)$, we get

$$I_n(p_1, \ldots, p_n) = \sum_{j=1}^{n-1} f(p_j) + h(p_n), \quad P \in \Gamma_n^\circ \ (n \geq 3). \qquad (2.3.2)$$

Until now, we have not used the full strength of the 4-symmetry. Now we do so, namely,

$$\sum_{j=1}^{3} f(p_j) + h(p_4) = I_4(p_1, p_2, p_3, p_4)$$

$$= I_4(p_1, p_2, p_4, p_3)$$

$$= \sum_{j=1}^{2} f(p_j) + f(p_4) + h(p_3).$$

In other words,

$$h(p_4) - f(p_4) = h(p_3) - f(p_3), \quad (p_3, p_4) \in D_2.$$

Thus, by Lemma 1.2.1, we have

$$h(p) = f(p) + c, \qquad p \in J,$$

for some constant c. With this, (2.3.2) takes the form

$$\begin{aligned}
I_n(p_1, \ldots, p_n) &= \sum_{j=1}^{n} f(p_j) + c \\
&= \sum_{j=1}^{n} [f(p_j) + c\pi_{1j}],
\end{aligned}$$

where π_{1j} is the first component of p_j ($j = 1, \ldots, n$). Defining now $g : J \to \mathbb{R}$ by

$$g(p) := f(p) + c\pi_1, \qquad p \in J, \tag{2.3.3}$$

where π_1 is the first component of p, we have therefore (2.3.1) for all $n \geq 3$.

It only remains to show that (2.3.1) holds also for $n = 2$. For this, we observe first that

$$G(x, y) = g(x) + g(y) - g(x + y), \quad (x, y) \in D_2, \tag{2.3.4}$$

follows from (2.2.6) and (2.3.3). (It can also be seen by comparing (2.3.1) with the definition (2.1.1) of the branching property.) Now let $(p_1, p_2) \in \Gamma_2^\circ$, and choose $x, y \in J$ so that $x + y = p_1$. So also $(x, y) \in D_2$ and $(x, y, p_2) \in \Gamma_3^\circ$. By (2.3.1) for $n = 3$, (2.3.4), and the branching property (2.1.1), we have

$$\begin{aligned}
I_2(p_1, p_2) &= I_2(x + y, p_2) \\
&= I_3(x, y, p_2) - G(x, y) \\
&= g(x) + g(y) + g(p_2) - [g(x) + g(y) - g(x + y)] \\
&= g(p_2) + g(p_1).
\end{aligned}$$

Thus (2.3.1) holds also for $n = 2$.

Conversely, it is easy to see that any $\{I_n\}$ of the form (2.3.1) is 4-symmetric and branching, with generating function G given by (2.3.4). This completes the proof of the theorem.

This is the cornerstone of our theory, and it (or Theorem 2.2.4, on which it is based) will be applied in every other chapter of this book.

2.4. Partially symmetric branching information measures

The theorem in the previous section is the principal result concerning branching measures of information, especially from the standpoint of applications. However, one could ask also what happens when the full strength of symmetry is not used. The results in this section show that one can get very nearly the sum property in such cases.

The problem we consider now is that of determining explicit forms of measures of information which satisfy the branching property and a weak symmetry property. The process of solving this problem is again complicated by the fact that we have no zeros in our domain to use for substitutions. Next we introduce the weak form of symmetry we shall use.

We say that $\{I_n\}$ is *4-weak-symmetric* (cf. (2.2.4)) if

$$I_4(p_1, p_2, p_3, p_4) = I_4(p_1, p_3, p_2, p_4), \quad P \in \Gamma_4^\circ. \tag{2.4.1}$$

Lemma 2.4.1. *Suppose $\{I_n\}$ has the branching property (2.1.1). Then $\{I_n\}$ is 4-weak-symmetric if and only if the generating function G satisfies the characteristic functional equation (cf. (2.2.2)) for branching measures*

$$G(x + y, z) + G(x, y) = G(x + z, y) + G(x, z), \quad (x, y, z) \in D_3. \tag{2.4.2}$$

Proof: Suppose $\{I_n\}$ is branching and 4-weak-symmetric. We have already observed in Remark 2.2.2 that only the 4-weak-symmetry is needed for (2.4.2). Conversely, the calculation made in the proof of Lemma 2.2.1 shows that (2.4.2) implies the 4-weak-symmetry. This completes the proof.

So an essential step in solving our problem is to find the solution of the characteristic equation (2.4.2). To do this we first prove a lemma which was proved in a more general setting in Ebanks (1989b).

Lemma 2.4.2. *A map $H : D_2 \to \mathbb{R}$ satisfies*

$$H(p, s) = H(p, q) + H(q, s), \quad (p, q, s) \in D_3, \tag{2.4.3}$$

if and only if there is a map $h : J \to \mathbb{R}$ for which

$$H(p, q) = h(p) - h(q), \quad (p, q) \in D_2. \tag{2.4.4}$$

Remark 2.4.3. On the closed domain

$$\overline{D}_3 = \left\{ P = (p_1, p_2, p_3) \mid p_1, p_2, p_3, \sum_{i=1}^{3} p_i \in \overline{J} \right\},$$

it is simple to find the solutions of (2.4.3). Putting $s = 0$ in (2.4.3), we get immediately $H(p, q) = H(p, 0) - H(q, 0)$. The absence of a zero makes the proof considerably more difficult on the open domain D_3. If we fix $t \in J$ and define

$$h(p) := H(p, t), \tag{2.4.5}$$

then there are two difficulties. First, this defines h only on a subset of J (since $p + t$ must be in J). Second, (2.4.3) with $s = t$ yields

$$H(p, q) = H(p, t) - H(q, t) = h(p) - h(q)$$

only when $p + q + t$ is in J. Since t is fixed, we cannot define h as in (2.4.5).

On the other hand, the function h used in representation (2.4.4) is determined only up to an additive constant. This suggests the idea of defining $h(p) := H(p, t) - H(c, t)$ for some appropriate constant c and "constant" t, depending on p.

Proof of Lemma 2.4.2. Let $p \in J$ and choose $t \in J$ so that

$$0 < t < \min\left\{ \frac{1}{2}(1 - p), \frac{3}{8} \right\}. \tag{2.4.6}$$

Then $(p, t), (\frac{1}{4}, t) \in D_2$, and we define $h : J \to \mathbb{R}$ by

$$h(p) := H(p, t) - H\left(\frac{1}{4}, t \right), \quad p \in J. \tag{2.4.7}$$

To show that h is well-defined, let us consider t_1 and t_2 satisfying (2.4.6). Then $t_1 + t_2 < 1 - p$ and $t_1 + t_2 < \frac{3}{4}$. Using these two facts and applying (2.4.3) twice, we get

$$H(p, t_1) - H\left(\frac{1}{4}, t_1 \right) = H(p, t_2) - H(t_1, t_2) - H\left(\frac{1}{4}, t_1 \right)$$

$$= H(p, t_2) - H\left(\frac{1}{4}, t_2 \right),$$

which shows that h is well-defined by (2.4.7).

Now, to show that (2.4.4) holds, we let $(p, q) \in D_2$ and fix t such that

$$0 < t < \min\left\{ \frac{1}{2}(1 - p), \frac{1}{2}(1 - q), \frac{3}{8}, 1 - p - q \right\}.$$

Then we calculate by (2.4.3) and (2.4.7) that

$$H(p,q) = H(p,t) - H(q,t)$$
$$= \left[H(p,t) - H\left(\frac{1}{4},t\right) \right] - \left[H(q,t) - H\left(\frac{1}{4},t\right) \right]$$
$$= h(p) - h(q),$$

which is (2.4.4).

Now we are ready to solve (2.4.2).

The next result, which appeared in Ebanks (1989), is the essential step in finding 4-weak-symmetric branching measures of information.

Theorem 2.4.4. *The general solution of the characteristic branching equation (2.4.2),*

$$G(x + y, z) + G(x, y) = G(x + z, y) + G(x, z), \quad (x, y, z) \in D_3,$$

for $G : D_2 \to \mathbb{R}$*, is given by*

$$G(x, y) = f(x) + g(y) - f(x + y), \quad (x, y) \in D_2, \qquad (2.4.8)$$

for arbitrary maps $f, g : J \to \mathbb{R}$*.*

Proof: Maps G of the form (2.4.8) clearly satisfy (2.2.2). We verify the converse.

Our first step is to create from G a symmetric function satisfying (2.2.2), since we can then apply Theorem 2.2.4. Interchanging x and y in (2.2.2) and subtracting the result from (2.2.2), we obtain

$$G(x, y) - G(y, x) = G(x, z) - G(y, z) + G(x + z, y) - G(y + z, x).$$

Alternatively, the substitution of (z, x, y) for (x, y, z) in (2.2.2) produces

$$G(z + x, y) - G(z + y, x) = G(z, y) - G(z, x).$$

Comparing these last two equations, we have

$$H(x, y) = H(x, z) + H(z, y), \quad (x, y, z) \in D_3,$$

for the map $H : D_2 \to \mathbb{R}$ defined by

$$H(x, y) := G(x, y) - G(y, x), \quad (x, y) \in D_2. \qquad (2.4.9)$$

That is, H satisfies (2.4.3). Thus, by Lemma 2.4.2, there exists an $h : J \to \mathbb{R}$ such that (2.4.4) holds, i.e.

$$H(x, y) = h(x) - h(y), \quad (x, y) \in D_2.$$

Now, defining the map $S : D_2 \to \mathbb{R}$ by

$$S(x, y) := G(x, y) + h(y), \quad (x, y) \in D_2, \tag{2.4.10}$$

we see from (2.4.9) and (2.4.4) that S is symmetric,

$$S(x, y) = S(y, x), \quad (x, y) \in D_2.$$

Moreover, (2.2.2) and (2.4.10) show that S also satisfies (2.2.2), viz.

$$\begin{aligned}
S(x + y, z) + S(x, y) &= G(x + y, z) + h(z) + G(x, y) + h(y) \\
&= G(x + z, y) + h(y) + G(x, z) + h(z) \\
&= S(x + z, y) + S(x, z).
\end{aligned}$$

According to Remark 2.2.3 and Theorem 2.2.4, there exists a map $f : J \to \mathbb{R}$ for which

$$S(x, y) = f(x) + f(y) - f(x + y), \quad (x, y) \in D_2. \tag{2.4.11}$$

Returning now to (2.4.10), we have by (2.4.11)

$$G(x, y) = f(x) + f(y) - h(y) - f(x + y), \quad (x, y) \in D_2.$$

Defining $g := f - h$, this is (2.4.8), and we are done.

We can now state our characterization theorem for branching, weakly symmetric measures of information.

Theorem 2.4.5. *A measure $\{I_n\}$ of information has the branching property (2.1.1) and is 4-weak-symmetric (2.4.1), if and only if there exist maps $f, g, h : J \to \mathbb{R}$ such that*

$$I_n(P) = f(p_1) + \sum_{j=2}^{n-1} g(p_j) + h(p_n), \quad P \in \Gamma_n^\circ. \tag{2.4.12}$$

In this case, the generating function G is given by (2.4.8):

$$G(x,y) = f(x) + g(y) - f(x+y), \quad (x,y) \in D_2.$$

Proof: $\{I_n\}$ of the form (2.4.8) obviously has the branching property with generating function G defined by (2.4.12). Moreover, $\{I_n\}$ is clearly 4-weak-symmetric.

For the converse, G must satisfy (2.4.2) by Lemma 2.4.1. Thus G is of the form (2.4.8) for some maps $f, g : J \to \mathbb{R}$, by Theorem 2.4.4. Now we apply Lemma 2.1.2 and substitute (2.4.8) into (2.1.2). This gives

$$
\begin{aligned}
I_n(P) &= I_2(1 - p_n, p_n) + \sum_{j=2}^{n-1} [f(p_1 + \ldots + p_{j-1}) \\
&\quad + g(p_j) - f(p_1 + \ldots + p_j)] \\
&= I_2(1 - p_n, p_n) + f(p_1) + \sum_{j=2}^{n-1} g(p_j) - f(1 - p_n),
\end{aligned}
$$

for all $P = (p_1, p_2, \cdots, p_n) \in \Gamma_n^\circ$, $n \geq 3$. Defining $h : J \to \mathbb{R}$ by

$$h(p) := I_2(1 - p, p) - f(1 - p), \quad p \in J,$$

we therefore have (2.4.12) for $n = 3, 4, \ldots$. The only thing left is to show that (2.4.12) holds for $n = 2$.

For this, we go back again to (2.1.1). Let $(p_1, p_2) \in \Gamma_2^\circ$, then choose $x, y \in J$ so that $x + y = p_1$. So $(x, y, p_2) \in \Gamma_3^\circ$, and we have

$$
\begin{aligned}
I_2(p_1, p_2) &= I_2(x + y, p_2) = I_3(x, y, p_2) - G(x, y) \\
&= f(x) + g(y) + h(p_2) - [f(x) + g(y) - f(x+y)] \\
&= f(p_1) + h(p_2),
\end{aligned}
$$

which is (2.4.12) for $n = 2$.

2.5 Historical remarks

We have called equation (2.2.5) the cocycle equation because in appearance it is the same as the equation of a cocycle in homological algebra (cf. Kurosh, 1956, pp. 48-50). In fact equation (2.2.5) first appeared as a factor system in the theory of group extensions (cf. Serre, 1959, p. 168). It is because of such connections that algebraic and set theoretical considerations were needed to establish Theorem 2.2.4. Both the history and bibliography of (2.2.5) are quite extensive.

Let us mention here the important papers of Schreier (1926a,b) and Baer (1934). S. Kurepa (1956) gave the differentiable solutions of (2.2.5) (among others) for functions $F : \mathbb{R}^2 \to \mathbb{R}$. Shortly after this, J. Erdős (1959) gave the general solution of the system of (2.2.5) and (2.2.3) among functions $F : \mathbb{R}^2 \to \mathbb{R}$.

In order to construct a simplified proof of the Dehn-Sydler Theorem on polyhedra, Jessen, Karpf and Thorup (1968) gave (among other things) the general solution of (2.2.5) and (2.2.3) for $F : G^2 \to X$ in the following cases: (a) G an abelian group, X a divisible abelian group; (b) G a free abelian group, X any abelian group; and (c) G the positive cone of an ordered abelian group, X a divisible abelian group. The proof of Erdős (1959) also works for case (a).

Concerning equation (2.2.5) alone, Erdős (1959) demonstrated the existence of solutions $F : \mathbb{R}^2 \to \mathbb{R}$ not of the form $F(x,y) = f(x+y) - f(x) - f(y)$. The general solution, containing an arbitrary skew-symmetric biadditive function, can be found in Hosszú (1963) and Aczél (1965).

Further results concerning the solution of (2.2.5) on certain classes of semigroups can be found in Ebanks (1979, 1981, 1982b, 1983b, 1984b) and in Davison and Ebanks (1996).

None of the proofs mention above works for maps F defined on such restricted domains as D_2. The result of Ng (1974) was the first in this direction, and it was motivated precisely by the branching property in the probabilistic theory of information. This result was proved with the aid of some boundary points, but we have seen now how it can be modified so that it is valid also on the open domain. Other extensions in the context of the probabilistic theory of information can be found in Forte and Bortone (1977), Ng (1979), and Ebanks (1979, 1989b).

For related results in the context of more general theories of information (with and without probability), see Forte and Ng (1974), Ebanks (1978, 1982a, 1990), Davidson and Ng (1981), Ebanks and Maksa (1986), Ebanks, Kannappan and Ng (1988) and Sander (1988).

CHAPTER 3

RECURSIVITY PROPERTIES

3.1. Introduction

In Chapter 2 we studied branching information measures, that is, measures $\{I_n\}$ satisfying

$$I_n(p_1, p_2, \ldots, p_n) = I_{n-1}(p_1 + p_2, p_3, \ldots, p_n) + G(p_1, p_2), \qquad P \in \Gamma_n^o,$$

for some function $G : D_2 \to \mathbb{R}$. It turned out that all 4-symmetric branching information measures $\{I_n\}$ have the sum form

$$I_n(P) = \sum_{j=1}^{n} g(p_j)$$

for some map $g : J \to \mathbb{R}$.

To arrive at some explicit information measures we need to know the possible form of the function g. On the other hand, the branching property gives us no further information about this function. With an arbitrary g, the branching property is satisfied.

Let us recall the motivation for the branching property which was presented in the introduction to Chapter 2. There we compared the respective information contents of two experiments A and B. Experiment A had outcomes $E_1, \ldots, E_n$, with corresponding probabilities $p_1, \ldots, p_n$. In B, we did not distinguish between events E_1 and E_2, but rather lumped them into a single event $E_1 \cup E_2$. So the outcomes of B are $E_1 \cup E_2, E_3, \ldots, E_n$, with corresponding probabilities $p_1 + p_2, p_3, \ldots, p_n$. Now, how much information is lost by performing B instead of A? The branching property makes the rather weak assumption that the information loss is some function of the probabilities of E_1 and E_2, but any function will do.

Now let us think a bit further. The information "lost" by performing B instead of A (if the outcome is $E_1 \cup E_2$) can be recaptured if we then perform a third experiment C to distinguish between the outcomes E_1 and E_2. This experiment C has only two outcomes, E_1 and E_2, with corresponding (conditional) probabilities $\frac{p_1}{p_1+p_2}$ and $\frac{p_2}{p_1+p_2}$. On the other hand, we only have to perform the additional experiment C if the outcome of B is $E_1 \cup E_2$. That is, the probability that we need to perform C is $p_1 + p_2$. Thus we arrive at the conclusion that the information lost by performing B instead of A is determined by the information content of C, weighted by the probability $p_1 + p_2$. The recursivity property which follows assumes that this weighting is accomplished by multiplying the information content of C by some function of $p_2 + p_2$.

Definition 3.1.1. A measure $\{I_n\}$ of information is said to be *recursive of type M*, for some function $M : J \to \mathbb{R}$, if

$$
\begin{aligned}
I_n(p_1, \ldots, p_n) = & I_{n-1}(p_1 + p_2, p_3, \ldots, p_n) \\
& + M(p_1 + p_2) I_2 \left(\frac{p_1}{p_1 + p_2}, \frac{p_2}{p_1 + p_2} \right)
\end{aligned}
\tag{3.1.1}
$$

for all $P \in \Gamma_n^o$ ($n = 3, 4, \ldots$). That is, $\{I_n\}$ is branching with generating function G given by

$$
G(x, y) = M(x + y) I_2 \left(\frac{x}{x+y}, \frac{y}{x+y} \right), \qquad (x, y) \in D_2.
$$

The next question is: What type of function should M be? Before giving our answer (in the next section), let us examine the types which have been proposed.

In a seminal paper in the characterization theory of information measures, Fadeev (1956) used (3.1.1) in the case $k = 1$ with $M(p) = p$ (along with several other properties of $\{I_n\}$) to characterize the Shannon entropy. The fact that Shannon's entropy has this property had been observed earlier by Shannon himself. So this is the "classical" property of recursivity, which may be considered as the linear case.

Havrda and Charvat (1967) introduced the family (again for $k = 1$) of recursivities obtained by taking $M(p) = p^\alpha$, for some real number α, in (3.1.1). They obtained the α-entropies, now called entropies of degree α, and proposed their usage in the measurement of classification schemes.

We formalize the types of recursivity described above in the following definition.

Definition 3.1.2. An information measure $\{I_n\}$ is *recursive of degree α*, $\alpha \in \mathbb{R}^k$, if it satisfies (3.1.1) with $M(x) = x^\alpha$, $x \in J$, for all $P \in \Gamma_n^o$ ($n = 3, 4, \ldots$).

Most of the information measures $\{I_n\}$ listed in Chapter 1 satisfy the α-recursivity condition. For example, the Shannon entropy is recursive of degree 1; the entropy of degree α ($\neq 1$) is recursive of degree α; Kullback's directed divergence is recursive of degree (1,0); Kerridge's inaccuracy is recursive of degree (1,0); and Theil's information improvement is recursive of degree (1,0,0). We have also seen error functions and directed divergences of degree α ($\neq 1$); these, too, are recursive of degree α.

Also, let us mention here that α-recursivity is connected to the Huffmann coding procedure (see for example Guiasu (1977)).

3.2. Recursivity of multiplicative type

Now we propose our answer to the question of what the proper type of function M should be for recursivity.

Definition 3.2.1. An information measure $\{I_n\}$ is *recursive of multiplicative type* M, if $\{I_n\}$ satisfies (3.1.1) for all $p \in \Gamma_n^o$, ($n = 3, 4, \ldots$), where the map $M : J \to \mathbb{R}$ is multiplicative.

There are two reasons for considering recursivity of multiplicative type. The first is one of convenience: For any $\alpha \in \mathbb{R}^k$, the map $M(p) = p^\alpha$ is multiplicative. Thus we can include α-recursivity for every α as a special case of recursivity of multiplicative type. Moreover in (3.1.1) we have no dependence on any parameter. But this is a superficial reason. There is a better and deeper reason to assume that M is multiplicative.

Let us suppose that $\{I_n\}$ satisfies (3.1.1) where $M : J \to \mathbb{R}$ now is an arbitrary function. If we assume, in addition to (3.1.1), the similar equation (for similar reasons)

$$I_4(p_1, p_2, p_3, p_4) = I_2(p_1 + p_2 + p_3, p_4)$$

$$+ M\left(\sum_{j=1}^{3} p_j\right) I_3\left(\frac{p_1}{\sum_{j=1}^{3} p_j}, \frac{p_2}{\sum_{j=1}^{3} p_j}, \frac{p_3}{\sum_{j=1}^{3} p_j}\right), \qquad (3.2.1)$$

then (3.1.1) and (3.2.1) imply that M is multiplicative, as we will now show. Without loss of generality, we may assume that $I_2(1 - a, a) \neq 0$ for some $a \in J$, since otherwise (3.1.1) implies that all I_n are 0. Let $p, q \in J$ and $a \in J$ such that $I_2(1-a, a)$ is not equal to zero. Then $(pq(1-a), pqa, p(1-q), 1-p) \in \Gamma_4^o$, $(q(1-a), qa, 1-q) \in \Gamma_3^o$ and $(pq, p(1-q), 1-p) \in \Gamma_3^o$.

Several applications of (3.1.1) and (3.2.1) imply

$$I_2(p,\, 1-p) + M(p)\, I_2(q,\, 1-q) + M(pq)\, I_2(1-a, a)$$
$$= I_3(pq,\, p(1-q), 1-p) + M(pq)\, I_2(1-a,a)$$
$$= I_4(pq(1-a),\, pqa,\, p(1-q), 1-p)$$
$$= I_2(p, 1-p) + M(p)\, I_3(q(1-a),\, qa,\, 1-q)$$
$$= I_2(p, 1-p) + M(p)\, I_2(q,\, 1-q) + M(p)\, M(q)\, I_2(1-a,\, a).$$

A comparison of the first member with the last member in the above sequence of equations yields

$$M(pq)I_2(1-a,a) = M(p)M(q)I_2(1-a,a)$$

which implies that M is multiplicative since $I_2(1-a,a) \neq 0$.

Henceforth, because of the considerations above, the only type of recursivity we consider is the multiplicative type. We shall present the explicit forms of all 3-symmetric (and, later, partially symmetric) measures $\{I_n\}$ of this type. Our first step toward this goal is the following.

Lemma 3.2.2. *Let $\{I_n\}$ be a 3-symmetric, recursive information measure of multiplicative type M. Then $\{I_n\}$ is 4-symmetric and branching, with generating function $G : D_2 \to \mathbb{R}$ defined by*

$$G(x,y) := M(x+y)\, I_2\left(\frac{x}{x+y}, \frac{y}{x+y}\right), \qquad (x,y) \in D_2, \qquad (3.2.2)$$

and satisfying the functional equations

$$G(x,y) = G(y,x), \qquad\qquad (3.2.3)$$

$$G(x,y) + G(x+y,z) = G(x,y+z) + G(y,z), \qquad\qquad (3.2.4)$$

$$G(tx,ty) = M(t)G(x,y), \qquad\qquad (3.2.5)$$

whenever $x, y, z, x+y+z, t \in J$.

Proof: First, we show that $\{I_n\}$ is 4-symmetric. For any $(p_1, p_2, p_3, p_4) \in \Gamma_4^o$, we have by (3.1.1)

$$I_4(p_1, p_2, p_3, p_4) = I_3(p_1 + p_2, p_3, p_4) + M(p_1 + p_2)I_2\left(\frac{p_1}{p_1 + p_2}, \frac{p_2}{p_1 + p_2}\right)$$

$$= I_2(1 - p_4, p_4) + M(p_1 + p_2 + p_3)I_2\left(\frac{p_1 + p_2}{p_1 + p_2 + p_3}, \frac{p_3}{p_1 + p_2 + p_3}\right)$$

$$+ M(p_1 + p_2)I_2\left(\frac{p_1}{p_1 + p_2}, \frac{p_2}{p_1 + p_2}\right)$$

$$= I_2(1 - p_4, p_4) + M(p_1 + p_2 + p_3)\left[I_2\left(\frac{p_1 + p_2}{p_1 + p_2 + p_3}, \frac{p_3}{p_1 + p_2 + p_3}\right)\right.$$

$$\left. + M\left(\frac{p_1 + p_2}{p_1 + p_2 + p_3}\right)I_2\left(\frac{p_1}{p_1 + p_2}, \frac{p_2}{p_1 + p_2}\right)\right]$$

$$= I_2(1 - p_4, p_4) + M(1 - p_4)I_3\left(\frac{p_1}{1 - p_4}, \frac{p_2}{1 - p_4}, \frac{p_3}{1 - p_4}\right).$$

By 3-symmetry, the top line shows that $I_4(p_1, p_2, p_3, p_4)$ is symmetric in p_3 and p_4, and the last line shows that it is symmetric in p_1, p_2, p_3. Hence I_4 is symmetric.

Now, defining G by (3.2.2), we have the branching property also for $\{I_n\}$, and so by Lemma 2.2.1 and Remark 2.2.3, we have (3.2.3) and (3.2.4).

Finally, (3.2.5) follows immediately from (3.2.2). Given any $t \in J$ and $(x, y) \in D_2$,

$$G(tx, ty) = M(tx + ty)I_2\left(\frac{tx}{tx + ty}, \frac{ty}{tx + ty}\right)$$

$$= M(t)M(x + y)I_2\left(\frac{x}{x + y}, \frac{y}{x + y}\right)$$

$$= M(t)G(x, y),$$

and this completes the proof.

We have found in Chapter 2 (cf. Theorem 2.2.4 and Remark 2.2.5) that the general solution of (3.2.3), (3.2.4) is given by (2.2.6)

$$G(x, y) = f(x) + f(y) - f(x + y), \qquad (x, y) \in D_2,$$

for an arbitrary function $f : J \to \mathbb{R}$. Now it turns out that with the additional condition (3.2.5), the function f must have a very special form depending on the multiplicative function M.

As it happens, the solution of our system (3.2.3) - (3.2.5) is much simpler in the case that M is not additive. We deal with that case first, then turn to the consideration of the additive case.

We shall find it convenient to work with the system (3.2.3) - (3.2.5) on the positive cone $\mathcal{P}$ of $\mathbb{R}^k$, and so we close this section with an extension theorem. Recall (Lemma 1.2.3) that any multiplicative M on J can be extended to a multiplicative $\overline{M}$ on $\mathcal{P}$.

Lemma 3.2.3. *Every solution $M : J \to \mathbb{R}$ (multiplicative), $G : D_2 \to \mathbb{R}$ of the system (3.2.3)–(3.2.5) for $x, y, z, x + y + z, t \in J$ can be extended to a solution $\overline{M} : \mathcal{P} \to \mathbb{R}$ (multiplicative), $\overline{G} : \mathcal{P} \times \mathcal{P} \to \mathbb{R}$ of (3.2.3)–(3.2.5) for $x, y, z, t \in \mathcal{P}$. If, moreover, M is additive on D_2, then $\overline{M}$ is additive on $\mathcal{P} \times \mathcal{P}$.*

Proof: We have already seen how to extend M on J to $\overline{M}$ on $\mathcal{P}$ in Lemma 1.2.3. There, we saw too that if M is also additive, then so is $\overline{M}$.

If $M \equiv 0$, then (3.2.5) shows that $G \equiv 0$. The extension $\overline{G} \equiv 0$ obviously satisfies (3.2.3)–(3.2.5) on $\mathcal{P}$.

If $M \not\equiv 0$, then (Lemma 1.2.2) M is nowhere zero, and the same is true of $\overline{M}$. Now, for any $x, y \in \mathcal{P}$, choose $t \in J$ such that $(tx, ty) \in D_2$ and define

$$\overline{G}(x, y) := M(t)^{-1} G(tx, ty). \tag{3.2.6}$$

We show that $\overline{G}$ is well-defined by (3.2.6). Suppose $s \in J$ with $(sx, sy) \in D_2$. Then $(tsx, tsy) \in D_2$ and we have by (3.2.5)

$$M(s)G(tx, ty) = G(stx, sty) = M(t)G(sx, sy).$$

Thus $M(t)^{-1}G(tx, ty) = M(s)^{-1}G(sx, sy)$, and (3.2.6) defines $\overline{G}$ on $\mathcal{P} \times \mathcal{P}$ unambiguously.

To verify (3.2.3) for $\overline{G}$, let $x, y \in \mathcal{P}$ and choose $t \in J$ such that $(tx, ty) \in D_2$. Then, by (3.2.3) for G and (3.2.6), we have

$$\overline{G}(x, y) = M(t)^{-1}G(tx, ty) = M(t)^{-1}G(ty, tx) = \overline{G}(y, x).$$

Similarly, for $x, y, z \in \mathcal{P}$, select any $t \in J$ for which $t(x + y + z) \in J$. Then

$$\begin{aligned}
\overline{G}(x, y) + \overline{G}(x + y, z) &= M(t)^{-1}[G(tx, ty) + G(tx + ty, tz)] \\
&= M(t)^{-1}[G(tx, ty + tz) + G(ty, tz)] \\
&= \overline{G}(x, y + z) + \overline{G}(y, z),
\end{aligned}$$

which demonstrates (3.2.4) for $\overline{G}$.

Finally, given any $t, x, y \in \mathcal{P}$, choose $p, q \in J$ such that $t = \frac{p}{q}$; then choose $s \in J$ with $(sx, sy), (stx, sty) \in D_2$. Now $(psx, psy) \in D_2$, and by (3.2.5) for G we get

$$
\begin{aligned}
M(q)\overline{G}(tx, ty) &= M(q)M(s)^{-1}G(stx, sty) \\
&= M(s)^{-1}G(psx, psy) \\
&= M(s)^{-1}M(p)G(sx, sy) \\
&= M(p)\overline{G}(x, y),
\end{aligned}
$$

Recalling that the extension of M to $\overline{M}$ was accomplished via $\overline{M}(t) = \overline{M}(\frac{p}{q}) = M(p)M(q)^{-1}$, we have

$$
\overline{G}(tx, ty) = \overline{M}(t)\overline{G}(x, y),
$$

which is (3.2.5) for $\overline{G}$ and $\overline{M}$. This completes the proof of the lemma.

3.3. Nonadditive multiplicative types

In this section, we first solve the system (3.2.3) - (3.2.5) in the case that M is not additive. We follow an idea of Sander (1988), which is similar to an argument of Aczél and Ng (1983), where we try to represent G through M in the same way that a symmetric bilinear form can be represented through its diagonal (a quadratic form).

Then immediately after, we give the form of all 3-symmetric $\{I_n\}$ which are recursive of nonadditive multiplicative type.

Theorem 3.3.1. *Let* $M : J \to \mathbb{R}$ *be multiplicative but not additive. Then the general solution of (3.2.3)–(3.2.5) for* $x, y, z, t, x + y + z \in J$ *is given by*

$$
G(x, y) = c[M(x) + M(y) - M(x + y)], \qquad (x, y) \in D_2. \qquad (3.3.1)
$$

for some constant $c \in \mathbb{R}$.

Proof: Evidently, any G of the form (3.3.1) satisfies (3.2.3)- (3.2.5). For the converse, let M be uniquely extended to a multiplicative function on P such that $M(1) = 1$. (See Lemma 1.2.3.) Since M is not additive, we can choose a constant $a \in J$ such that

$$
b := M(a) + M(1 - a) - 1 \neq 0.
$$

Let $(x, y) \in D_2$. Since $(y, xa, x(1 - a)) \in D_3$, several applications of (3.2.3) and (3.2.4) lead to

$$
\begin{aligned}
&G(x, y) + G(xa, x(1 - a)) \\
&= G(y, xa + x(1 - a)) + G(xa, x(1 - a)) \\
&= G(y, xa) + G(y + xa, x(1 - a)) \\
&= G(y(1 - a) + ya, xa) + G((x + y)a + y(1 - a), x(1 - a)) \\
&= G(y(1 - a), (y + x)a) + G(ya, xa) - G(y(1 - a), ya) \\
&\quad + G(y(1 - a), x(1 - a)) - G((x + y)a, y(1 - a)) \\
&\quad + G((x + y)a, (x + y)(1 - a)) \\
&= G(ax, ay) - G(ya, y(1 - a)) \\
&\quad + G((x + y)a, (x + y)(1 - a)) + G(y(1 - a), x(1 - a)).
\end{aligned}
$$

Using (3.2.5) in the top line and the last two lines we arrive at

$$
-b \cdot G(x, y) = -G(a, 1 - a)(M(x) + M(y) - M(x + y)). \qquad (3.3.2)
$$

Thus we get (3.3.1) with $c = G(a, 1 - a)b^{-1}$ and the theorem is established.

Theorem 3.3.2. *Let $M : J \to \mathbb{R}$ be multiplicative but not additive. Then a measure $\{I_n\}$ of information is 3-symmetric and recursive of type M, if and only if there exists a constant $c \in \mathbb{R}$ such that*

$$
I_n(p_1, \ldots, p_n) = c \left[\sum_{i=1}^{n} M(p_i) - 1 \right], \qquad P \in \Gamma_n^o, \qquad (3.3.3)
$$

for all $n = 2, 3, \ldots$.

Proof: Clearly, any $\{I_n\}$ of the form (3.3.3) is both 3-symmetric and recursive of (multiplicative) type M.

For the converse, by Lemma 3.2.2 and Theorem 3.3.1, we have (3.3.1) for the map G defined by (3.3.2). That is,

$$
M(x + y)I_2 \left(\frac{x}{x + y}, \frac{y}{x + y} \right) = c[M(x) + M(y) - M(x + y)]
$$

for all $(x, y) \in D_2$. Since M is not additive, certainly $M \not\equiv 0$, hence (Lemma 1.2.2 again) M is nowhere zero. Thus,

$$I_2 \left(\frac{x}{x+y}, \frac{y}{x+y} \right)$$

$$= \frac{c[M(x) + M(y) - M(x+y)]}{M(x+y)}$$

$$= c \left[M \left(\frac{x}{x+y} \right) + M \left(\frac{y}{x+y} \right) - 1 \right],$$

which is (3.3.3) for $n = 2$.

Now we proceed by induction. Suppose (3.3.3) holds for some $n \geq 2$. Then, using the recursivity once more, we have

$$I_{n+1}(p_1, \ldots, p_{n+1})$$

$$= I_n(p_1 + p_2, p_3, \ldots, p_{n+1}) + M(p_1 + p_2)I_2 \left(\frac{p_1}{p_1 + p_2}, \frac{p_2}{p_1 + p_2} \right)$$

$$= c[M(p_1 + p_2) + \sum_{i=3}^{n+1} M(p_i) - 1] + M(p_1 + p_2)c \left[M \left(\frac{p_1}{p_1 + p_2} \right) \right.$$

$$\left. + M \left(\frac{p_2}{p_1 + p_2} \right) - 1 \right]$$

$$= cM(p_1 + p_2) + c \left[\sum_{i=3}^{n+1} M(p_i) - 1 \right]$$

$$+ c[M(p_1) + M(p_2) - M(p_1 + p_2)]$$

$$= c \left[\sum_{i=1}^{n+1} M(p_i) - 1 \right],$$

for all $P \in \Gamma_{n+1}^o$. This completes the proof.

So we see that if, for example, $M(p) = p^\alpha$, where $\alpha \in \mathbb{R}^k$ is *not* a basic unit vector (that is, a single 1 and the rest 0's), then all 3-symmetric information measures $\{I_n\}$ which are recursive of degree α are of this form

$$I_n(P) = c \left[\sum_{i=1}^{n} p_i^\alpha - 1 \right], \qquad P \in \Gamma_n^o.$$

It is not necessary to make any "regularity" assumption (say, continuity or measurability) about I_n in order to achieve this result. This was first realized (for $k = 1$ and $\alpha \neq 1$) by Daróczy (1970).

3.4. Additive multiplicative types

Now we consider the system (3.2.3)–(3.2.5) in the case that M is additive (in addition to being multiplicative). The original published solution in this case is a very long one which is spread over a couple of papers by different authors. (See historical remarks in the next section.) We present here a newer and shorter proof which appeared first in Ebanks (1995). Both this proof and the original one are based on the following result of Jessen, Karpf, and Thorup (1968), which handles dimension $k = 1$.

Theorem 3.4.1. *Let $k = 1$. The general solution $G : \mathcal{P}_1 \times \mathcal{P}_1 \to \mathbb{R}$ of the system (3.2.3), (3.2.4), and*

$$G(tx, ty) = tG(x, y), \tag{3.4.1}$$

for all $x, y, z, t \in \mathcal{P}_1 =]0, \infty[$, is given by

$$G(x, y) = h(x) + h(y) - h(x + y), \qquad x, y > 0, \tag{3.4.2}$$

where $h : \mathcal{P}_1 \to \mathbb{R}$ satisfies

$$h(uv) = uh(v) + vh(u), \qquad u, v > 0. \tag{3.4.3}$$

Proof: If G has the above representation (3.4.2) with some map $h : \mathcal{P}_1 \to \mathbb{R}$ satisfying (3.4.3) then straightforward calculations show that G satisfies (3.2.3), (3.2.4) and is t-homogeneous (3.4.1). To prove the converse we first give an outline of the steps involved.

Step 1. The function $G : \mathcal{P}_1^2 \to \mathbb{R}$ can be extended to a map $\overline{G} : \mathbb{R}^2 \to \mathbb{R}$ satisfying (3.2.3), (3.2.4) and (3.4.1) for all $x, y, z, t \in \mathbb{R}$.

Step 2. The triple $T := (\mathbb{R}^2, \oplus, \otimes)$ forms a commutative ring (with zero and unit elements $0_T = (0, 0)$ and $1_T = (1, 0)$, respectively) where the operations $\oplus$ and $\otimes$ on $\mathbb{R}^2$ are defined by $(x, y, u, v \in \mathbb{R})$

$$(x, u) \oplus (y, v) := (x + y, \ u + v - G(x, y)) \tag{3.4.4}$$

$$(x, u) \otimes (y, v) := (xy, \ vx + uy). \tag{3.4.5}$$

(The extension $\overline{G}$ will be denoted again by G). Moreover, if we define $n \cdot 1_T$, $n \in \mathbb{N}$, by

$$n \cdot (1,0) := (1,0) \oplus (1,0) \oplus \cdots \oplus (1,0) \quad (n - \text{fold})$$
$$(-n) \cdot (1,0) := \ominus (n \cdot (1,0)).$$
(3.4.6)

then $\mathbf{Z}^* := \{z \cdot (1,0) : z \in \mathbf{Z}\}$ is a subring of $T = (\mathbb{R}^2, \oplus, \otimes)$.

Step 3. The class $\mathcal{O} = \{(U,S)\}$ of pairs (U,S) satisfying
    01. $\mathbf{Z} \subset U \subset \mathbb{R}$, U is a subring of $(\mathbb{R}, +, \cdot)$,
    02. $\mathbf{Z}^* \subset S \subset T$, S is a subring of T,
    03. The projection $\phi : \mathbb{R}^2 \to \mathbb{R}$ given by

$$\phi(x,u) = x, \qquad x, u \in \mathbb{R}, \tag{3.4.7}$$

is a ring isomorphism of S onto U,

is nonempty and complete under the partial ordering $\prec$ defined by

$$(U,S) \prec (V,T) \iff U \subset V \text{ and } S \subset T.$$

By Zorn's Lemma there exists a maximal element (U_0, S_0) of $\mathcal{O}$.

Step 4. We show that $U_0 = \mathbb{R}$. Since $\varphi|_{S_0} : S_0 \to \mathbb{R} = U_0$ is bijective, we know that for all $x \in \mathbb{R}$ there exists a unique $u_x \in \mathbb{R}$ such that $(x, u_x) \in S_0$ and $\varphi(x, u_x) = x$. This means that there is a map $h : \mathbb{R} \to \mathbb{R}$ such that $h(x) = u_x$, $x \in \mathbb{R}$ and
$$S_0 = \{(x, h(x)) : x \in \mathbb{R}\}.$$

Using the fact that S_0 is a subring of T, we get for all $x, y \in \mathbb{R}$

$$(x + y, h(x) + h(y) - G(x,y)) = (x, h(x)) \oplus (y, h(y)) = (x + y, h(x + y))$$

$$(x \cdot y, xh(y) + h(x)y) = (x, h(x)) \otimes (y, h(y)) = (x \cdot y, h(x \cdot y))$$

which means that G and h satisfy (3.4.2) and (3.4.3).

Verification of Step 1. The desired extension $\overline{G}$ of G can be given explicitly (cf. Lemma 1 in Jessen, Karpf, Thorup (1968) or Lemma (3.5.10) in Aczél and Daróczy (1975)) but the verification requires the investigation of seven subcases. Therefore we present an algebraic proof which indicates why the cocycle equation (3.2.4) plays an important role in the theory of group extensions.

In the product set $S := \mathcal{P}_1 \times \mathbb{R}$ we define two operations $\oplus, \otimes : S \times S \to \mathbb{R}$ by (3.4.4) and (3.4.5) for all $x, y \in \mathcal{P}_1$ and $u, v \in \mathbb{R}$ (cf. equation (2.2.7)). As in Step 2 of the proof of Theorem 2.2.4 we find that $(S, \oplus)$ is an abelian semigroup in which the cancellation law holds. Then an easy calculation shows that also $(S, \otimes)$ is an abelian semigroup. Moreover, the two distributive laws are valid. We prove one distributive law using (3.4.1):

$$
\begin{aligned}
(x, u) \otimes \big[(y, v) \oplus (z, w)\big] &= (x, u) \otimes (y + z, v + w - G(y, z)) \\
&= (xy + xz, xv + xw + uy + uz - xG(y, z)) \\
&= (xy + xz, xv + uy + xw + uz - G(xy, xz)) \\
&= (xy, xv + uy) \oplus (xz, xw + uz) \\
&= \big[(x, u) \otimes (y, v)\big] \oplus \big[(x, u) \otimes (z, w)\big].
\end{aligned}
$$

The other distributive law can be proven in the same manner.

Now we want to show that $(S, \oplus, \otimes)$ can be embedded in a ring. Since every cancellative semigroup can be embedded uniquely in a group, let $W = S \ominus S$ be the group generated by $(S, \oplus)$ (cf. Kuczma (1985), p. 90 or Lam (1973), p. 34). Thus every element of W is a difference of elements of S. (Note that we don't know explicitly the operation $\ominus$.) The extension of $\oplus$ on S to an addition on W will be denoted again by $\oplus$. We extend the multiplication $\otimes$ on S in the natural way, namely

$$
(s \ominus t) \otimes (s' \ominus t') := \big[(s \otimes s') \oplus (t \otimes t')\big] \ominus \big[(s \otimes t') \oplus (t \otimes s')\big], \quad (3.4.8)
$$

$(s, t, s', t' \in S)$ to a multiplication on W (which is denoted again by $\otimes$). Using the above rules for $\oplus, \otimes$ on S we get immediately that $(W, \oplus, \otimes)$ is a ring.

The projection $\varphi : S \to \mathcal{P}_1$ defined by

$$
\varphi(x, u) := x, \quad x \in \mathcal{P}_1, \quad u \in \mathbb{R}, \quad (3.4.9)
$$

is an additive homomorphism (by the definition of φ and the operation $\oplus$ in (3.4.4)) of S onto $\mathcal{P}_1$. Applying Lemma 1.2.5 and (1.2.11) we get that $\overline{\varphi} : W \to \mathbb{R}$ given by

$$
\overline{\varphi}((x, u) \ominus (y, v)) = \overline{\varphi}(x, u) - \overline{\varphi}(y, v) = x - y \quad (3.4.10)
$$

$(x, y \in \mathcal{P}_1, u, v \in \mathbb{R})$ is a (uniquely determined) surjective additive homomorphism of W onto $\mathbb{R}$ which extends φ. Let us prove that $\overline{\varphi}$ is also a ring homomorphism. Putting $s = (x, u)$, $t = (y, v)$, $s' = (x', u')$, $t' = (y', v')$ into (3.4.8) (where $x, y, x', y' \in \mathcal{P}_1; u, v, u', v' \in \mathbb{R}$) we arrive at (using (3.4.4) and (3.4.5))

$$\left. \begin{aligned} &((x, u) \ominus (y, v)) \otimes ((x', u') \ominus (y', v')) \\ &\quad = (xx' + y'y, xu' + ux' + yv' + vy' - G(xx', yy')) \\ &\qquad \ominus (xy' + x'y, xv' + vx' + yu' + uy' - G(xy', yx')). \end{aligned} \right\} \quad (3.4.11)$$

Thus we have

$$\left. \begin{aligned} &\overline{\varphi}\big((s \ominus t) \otimes (s' \ominus t')\big) \\ &\quad = \overline{\varphi}\big[(s \otimes s') \oplus (t \otimes t')\big] - \overline{\varphi}\big[(s \otimes t') \oplus (t \otimes s')\big] \\ &\quad = xx' + y'y - (xy' + x'y) = (x - y)(x' - y') \\ &\quad = \big((\overline{\varphi}(s) - \overline{\varphi}(t))(\overline{\varphi}(s') - \overline{\varphi}(t'))\big) \\ &\quad = \overline{\varphi}(s \ominus t)\,\overline{\varphi}(s' \ominus t'). \end{aligned} \right\} \quad (3.4.12)$$

But this implies that the subgroup

$$W_0 := \overline{\varphi}^{-1}(0) = \{(x, u) \ominus (x, v) : x \in \mathcal{P}_1, u, v \in \mathbb{R}\}$$

is an ideal in W. We observe that two elements $(x, u) \ominus (x, v)$ and $(x', u') \ominus (x', v')$ of W_0 are equal iff $u - v = u' - v'$, $(x, x' \in \mathcal{P}_1, u, v, u', v' \in \mathbb{R})$. Indeed,

$$(x, u) \ominus (x, v) = (x', u') \ominus (x', v') \Leftrightarrow$$
$$(x, u) \oplus (x', v') = (x', u') \oplus (x, v) \Leftrightarrow$$
$$(x + x', u + v' - G(x, x')) = (x + x', u' + v - G(x, x')) \Leftrightarrow$$
$$u + v' = u' + v \Leftrightarrow u - v = u' - v'.$$

This calculation shows that the mapping $\psi : W_0 \to \mathbb{R}$ given by

$$\psi((x, u) \ominus (x, v)) := u - v \quad (3.4.13)$$

is well-defined and bijective.

To prove that ψ is an additive group isomorphism of W_0 onto $\mathbb{R}$, we show that ψ is additive by the calculation

$$\begin{aligned} \psi\big[((x, u) &\ominus (x, v)) \oplus ((x', u') \ominus (x', v'))\big] \\ &= \psi\big[((x, u) \oplus (x', u')) \ominus ((x, v) \oplus (x', v'))\big] \\ &= \psi\big[(x + x', u + u' - G(x, x')) \ominus (x + x', v + v' - G(x, x'))\big] \\ &= u + u' - G(xx') - v - v' + G(xx') = (u - v) + (u' - v') \\ &= \psi\big[(x, u) \ominus (x, v)\big] + \psi\big[(x', u') \ominus (x', v')\big]. \end{aligned}$$

Concerning multiplication we prove the formula

$$\psi(g \otimes g_o) = \overline{\varphi}(g)\psi(g_o), \quad g \in W, g_o \in W_0. \tag{3.4.14}$$

Putting $g = (x, u) \ominus (y, v) \in W$ and $g_o = (x', u') \ominus (x', v') \in W_0$ (where $x, y, x' \in \mathcal{P}_1$, $u, v, u', v' \in \mathbb{R}$) we get (using (3.4.11) with $x' = y'$) the desired result

$$\begin{aligned}
\psi(g \otimes g_o) &= \psi\big[((x+y)x', xu' + ux' + yv' + vx' - G(xx', yx')) \\
&\quad \ominus ((x+y)x', xv' + vx' + yu' + ux' - G(xx', yx'))\big] \\
&= xu' + yv' - (xv' + yu') = (x-y)(u' - v') = \overline{\varphi}(g)\psi(g_o).
\end{aligned}$$

Next we are ready to define the extension $\overline{G}$ of G. Since $\overline{\varphi}$ is surjective we get for every $x \in \mathbb{R}$ an element $s_x \in W$ such that $\overline{\varphi}(s_x) = x$. If $x \in \mathcal{P}_1$ we may choose $s_x := (x, 0)$ since $\overline{\varphi}(s_x) = x = \varphi(x, 0)$ (see (3.4.9)). For arbitrary $x, y \in \mathbb{R}$ we have

$$\overline{\varphi}(s_x \oplus s_y \ominus s_{x+y}) = \overline{\varphi}(s_x) + \overline{\varphi}(s_y) - \overline{\varphi}(s_{x+y}) = x + y - (x+y) = 0$$

that is

$$s_x \oplus s_y \ominus s_{x+y} \in W_0. \tag{3.4.15}$$

Defining

$$\overline{G}(x, y) := -\psi(s_x \oplus s_y \ominus s_{x+y}), \quad x, y \in \mathbb{R} \tag{3.4.16}$$

we see that $\overline{G}$ is an extension of G since

$$-\psi((x, 0) \oplus (y, 0) \ominus (x+y, 0)) = -\psi((x+y, G(x, y)) \ominus (x+y, 0)) = G(x, y)$$

for $x, y \in \mathcal{P}_1$. By definition $\overline{G}$ is symmetric; moreover, $\overline{G}$ satisfies (3.2.4):

$$\begin{aligned}
\overline{G}(x, y) + \overline{G}(x+y, z) &= -\psi(s_x \oplus s_y \ominus s_{x+y} \oplus s_{x+y} \oplus s_z \ominus s_{x+y+z}) \\
&= -\psi(s_x \oplus s_y \oplus s_z \ominus s_{x+y+z}) \\
&= -\psi(s_x \oplus s_{y+z} \oplus s_{x+y+z} \oplus s_y \oplus s_z \ominus s_{y+z}) \\
&= \overline{G}(x, y+z) + \overline{G}(y, z)
\end{aligned}$$

for all $x, y, z \in \mathbb{R}$.

Concerning equation (3.4.1) we begin by showing that

$$s_{xy} = s_x \otimes s_y \tag{3.4.17}$$

for all $x, y \in \mathbb{R}$. Writing $x = x_1 - x_2$ and $y = y_1 - y_2$ for $x_1, x_2, y_1, y_2 \in \mathcal{P}_1$, we may choose

$$s_x = (x_1, 0) \ominus (x_2, 0) = s_{x_1} \ominus s_{x_2}, \quad s_y = s_{y_1} \ominus s_{y_2},$$
$$s_{xy} = (s_{x_1 y_1} \oplus s_{x_2 y_2}) \ominus (s_{x_1 y_2} \oplus s_{x_2 y_1}),$$

since $\overline{\phi}(s_{xy}) = \phi(s_{x_1}) - \phi(s_{x_2}) = x_1 - x_2 = x$, $\overline{\phi}(s_y) = y$, and

$$\overline{\phi}(s_{xy}) = \phi(s_{x_1 y_1}) + \phi(s_{x_2 y_2}) - \phi(s_{x_1 y_2}) - \phi(s_{x_2 y_1})$$
$$= (x_1 - x_2)(y_1 - y_2) = xy.$$

Now an application of (3.4.11) gives

$$s_y \otimes s_y$$
$$= ((x_1, 0) \ominus (x_2, 0)) \otimes ((y_1, 0) \ominus (y_2, 0))$$
$$= (x_1 y_1 + x_2 y_2, -G(x_1 y_1, x_2 y_2)) \ominus (x_1 y_2 + x_2 y_1, -G(x_1 y_2, x_2 y_1))$$
$$= ((x_1 y_1, 0) \oplus (x_2 y_2, 0)) \ominus ((x_1 y_2, 0) \oplus (x_2 y_1, 0))$$
$$= s_{xy},$$

which establishes (3.4.17).

Next we observe that

$$(s_x \otimes s_y) \ominus s_{xy} \in W_0$$

for all $x, y \in \mathbb{R}$. Indeed, using (3.4.10) and (3.4.12) we have

$$\overline{\varphi}((s_x \otimes s_y) \ominus s_{xy}) = \overline{\varphi}(s_x)\overline{\varphi}(s_y) - \overline{\varphi}(s_{xy}) = x \cdot y - x \cdot y = 0.$$

Now we prove that $\overline{G}$ satisfies (3.4.1). Indeed, for $t, x, y \in \mathbb{R}$ (using (3.4.14)-(3.4.17) and the ring structure)

$$\overline{G}(tx, ty) = -\psi(s_{tx} \oplus s_{ty} \ominus s_{t(x+y)})$$
$$= -\psi(s_t \otimes (s_x \oplus s_y \ominus s_{x+y}))$$
$$= -\overline{\varphi}(s_t)\psi(s_x \oplus s_y \ominus s_{x+y})$$
$$= t\overline{G}(x, y).$$

Verification of Step 2. By Step 1 we may assume that $G : \mathbb{R}^2 \to \mathbb{R}$ satisfies (3.2.3), (3.2.4) and (3.4.1). Putting $t = 0$ in (3.4.1) we have $G(0, 0) = 0$.

Substituting $y = 0$ into (3.2.4) we arrive at $G(x,0) = G(0,z)$ for all $x, z \in \mathbb{R}$. Moreover we obtain from (3.4.1) and (3.2.3) that $G(x,-x) = -G(-x,x) = -G(x,-x)$ for all $x \in \mathbb{R}$, so

$$G(x,0) = G(0,x) = G(x,-x) = G(0,0) = 0 \qquad (3.4.18)$$

for all $x \in \mathbb{R}$. Using (3.2.3), (3.2.4), (3.4.1) and (3.4.18) we get, as in Step 1, that $T = (\mathbb{R}^2, \oplus, \otimes)$ is a commutative ring with zero element $0_T = (0,0)$ and unit element $1_T = (1,0)$. The last statement is valid because of

$$\left. \begin{aligned}
(x,u) \oplus (-x,-u) &= (0, 0 - G(-x,x)) = (0,0), \\
(x,u) \oplus (0,0) &= (x+0, u+0 - G(x,0)) = (x,u), \\
(x,u) \otimes (1,0) &= (x \cdot 1, 0 + u) = (x,u).
\end{aligned} \right\} \qquad (3.4.19)$$

By induction we find that, for $n \in \mathbb{N}$,

$$n \cdot 1_T = \left(n, -\sum_{j=1}^{n-1} G(j,1) \right).$$

Since it can be verified also that $(n \cdot 1_T) \oplus (m \cdot 1_T) = (n+m) \cdot 1_T$ and $(n \cdot 1_T) \otimes (m \cdot 1_T) = (n \cdot m) \cdot 1_T$ for all $n, m \in \mathbb{Z}$, it follows that $\mathbb{Z}^*$ is a subring of T.

Verification of Step 3. Since the projection in (3.4.7) is obviously a ring isomorphism of $\mathbb{Z}^*$ onto $\mathbb{Z}$, the element $(\mathbb{Z}, \mathbb{Z}^*)$ is in $\mathcal{O}$, and thus $\mathcal{O}$ is nonempty. If $\mathcal{C} = \{(U_i, S_i) \mid i \in I\}$ is a nonempty chain in $\mathcal{O}$ then $\left(\bigcup_{i \in I} U_i, \bigcup_{i \in I} S_i \right)$ is clearly an upper bound for $\mathcal{C}$ and belongs to $\mathcal{O}$. By Zorn's Lemma there exists (U_0, S_0) which is maximal in $\mathcal{O}$.

Verification of Step 4. To show that $U_0 = \mathbb{R}$ we suppose that U_0 is a proper subset of $\mathbb{R}$. Hence there exists an element $a \in \mathbb{R}$ such that $a \notin U_0$. Consider the subring generated by U_0 and a, namely

$$U_0^* = U_0[a] = \bigcup_{n \in \mathbb{N}} \{ P(a) = \sum_{\nu=0}^{n} x_\nu a^\nu \mid x_\nu \in U_0, 0 \le \nu \le n \}.$$

Note that every polynomial $P(a) \in U_0^*$ can be written as

$$P(a) = \varphi \left[\bigoplus_{\nu=0}^{n} ((x_\nu, u_\nu) \otimes (a,u)^\nu) \right] = \sum_{\nu=0}^{n} \varphi(x_\nu, u_\nu) \varphi((a,u)^\nu) \qquad (3.4.20)$$

where $(x_\nu, u_\nu) \in S_0$, $0 \le \nu \le n$, u is any $u \in \mathbb{R}$ and the projection $\varphi : \mathbb{R}^2 \to \mathbb{R}$, given by (3.4.7), is a ring isomorphism of S_0 onto U_0.

Now consider, for $u \in \mathbb{R}$, the subrings

$$S_0^*(u) := S_0[(a,u)] =$$

$$\bigcup_{n \in \mathbb{N}} \left\{ \overline{P}(a,u) = \bigoplus_{\nu=1}^{n} ((x_\nu, u_\nu) \otimes (a,u)^\nu) \mid (x_\nu, u_\nu) \in S_0, 0 \le \nu \le n \right\}.$$

$$(3.4.21)$$

Then (U_0, S_0) is a proper subset of $(U_0^*, S_0^*(u))$ since $\varphi(a,u) = a \notin U_0$ implies $(a,u) \notin \varphi^{-1}(U_0) = S_0$ whereas $(a,u) \in S_0^*(u)$. Our final aim now is to show that there exists $u \in \mathbb{R}$ such that $(U_0^*, S_0^*(u)) \in \mathcal{O}$, which contradicts the maximality of (U_0, S_0).

Case 1. Let a be transcendental over U_0. Then we can take any $u \in \mathbb{R}$ in (3.4.21). The ring isomorphism $\varphi : S_0 \to U_0$ extends to a ring isomorphism of $S_0^*(u)$ onto U_0^* since by comparison of (3.4.20) and (3.4.21) we see that $\varphi(\overline{P}(a,u)) = P(a)$ and that φ is surjective. Moreover, since a is (in Case 1) not a zero of any polynomial in $U_0[a]$ the equation $\varphi(\overline{P}(a,u)) = 0$ implies $\overline{P}(a,u) = (0,0)$. Thus φ is injective and $(U_0^*, S_0^*(u)) \in \mathcal{O}$.

Case 2. If a is algebraic over U_0, then let $P(\neq 0) \in U_0[a]$ be of minimal degree (say n) such that $P(a) = 0$. We show now that the equation $\overline{P}(a,u) = (0,0)$ has a unique solution $u \in \mathbb{R}$. Because of (3.4.18) we have for all $u \in \mathbb{R}$

$$(a,0) \oplus (0,u) = (a, u - G(a,0)) = (a,u),$$

and moreover $(0,u)^\nu = (0,0)$ for all $\nu \in \mathbb{N}$, $\nu \ge 2$. Hence, by Taylor's formula

$$\overline{P}(a,u) = \overline{P}((a,0) \oplus (0,u)) = \bigoplus_{\nu=0}^{n} \frac{1}{\nu!} \left[\overline{P}^{(\nu)}(a,0) \otimes (0,u)^\nu \right]$$

$$= \overline{P}(a,0) \oplus \left(\overline{P}'((a,0) \otimes (0,u)) \right).$$

Because of $\varphi(\overline{P}(a,0)) = P(a) = 0$ and $P \neq 0$ we obtain $\overline{P}(a,0) = (0,w)$ for some $w \in \mathbb{R}$. Furthermore, since P' is of lower degree than P, and $P' \neq 0$, we have $\overline{P}'(a,0) = (b,v)$ for some $v \in \mathbb{R}$, where $b := P'(a) \neq 0$. Thus $\overline{P}(a,u) = (0,0)$ is equivalent to

$$\ominus(0,w) = \ominus\overline{P}(a,0) = \overline{P}'(a,0) \otimes (0,u) = (b,v) \otimes (0,u) = (0,bu).$$

Putting $u := -wb^{-1}$ in (3.4.21) we find, as in Case 1, that $\varphi : S_0^*(-wb^{-1}) \to$ U_0^* is a surjective ring homomorphism. To prove that φ is injective we show that $\overline{Q}(a, u) = (0, 0)$ for every $Q \in U_0[a]$ for which $Q(a) = 0$. The choice of P shows that $Q = P \cdot R$ for some polynomial $R(\neq 0) \in U_0[a]$. But this implies

$$\overline{Q}(a, u) = \overline{P}(a, u) \otimes \overline{R}(a, u) = (0, 0) \otimes \overline{R}(a, u) = (0, 0).$$

Thus again $(U_0^*, S_0^*(-wb^{-1})) \in \mathcal{O}$, contradicting the maximality of (U_0, S_0). This finishes the verification of Step 4, and the theorem is established.

In order to solve the system (3.2.3)-(3.2.5) for additive and multiplicative functions $M : J \to \mathbb{R}$, we need also the following result.

Lemma 3.4.2. *Let j be a positive integer, $\mathcal{P}_j$ the positive cone of $\mathbb{R}^j$. Suppose $f : \mathcal{P}_1 \times \mathcal{P}_j \to \mathbb{R}$ satisfies*

$$f(\tau\xi, u) = \tau f(\xi, u) + a(\tau, \xi) + A(\tau, u), \quad \tau, \xi \in \mathcal{P}_1, u \in \mathcal{P}_j, \qquad (3.4.22)$$

where $u \mapsto A(\tau, u)$ is additive on $\mathbb{R}^j$ and $\xi \mapsto a(\tau, \xi)$ is additive on $\mathbb{R}$ for each $\tau \in \mathcal{P}_1$, but otherwise $a : \mathcal{P}_1 \times \mathbb{R} \to \mathbb{R}$ and $A : \mathcal{P}_1 \times \mathbb{R}^j \to \mathbb{R}$ are arbitrary. Then there exist functions $g : \mathcal{P}_j \to \mathbb{R}, b : \mathcal{P}_1 \to \mathbb{R}$ and additive $B : \mathbb{R}^j \to \mathbb{R}$ for which

$$f(\tau, u) = \tau g(u) + b(\tau) - B(u), \quad \tau \in \mathcal{P}_1, u \in \mathcal{P}_j, \qquad (3.4.23)$$

$$A(\tau, u) = (\tau - 1)B(u), \qquad u \in \mathcal{P}_j, \qquad (3.4.24)$$

$$a(\tau, \xi) = b(\tau\xi) - \tau b(\xi), \quad \tau, \xi \in \mathcal{P}_1. \qquad (3.4.25)$$

Proof: Putting $\xi = 1$ in (3.4.22), we get immediately

$$f(\tau, u) = \tau g_1(u) + b(\tau) + A(\tau, u), \quad \tau \in \mathcal{P}_1, u \in \mathcal{P}_j, \qquad (3.4.26)$$

for $g_1(u) := f(1, u)$ and $b(\tau) := a(\tau, 1)$. Substituting (3.4.26) back into (3.4.22), we see that

$$b(\tau\xi) + A(\tau\xi, u) = \tau[b(\xi) + A(\xi, u)] + a(\tau, \xi) + A(\tau, u),$$

or

$$A(\tau\xi, u) - \tau A(\xi, u) - A(\tau, u) = a(\tau, \xi) - b(\tau\xi) + \tau b(\xi), \qquad (3.4.27)$$

for all $\tau, \xi \in \mathcal{P}_1$ and $u \in \mathcal{P}_j$.

Observe that the left-hand side of (3.4.27) is additive in u, by our hypothesis about A. On the other hand, the right-hand side of (3.4.27) is constant with respect to u. But the only constant (even on $\mathcal{P}_j$) additive function is the zero function, so (3.4.27) yields

$$A(\tau\xi, u) = \tau A(\xi, u) + A(\tau, u), \quad \tau, \xi \in \mathcal{P}_1, u \in \mathcal{P}_j. \tag{3.4.28}$$

Equations (3.4.27) and (3.4.28) give (3.4.25) immediately.

Furthermore, the left hand side of (3.4.28) is symmetric in τ and ξ, so we also have

$$\tau A(\xi, u) + A(\tau, u) = \xi A(\tau, u) + A(\xi, u),$$

or

$$(\tau - 1)A(\xi, u) = (\xi - 1)A(\tau, u).$$

Putting $\xi = 2$ in this equation and defining $B : \mathbb{R}^j \to \mathbb{R}$ by

$$B(u) := A(2, u), \qquad u \in \mathbb{R}^j,$$

we get (3.4.24) together with the additivity of B.

Finally, (3.4.24) and (3.4.26) yield

$$f(\tau, u) = \tau g_1(u) + b(\tau) + (\tau - 1)B(u), \quad \tau \in \mathcal{P}_1, u \in \mathcal{P}_j.$$

Defining $g : \mathcal{P}_j \to \mathbb{R}$ by

$$g(u) := g_1(u) + B(u), \qquad u \in \mathcal{P}_j,$$

we have (3.4.23), and this finishes the proof of the lemma.

Now we are ready to solve (3.2.3) - (3.2.5) in higher dimensions. As we saw in Lemma 1.2.10, whenever M is both additive and multiplicative on $\mathcal{P}$ (or even just on D_2, respectively J), then M must be either the zero map or a *projection*:

$$M(x) = M(\xi_1, \ldots, \xi_k) = \xi_j, \qquad x \in \mathcal{P},$$

for some $j \in \{1, \ldots, k\}$. We use this fact in the next theorem.

Theorem 3.4.3. *Let $M : J \to \mathbb{R}$ be multiplicative on J and additive on D_2. Then the general solution of the system (3.2.3)–(3.2.5) for $x, y, z, t, x+y+z \in J$ is given by*

$$G(x, y) = M(x)L(x) + M(y)L(y) - M(x + y)L(x + y) \tag{3.4.29}$$

for all $(x, y) \in D_2$, *where* $L : \mathcal{P} \to \mathbb{R}$ *is a logarithmic function*

$$L(st) = L(s) + L(t), \qquad s, t, \in \mathcal{P}, \tag{3.4.30}$$

and either $M \equiv 0$ *or* M *is a projection.*

Proof: Any G of the form (3.4.29) satisfies (3.2.3) and (3.2.4). If $M \equiv 0$, then (3.4.29) gives $G \equiv 0$ on D_2, and this certainly satisfies (3.2.5). If M is a projection and L satisfies (3.4.30), then a straightforward calculation verifies that G of the form (3.4.29) satisfies (3.2.5). One needs only (3.4.30) and the additivity and multiplicativity of M.

For the converse, we first apply Lemma 3.2.3 to extend M, G to a solution pair $\overline{M}$ (on $\mathcal{P}$), $\overline{G}$ (on $\mathcal{P} \times \mathcal{P}$), with $\overline{M}$ additive and multiplicative on $\mathcal{P}$. We drop the bars and identify $\overline{M}$ with M and $\overline{G}$ with G.

Now we also apply Lemma 1.2.10, which shows that either $M \equiv 0$ or M is a projection. If $M = 0$ on $\mathcal{P}$, then (3.2.5) shows that also $G = 0$ on $\mathcal{P} \times \mathcal{P}$, so (3.4.29) holds in this case.

In the case that M is a projection, let us assume temporarily (for ease of notation) that

$$M(t) = M(\tau_1, \dots, \tau_k) = \tau_1, \qquad t \in \mathcal{P}. \tag{3.4.31}$$

If $k = 1$, then Theorem 3.4.1 yields the desired result, with $L(u) := \frac{h(u)}{u}$, $u > 0$. Henceforth, we assume $k \geq 2$. Applying Theorem 2.2.4 (see also Remark 2.2.5) to the subsystem (3.2.3) - (3.2.4), we obtain

$$G(x, y) = f(x) + f(y) - f(x + y), \qquad x, y \in \mathcal{P}. \tag{3.4.32}$$

And now for convenience we write $t = (\tau, s)$, $x = (\xi, u)$, $y = (\eta, v)$ for $\tau, \xi, \eta \in \mathcal{P}_1$ and $s, u, v \in \mathcal{P}_{k-1}$. Inserting (3.4.31) and (3.4.32) into (3.2.5), we have with this notation

$$\begin{aligned} f(\tau\xi, su) + f(\tau\eta, sv) &- f(\tau(\xi + \eta), s(u + v)) \\ &= \tau[f(\xi, u) + f(\eta, v) - f(\xi + \eta, u + v)], \end{aligned} \tag{3.4.33}$$

for all $\tau, \xi, \eta \in \mathcal{P}_1$ and $u, v, s \in \mathcal{P}_{k-1}$.

Put first $s = 1$ into (3.4.33). Then, defining $D : \mathcal{P}_1 \times \mathcal{P}_1 \times \mathcal{P}_{k-1} \to \mathbb{R}$ by

$$D(\tau, \xi, u) := f(\tau\xi, u) - \tau f(\xi, u), \qquad \tau, \xi \in \mathcal{P}_1, u \in \mathcal{P}_{k-1}, \tag{3.4.34}$$

we get

$$D(\tau, \xi, u) + D(\tau, \eta, v) = D(\tau, \xi + \eta, u + v).$$

That is, $x \mapsto D(\tau, x)$ is additive on $\mathcal{P}$ for each fixed $\tau \in \mathcal{P}_1$. By Lemma 1.2.5, this means that for each $\tau \in \mathcal{P}_1$ there exist additive maps $a(\tau, \cdot), a_i(\tau, \cdot)$ on $\mathbb{R}$ ($i = 1, \dots k - 1$) such that

$$D(\tau, \xi, u) = D(\tau, \xi, \omega_1, \dots, \omega_{k-1})$$
$$= a(\tau, \xi) + \sum_{i=1}^{k-1} a_i(\tau, \omega_i).$$

Letting $A(\tau, u) = \sum_{i=1}^{k-1} a_i(\tau, \omega_i)$, we have

$$D(\tau, \xi, u) = a(\tau, \xi) + A(\tau, u), \quad \tau, \xi \in \mathcal{P}_1, u \in \mathcal{P}_{k-1},$$

where $a(\tau, \cdot)$, $A(\tau, \cdot)$ are additive on $\mathbb{R}$, respectively $\mathbb{R}^{k-1}$, for each fixed $\tau \in \mathcal{P}_1$. Substituting this back into (3.4.34), we have

$$f(\tau\xi, u) = \tau f(\xi, u) + a(\tau, \xi) + A(\tau, u),$$

valid for $\tau, \xi \in \mathcal{P}_1$ and $u \in \mathcal{P}_{k-1}$.

Here, we apply Lemma 3.4.2 (with $j = k - 1 \geq 1$). Thereby, we obtain

$$f(\tau, u) = \tau g(u) + b(\tau) - B(u), \quad \tau \in \mathcal{P}_1, u \in \mathcal{P}_{k-1}, \tag{3.4.23}$$

$$a(\tau, \xi) = b(\tau\xi) - \tau b(\xi), \quad \tau, \xi \in \mathcal{P}_1, \tag{3.4.25}$$

where $B : \mathbb{R}^{k-1} \to \mathbb{R}$ is additive. Putting (3.4.23) back into (3.4.33) and using the additivity of B, we have now

$$\tau[\xi g(su) + \eta g(sv) - (\xi + \eta)g(s(u + v))] + b(\tau\xi)$$
$$+ b(\tau\eta) - b(\tau(\xi + \eta))$$
$$= \tau[\xi g(u) + \eta g(v) - (\xi + \eta)g(u + v)$$
$$+ b(\xi) + b(\eta) - b(\xi + \eta)].$$

But also (3.4.25) together with the additivity of $a(\tau, \cdot)$ reduces this to

$$\xi[g(su) - g(u)] + \eta[g(sv) - g(v)] = (\xi + \eta)[g(s(u + v)) - g(u + v)].$$

Letting first $\xi = \eta = 1$, then $\xi = 1$ and $\eta = 2$, the difference is

$$g(sv) - g(v) = g(s(u + v)) - g(u + v).$$

Since the right hand side is symmetric in u and v, so is the left. It follows that

$$g(sv) - g(v) = g(su) - g(u), \qquad s, v, u \in \mathcal{P}_{k-1}.$$

In particular, with $u = 1$,

$$g(sv) - g(v) = g(s) - g(1) =: \ell(s), \qquad s \in \mathcal{P}_{k-1}.$$

That is, with $c := g(1)$, we get from the last equation both

$$g(s) = \ell(s) + c, \qquad s \in \mathcal{P}_{k-1}, \tag{3.4.35}$$

and

$$\ell(sv) = \ell(v) + \ell(s), \qquad s, v \in \mathcal{P}_{k-1}. \tag{3.4.36}$$

Therefore, with (3.4.35), (3.4.23) and the additivity of B, (3.4.32) becomes

$$\begin{aligned}
G((\xi, u), (\eta, v)) &= \xi[\ell(u) + c] + b(\xi) + \eta[\ell(v) + c] + b(\eta) \\
&\quad - (\xi + \eta)[\ell(u + v) + c] - b(\xi + \eta) \\
&= \xi\ell(u) + \eta\ell(v) - (\xi + \eta)\ell(u + v) \\
&\quad + b(\xi) + b(\eta) - b(\xi + \eta).
\end{aligned} \tag{3.4.37}$$

For the final step, we return to the fact that $a(\tau, \cdot)$ is additive. This carries crucial implications for our function b. We define $F : \mathcal{P}_1 \times \mathcal{P}_1 \to \mathbb{R}$ by

$$F(\xi, \eta) := b(\xi) + b(\eta) - b(\xi + \eta), \qquad \xi, \eta \in \mathcal{P}_1. \tag{3.4.38}$$

Then obviously F is symmetric (3.2.3) and satisfies the cocycle equation (3.2.4). Moreover, (3.4.25) and the additivity of $a(\tau, \cdot)$ yield also

$$\begin{aligned}
F(\tau\xi, \tau\eta) - \tau F(\xi, \eta) &= b(\tau\xi) + b(\tau\eta) - b(\tau(\xi + \eta)) \\
&\quad - \tau[b(\xi) + b(\eta) - b(\xi + \eta)] \\
&= a(\tau, \xi) + a(\tau, \eta) - a(\tau, \xi + \eta) \\
&= 0.
\end{aligned}$$

That is, F satisfies the homogeneity equation (3.2.5) with $M(t) = t$ (and $k = 1$). Hence, by Theorem 3.4.1, we find that

$$F(\xi, \eta) = h(\xi) + h(\eta) - h(\xi + \eta), \qquad \xi, \eta \in \mathcal{P}_1, \tag{3.4.39}$$

where h satisfies (3.4.1):

$$h(\xi\eta) = \xi h(\eta) + \eta h(\xi), \qquad \xi, \eta \in \mathcal{P}_1.$$

Let $\ell_1 : \mathcal{P}_1 \to \mathbb{R}$ be defined by

$$\ell_1(\xi) := \frac{h(\xi)}{\xi}, \qquad \xi > 0. \tag{3.4.40}$$

Then (3.4.1) shows that ℓ_1 is logarithmic:

$$\ell_1(\xi\eta) = \ell_1(\xi) + \ell_1(\eta), \qquad \xi, \eta \in \mathcal{P}_1, \tag{3.4.41}$$

Finally, by (3.4.37) - (3.4.40), we have

$$G((\xi, u), (\eta, v)) = \xi\ell(u) + \eta\ell(v) - (\xi + \eta)\ell(u + v)$$
$$+ \xi\ell_1(\xi) + \eta\ell_1(\eta) - (\xi + \eta)\ell_1(\xi + \eta).$$

Thus, defining $L : \mathcal{P} \to \mathbb{R}$ by

$$L(\xi, u) := \ell(u) + \ell_1(\xi), \quad \xi \in \mathcal{P}_1, u \in \mathcal{P}_{k-1}, \tag{3.4.42}$$

we have on the one hand

$$G((\xi, u), (\eta, v)) = \xi L(\xi, u) + \eta L(\eta, v) - (\xi + \eta) L(\xi + \eta, u + v), \tag{3.4.43}$$

for all $(\xi, u), (\eta, v) \in \mathcal{P}$. And on the other hand, by (3.4.36), (3.4.41) and (3.4.42), we see that L is (3.4.30) logarithmic on $\mathcal{P}$. But (3.4.43) was obtained under the assumption (3.4.31) that M is the projection onto the first component, whereas, in the general case, M is a projection onto any component. We recapture full generality by writing (3.4.43) in the form (3.4.29).

This completes the proof of the theorem.

Now we can give the explicit forms of all 3-symmetric measures of information which are recursive of additive multiplicative type.

Theorem 3.4.4. *Let* $M : J \to \mathbb{R}$ *be multiplicative and additive. Then a measure of information* $\{I_n\}$ *is 3-symmetric and recursive of type* M, *if and only if either* M *is a projection and*

$$I_n(p_1, \ldots, p_n) = \sum_{i=1}^{n} M(p_i)L(p_i), \qquad P \in \Gamma_n^o, \qquad (3.4.44)$$

$(n = 2, 3, \ldots)$ *for some* $L : \mathcal{P} \to \mathbb{R}$ *satisfying (3.4.30), or* $M \equiv 0$ *and*

$$I_n(p_1, \ldots, p_n) = c, \qquad P \in \Gamma_n^o, \qquad (3.4.45)$$

$(n = 2, 3, \ldots)$ *for some constant c.*

Proof: Both (3.4.44) and (3.4.45) are obviously 3-symmetric. It is also clear that (3.4.45) is recursive of type $M \equiv 0$. To verify that (3.4.44) is recursive of type M = projection, one needs also the fact that

$$L\left(\frac{x}{x+y}\right) = L(x) - L(x+y), \qquad x, y \in \mathcal{P}, \qquad (3.4.46)$$

which follows from (3.4.30) with $s = x + y, t = \frac{x}{x+y}$.

For the converse, we apply first Lemma 3.2.2 and then Theorem 3.4.3. We have two cases to consider.

First, suppose that M is a projection. By (3.2.2) and (3.4.29), we have then

$$I_2\left(\frac{x}{x+y}, \frac{y}{x+y}\right) = \frac{M(x)L(x) + M(y)L(y) - M(x+y)L(x+y)}{M(x+y)},$$

for all $(x, y) \in D_2$. Using the properties of M along with (3.4.46), we calculate that

$$\begin{aligned} I_2\left(\frac{x}{x+y}, \frac{y}{x+y}\right) &= M\left(\frac{x}{x+y}\right)L(x) + M\left(\frac{y}{x+y}\right)L(y) \\ &\quad - M\left(\frac{x}{x+y} + \frac{y}{x+y}\right)L(x+y) \\ &= M\left(\frac{x}{x+y}\right)[L(x) - L(x+y)] + M\left(\frac{y}{x+y}\right)[L(y) - L(x+y)] \\ &= M\left(\frac{x}{x+y}\right)L\left(\frac{x}{x+y}\right) + M\left(\frac{y}{x+y}\right)L\left(\frac{y}{x+y}\right), \end{aligned}$$

which is (3.4.44) for $n = 2$. Now we proceed by induction. Supposing that (3.4.44) holds for some $n \geq 2$, we find by recursivity and (3.4.46) that

$$I_{n+1}(p_1, \ldots, p_{n+1})$$

$$= I_n(p_1 + p_2, p_3, \ldots, p_{n+1}) + M(p_1 + p_2) \left[M \left(\frac{p_1}{p_1 + p_2} \right) L \left(\frac{p_1}{p_1 + p_2} \right) \right.$$

$$+ M \left(\frac{p_2}{p_1 + p_2} \right) L \left(\frac{p_2}{p_1 + p_2} \right) \right]$$

$$= M(p_1 + p_2) L(p_1 + p_2) + \sum_{j=3}^{n+1} M(p_j) L(p_j)$$

$$+ M(p_1)[L(p_1) - L(p_1 + p_2)] + M(p_2)[L(p_2) - L(p_1 + p_2)]$$

$$= \sum_{j=1}^{n+1} M(p_j) L(p_j) + [M(p_1 + p_2) - M(p_1) + M(p_2)] L(p_1 + p_2),$$

$$= \sum_{j=1}^{n+1} M(p_j) L(p_j),$$

which is (3.4.44) for $P \in \Gamma_{n+1}^o$.

The only other case to consider is $M \equiv 0$. In this case, (3.1.1) with $M \equiv 0$ gives

$$I_n(p_1, \ldots, p_n) = I_{n-1}(p_1 + p_2, p_3, \ldots, p_n), \quad P \in \Gamma_n^o, \tag{3.4.47}$$

$n = 3, 4, \ldots$. For $n = 3$, this means that

$$I_3(p_1, p_2, p_3) = I_2(1 - p_3, p_3) =: \varphi(p_3), \quad P \in \Gamma_3^o. \tag{3.4.48}$$

But I_3 is symmetric, so (3.4.48) implies in particular

$$\varphi(p_3) = \varphi(p_2), \quad (p_2, p_3) \in D_2.$$

By Lemma 1.2.1, φ must be constant, say

$$\varphi(p) = c, \quad p \in J.$$

Then (3.4.48) yields $I_3(P) = I_2(Q) = c$, for all $P \in \Gamma_3^o$ and $Q \in \Gamma_2^o$. By (3.4.47) this extends to all $n \geq 2$, and we conclude that (3.4.45) holds.

This concludes the proof of the theorem.

Observe that solution (3.4.45) ties in very well with solution (3.3.3), in the previous section. For, taking $M \equiv 0$ in (3.3.3) one gets $I_n(P) = -c$ for all $P \in \Gamma_n^o$ $(n = 2, 3, \ldots)$.

We would like to point out that, in contrast to the situation in the previous section, it is not an easy task to select the "regular" (say, continuous or measurable) solutions from among the general solutions in Theorems 3.4.3 and 3.4.4. Indeed, when M is a projection, it is already *very* regular, but there exist very *irregular* solutions of (3.4.30). The general solution of (3.4.30) is of the form (see Lemma 1.2.12)

$$L(t) = L(\tau_1, \ldots, \tau_k) = \sum_{i=1}^{k} \ell_i(\tau_i), \quad t = (\tau_1, \ldots, \tau_k) \in \mathcal{P},$$

where each $\ell_i : \mathcal{P}_1 \to \mathbb{R}$ $(i = 1, \ldots, k)$ is again logarithmic:

$$\ell_i(\sigma\tau) = \ell_i(\sigma) + \ell_i(\tau), \quad \sigma, \tau > 0.$$

Of course, if one assumes that L satisfies some regularity condition (say, measurability), then each ℓ_i possesses the same type of regularity, and is therefore of the form

$$\ell_i(\tau) = c_i \log \tau, \quad \tau > 0,$$

and so we have

$$L(t) = \sum_{i=1}^{k} c_i \log \tau_i, \quad t = (\tau_1, \ldots, \tau_k) \in \mathcal{P}.$$

But the situation is worse than that. Unfortunately, the assumption that I_n is regular does not lead to the conclusion that L is regular. For instance, taking $k = 1$ in (3.4.44), we have

$$I_n(p_1, \ldots, p_n) = \sum_{i=1}^{n} p_i \ell(p_i), \quad P \in \Gamma_n^o.$$

We can write this in the form

$$I_n(P) = \sum_{i=1}^{n} h(p_i), \quad P \in \Gamma_n^o, \tag{3.4.49}$$

where $h(p) := p\ell(p)$ satisfies (3.4.3):

$$h(uv) = uh(v) + vh(u), \qquad u, v \in \,]0, 1[.$$

The function h appearing in (3.4.49) is determined only up to an added *real derivation* d (that is, d is additive and satisfies (3.4.3) on $\mathbb{R}$) restricted to $]0, 1[$, since every derivation fulfills

$$\sum_{i=1}^{n} d(p_i) = d(1) = 0.$$

(Put $u = v = 1$ into (3.4.3) to see that $d(1) = 0$. Then use induction). But it is known that there exist derivations which are not identically zero (cf. Kuczma (1985), p. 352). These derivations d are automatically very irregular: If such a derivation d would be measurable or bounded from below on an open interval, then d, being an additive function, would be continuous and thus linear, hence $d(x) = x \cdot d(1) = 0$ for all $x \in \mathbb{R}$. Thus we see that even a very regular I_n can be represented through (3.4.49) (and so also (3.4.44)) by a highly irregular h. The relevant question, though, is whether *there exists* a regular h representing I_n in (3.4.49), if we know that I_n is regular. We shall return to this question in Chapter 4.

3.5. Summary and further remarks

In the previous two sections, we have determined all recursive $\{I_n\}$ of multiplicative type which are 3-symmetric. We can summarize the main results in the following two corollaries.

Corollary 3.5.1. *Let $M : J \to \mathbb{R}$ be multiplicative. The general solution of the system of functional equations (3.2.3) - (3.2.5), for $x, y, z, t, x+y+z \in J$, is given either by (3.4.29):*

$$G(x,y) = M(x)L(x) + M(y)L(y) - M(x+y)L(x+y), \quad (x,y) \in D_2,$$

with M a projection and $L : \mathcal{P} \to \mathbb{R}$ any (3.4.30) logarithmic map, or by (3.3.1):

$$G(x,y) = c[M(x) + M(y) - M(x+y)], \quad (x,y) \in D_2,$$

with $M \equiv 0$ or M not additive and with arbitrary $c \in \mathbb{R}$.

Proof: Theorems 3.3.1 and 3.4.3.

Remark 3.5.2. Since we have proved Theorems 3.3.1 and 3.4.3 by extension of our functions and equations to $\mathcal{P}$, both those theorems and Corollary 3.5.1 are also valid on $\mathcal{P}$.

Corollary 3.5.3. *A measure $\{I_n\}$ of information on the open domain is 3-symmetric and recursive of multiplicative type M, if and only if I_n has one of the following two forms. Either M is a projection and*

$$I_n(p_1,\ldots,p_n) = \sum_{i=1}^{n} M(p_i)L(p_i), \qquad P \in \Gamma_n^o \qquad (3.4.44)$$

$(n = 2,3,\ldots)$ holds for some logarithmic $L : \mathcal{P} \to \mathbb{R}$ (see (3.4.30)) , or

$$I_n(p_1,\ldots,p_n) = c\left[\sum_{i=1}^{n} M(p_i) - 1\right], \qquad P \in \Gamma_n^o \qquad (3.3.3)$$

$(n = 2,3,\ldots)$ holds for some constant $c \in \mathbb{R}$ and for $M \equiv 0$ or M not additive.

Proof: Theorems 3.3.2 and 3.4.4.

Remark 3.5.4. Note that we could also have proved Theorems 3.3.2, 3.4.4, and Corollary 3.5.3 with the aid of Theorem 2.3.1. Any branching and 4-symmetric $\{I_n\}$ has the sum form (2.3.1):

$$I_n(p_1,\ldots,p_n) = \sum_{i=1}^{n} g(p_i), \qquad P \in \Gamma_n^o$$

$(n = 2,3,\ldots)$, for some $g : J \to \mathbb{R}$, and the generating function G has the form (2.3.4):

$$G(x,y) = g(x) + g(y) - g(x+y), \quad (x,y) \in D_2.$$

Moreover G satisfies the characteristic functional equation for branching measures of information (2.2.2) and G is symmetric (2.2.3). But here we have the additional property that G is M-homogeneous (3.2.5). By this property we can explicitly determine g up to an added additive function.

Indeed, subtracting (3.4.29), respectively (3.3.1) from (2.3.4) we arrive at

$$(g - ML)(x) + (g - ML)(y) = (g - ML)(x + y)$$
$$(g - cM)(x) + (g - cM)(y) = (g - cM)(x + y)$$

which is

$$a(x) + a(y) = a(x + y),$$

where

$$a(x) = (g - ML)(x)$$

or

$$a(x) = (g - cM)(x)$$

respectively (for some additive function $a : \mathbb{R}^k \to \mathbb{R}$). We may take $a \equiv 0$ (or more generally $a(1) = 0$) in the first case and $a(1) = -c$ (for instance, $g(p) = g(\pi_1, \ldots, \pi_k) = c[M(p) - \pi_1)]$ in the second. Thus, using (2.3.1), we arrive at (3.4.44), respectively (3.3.3).

Remark 3.5.5. The proof of Theorem 3.4.3 presented here (including Lemma 3.4.2) is a new one found recently by the first author. The original proof was proposed in Aczél (1980e) and carried out in Aczél and Ng (1983), based on Maksa (1982). Together with other things, we give a brief outline of this proof near the end of Chapter 4.

CHAPTER 4

THE FUNDAMENTAL EQUATION OF INFORMATION
AND REGULAR RECURSIVE MEASURES

4.1. Introduction

This chapter deals again with the combination of recursivity and symmetry. Here we shall need only a partial symmetry of I_3 in order to obtain the fundamental equation of information (of multiplicative type). There are at least two reasons for studying this equation. One is that it has played an important historical role in the development of a characterization theory for measures of information. The second reason is that the results obtained show, similar to the situation with branching in Chapter 2, that the recursivity gives almost the same form of solution with partial symmetry as with full symmetry.

We begin with the assumption that $\{I_n\}$ is recursive of multiplicative type M. That is, again, (3.1.1):

$$I_n(p_1, \ldots, p_n) = I_{n-1}(p_1 + p_2, p_3, \ldots, p_n)$$
$$+ M(p_1 + p_2)I_2\left(\frac{p_1}{p_1 + p_2}, \frac{p_2}{p_1 + p_2}\right)$$

with $M : J \to \mathbb{R}$ multiplicative. The only other assumption we make is that $\{I_n\}$ is *3-semisymmetric*:

$$I_3(p_1, p_2, p_3) = I_3(p_1, p_3, p_2), \qquad P \in \Gamma_3^o. \tag{4.1.1}$$

Lemma 4.1.1. *Suppose $\{I_n\}$ is recursive of multiplicative type M and 3-semisymmetric. Then the map $\varphi : J \to \mathbb{R}$ defined by*

$$\varphi(p) := I_2(1 - p, p), \qquad p \in J, \tag{4.1.2}$$

satisfies the fundamental equation of information (of multiplicative type)

$$\varphi(x) + M(1 - x)\varphi\left(\frac{y}{1 - x}\right) = \varphi(y) + M(1 - y)\varphi\left(\frac{x}{1 - y}\right) \tag{4.1.3}$$

for all $(x, y) \in D_2$. Such φ are called information functions of type M.

Proof: Into (3.1.1) for $n = 3$, we put $x = p_3$ and $y = p_2$. Using (4.1.1), we compute that

$$I_2(1 - x, x) + M(1 - x)I_2\left(\frac{1 - x - y}{1 - x}, \frac{y}{1 - x}\right)$$
$$= I_3(1 - x - y, y, x)$$
$$= I_3(1 - y - x, x, y)$$
$$= I_2(1 - y, y) + M(1 - y)I_2\left(\frac{1 - y - x}{1 - y}, \frac{x}{1 - y}\right),$$

for all $(x, y) = (p_3, p_2) \in D_2$. Making the definition (4.1.2), this is (4.1.3).

Remark 4.1.2. Note that if we were to assume full 3-symmetry, then (applying Lemma 3.2.2) by (3.2.2) and (3.2.3) we would get also $\varphi(x) = \varphi(1-x), x \in J$. Such φ are called *symmetric* information functions.

To call (4.1.3) the fundamental equation *of multiplicative type* might give the impression that this equation is a special case of something called the fundamental equation. In fact, just the opposite is true. In most of the literature, one finds the title "fundamental equation of information" used to refer to a special case of (4.1.3). In fact, it was originally the special case $k = 1$ and $M(x) = x$ of (4.1.3), the case satisfied by Shannon's entropy. Later, the same title was sometimes used for the higher dimensional special case of (4.1.3) in which M is a projection. Our equation includes all of these "fundamental equations" and more. One would be tempted, therefore, to call it the "generalized" fundamental equation of information. Unfortunately, that title has already been given (with reason) to equation (3.5.1). So we have settled on the name given here, also because it has been used in some of the literature.

A different justification of the fundamental equation is given in Aczél and Daróczy (1975) [see sections 3.1, 6.1, 6.2], where they consider the information contained in a random event.

In the next lemma, we see that we shall revisit some familiar territory. We take the following result from Aczél and Ng (1983).

Lemma 4.1.3. *Suppose $M : J \to \mathbb{R}$ is multiplicative, and that $\varphi : J \to \mathbb{R}$ satisfies (4.1.3). Then the map $G : D_2 \to \mathbb{R}$ defined by*

$$G(x, y) := \varphi(x) + M(1 - x)\varphi\left(\frac{y}{1 - x}\right) - \varphi(x + y), \quad (x, y) \in D_2, \quad (4.1.4)$$

satisfies the system (3.2.3) - (3.2.5) for all $x, y, z, t,\ x + y + z \in J$.

Proof: The (3.2.3) symmetry of G [i.e. $G(x, y) = G(y, x)$] follows immediately from (4.1.3) and the definition (4.1.4).

To get (3.2.4) and (3.2.5), we use (4.1.3) several times, and the multiplicativity of M once. For all $x, y, z,\ x + y + z \in J$, we have

$$G(x, y) + G(x + y, z)$$
$$= \left[\varphi(x) + M(1 - x)\varphi\left(\frac{y}{1 - x}\right) - \varphi(x + y)\right] + \left[\varphi(x + y)\right.$$
$$\left. + M(1 - x - y)\varphi\left(\frac{z}{1 - x - y}\right) - \varphi(x + y + z)\right]$$
$$= M(1 - x)\left[\varphi\left(\frac{y}{1 - x}\right) + M\left(1 - \frac{y}{1 - x}\right)\varphi\left(\frac{z/(1 - x)}{1 - y/(1 - x)}\right)\right]$$
$$+ \varphi(x) - \varphi(x + y + z)$$
$$= \varphi(x) + M(1 - x)\left[G\left(\frac{y}{1 - x}, \frac{z}{1 - x}\right) + \varphi\left(\frac{y + z}{1 - x}\right)\right] - \varphi(x + y + z)$$
$$= G(x, y + z) + M(1 - x)G\left(\frac{y}{1 - x}, \frac{z}{1 - x}\right).$$

$$(4.1.5)$$

Now the left-hand side of (4.1.5) is symmetric in x and y (by (3.2.3), already proved), while the right-hand side is symmetric in y and z. Hence both sides are fully symmetric in x, y and z. Applying this symmetry to the left-hand side, and using (3.2.3) again, we get (3.2.4):

$$G(x, y) + G(x + y, z) = G(x, y + z) + G(y, z).$$

Finally, comparing this with (4.1.5), we obtain (for $x, y, z,\ x + y + z \in J$)

$$G(y, z) = M(1 - x)G\left(\frac{y}{1 - x}, \frac{z}{1 - x}\right). \qquad (4.1.6)$$

Given any $t \in J$, $(u, v) \in D_2$, let $x = 1 - t$, $y = tu$, and $z = tv$. Then $x + y + z = 1 - t(1 - u - v) \in J$, and (4.1.6) yields

$$G(tu, tv) = M(t)G(u, v),$$

which is (3.2.5). This finishes the proof.

So we see that we again arrive at the system (3.2.3)–(3.2.5), which was solved in the last chapter, even though we don't have the full 3-symmetry.

The difference here is that we have to define our G more cleverly. The advantage is that we can again use the results of Section 3. But on the other hand, the function G defined by (4.1.4) is not the same as the generating function defined previously (see Definition 3.1.1) as

$$M(x+y)I_2\left(\frac{x}{x+y},\frac{y}{x+y}\right),\qquad (x,y)\in D_2. \qquad (4.1.7)$$

The next task is to reinterpret our results about (3.2.3)–(3.2.5) in light of equation (4.1.4).

4.2. General solution of the fundamental equation of information

In this section we shall solve (4.1.3) and determine all 3-semisymmetric information measures which are recursive of multiplicative type. But first we isolate one additional helpful fact: We solve the special case of (4.1.4) in which $G \equiv 0$.

Lemma 4.2.1. *Let* $M : J \to \mathbb{R}$ *be multiplicative. Then a map* $d : J \to \mathbb{R}$ *satisfies*

$$d(x) + M(1-x)d\left(\frac{y}{1-x}\right) - d(x+y) = 0,\quad (x,y)\in D_2, \qquad (4.2.1)$$

if and only if either

$$M \not\equiv 1 \text{ and } d(x) = b[M(1-x)-1],\qquad x\in J, \qquad (4.2.2)$$

for some constant $c \in \mathbb{R}$*, or*

$$M \equiv 1 \text{ and } d(x) = L(1-x),\qquad x\in J, \qquad (4.2.3)$$

for some $L : J \to \mathbb{R}$ *satisfying the logarithm equation*

$$L(xy) = L(x) + L(y),\qquad x,y\in J. \qquad (4.2.4)$$

Proof: Suppose d satisfies (4.2.1). Put $y = (1-x)z$ with arbitrary $z \in J$; then

$$d(x) + M(1-x)d(z) = d(x+z-xz),\qquad x,z\in J. \qquad (4.2.5)$$

If $M \not\equiv 1$, then we use the symmetry of the right-hand side of (4.2.5) to get

$$d(x) + M(1-x)d(z) = d(z) + M(1-z)d(x).$$

Inserting $z = z_0$ such that $M(1 - z_0) \neq 1$, we get (4.2.2) with constant $b = d(z_0)[M(1 - z_0) - 1]^{-1}$.

On the other hand, if $M \equiv 1$, then (4.2.5) reduces to

$$d(x) + d(z) = d(x + z - xz),$$

or

$$d(1 - (1 - x)) + d(1 - (1 - z)) = d(1 - (1 - x)(1 - z)),$$

which means that the function $L : J \to \mathbb{R}$ defined by $L(x) := d(1-x)$ satisfies

$$L(1 - x) + L(1 - z) = L((1 - x)(1 - z)).$$

Thus we have (4.2.3) with logarithmic L.

The converse is easy to verify.

Now we are ready to present the main results. The proof in the case that M is a projection is a new one.

Theorem 4.2.2. *Let $M : J \to \mathbb{R}$ be multiplicative. Then the general solution $\varphi : J \to \mathbb{R}$ of (4.1.3) is given by*

$$\varphi(x) = M(1 - x)L(1 - x) + M(x)[L(x) + c], \qquad x \in J, \qquad (4.2.6)$$

in case M is a projection; by

$$\varphi(x) = L(1 - x) + c, \qquad x \in J, \qquad (4.2.7)$$

in case $M \equiv 1$; and finally by

$$\varphi(x) = bM(1 - x) + cM(x) - b, \qquad x \in J, \qquad (4.2.8)$$

in all other cases. Here $L : J \to \mathbb{R}$ is an arbitrary solution of (4.2.4), and b, c are arbitrary constants.

Proof: Starting with (4.1.3), we apply first Lemma 4.1.3 to obtain the system (3.2.3) - (3.2.5) for $G : D_2 \to \mathbb{R}$ defined by (4.1.4). Next, we look to Corollary 3.5.1 for the solutions of (3.2.3) - (3.2.5).

If M is a projection, then G is given by (3.4.9) for logarithmic L. Inserting this in (4.1.4), we have

$$M(x)L(x) + M(y)L(y) - M(x + y)L(x + y)$$
$$= \varphi(x) + M(1 - x)\varphi\left(\frac{y}{1 - x}\right) - \varphi(x + y), \qquad (x, y) \in D_2. \qquad (4.2.9)$$

Now we show that the function ψ defined by

$$\psi(x) = M(x)L(x) + M(1-x)L(1-x) \qquad (4.2.10)$$

also satisfies the functional equation (compare (4.2.9))

$$M(x)L(x) + M(y)L(y) - M(x+y)L(x+y)$$
$$= \psi(x) + M(1-x)\psi\left(\frac{y}{1-x}\right) - \psi(x+y), \ (x,y) \in D_2. \qquad (4.2.11)$$

Bearing in mind that M is both additive and multiplicative we get

$$M(x)L(x) + M(1-x)L(1-x) + M(1-x)\left[M\left(\frac{y}{1-x}\right)(L(y)\right.$$
$$\left. -L(1-x)) + M\left(1 - \frac{y}{1-x}\right)(L(1-x-y) - L(1-x))\right]$$
$$- M(x+y)L(x+y) - M(1-(x+y))L(1-(x+y))$$
$$= M(x)L(x) + M(1-x)L(1-x) + M(y)L(y) - M(y)L(1-x)$$
$$+ M(1-x-y)L(1-x-y) - M(1-x-y)L(1-x)$$
$$- M(x+y)L(x+y) - M(1-x-y)L(1-x-y)$$
$$= M(x)L(x) + M(y)L(y) - M(x+y)L(x+y).$$

By (4.2.4) and (4.2.10), this is equivalent to (4.2.11). Subtracting (4.2.11) from (4.2.9) we see that the function $d : J \to \mathbb{R}$ defined by

$$d(x) = \varphi(x) - \psi(x)$$
$$= \varphi(x) - M(x)L(x) - M(1-x)L(1-x), x \in J, \qquad (4.2.12)$$

satisfies (4.2.1). Here $M \not\equiv 1$ (since M is additive), so we get from Lemma 4.2.1 that

$$d(x) = b[M(1-x) - 1], \qquad x \in J,$$

or, since M is a projection,

$$d(x) = -bM(x), \qquad x \in J.$$

Together with (4.2.12), this leads to (4.2.6) with $c = -b$.

On the other hand, if M is not a projection, then G is given by (3.3.1) with either $M \equiv 0$ or M not additive. Equation (4.1.4) now becomes

$$cM(x) + cM(y) - cM(x+y)$$
$$= \varphi(x) + M(1-x)\varphi\left(\frac{y}{1-x}\right) - \varphi(x+y), \; (x,y) \in D_2.$$

Because of $M(y) = M(1-x)M\left(\frac{y}{1-x}\right)$, the last equation is also valid when $\phi(x)$ is replaced by $cM(x)$. Forming the difference we find that the function $d : J \to \mathbb{R}$ defined by

$$d(x) = \varphi(x) - cM(x), \quad x \in J, \tag{4.2.13}$$

satisfies (4.2.1). There are two cases to consider.

If $M \equiv 1$, then Lemma 4.2.1 yields

$$d(x) = L(1-x), \quad x \in J,$$

with $L : J \to \mathbb{R}$ (4.2.4) logarithmic. Comparing this with (4.2.13) for $M \equiv 1$, we have (4.2.7).

Finally, if $M \not\equiv 1$ (and either $M \equiv 0$ or M not additive), then Lemma 4.2.1 provides

$$d(x) = b[M(1-x) - 1], \quad x \in J.$$

Inserting this into (4.2.13), we obtain (4.2.8).

Conversely, in each case it can be checked by a direct calculation that the given φ and M satisfy (4.1.3). This finishes the proof.

So we have determined all information functions of multiplicative type, and finally we can give the main result concerning weakly symmetric and generalized recursive information measures.

Theorem 4.2.3. *Let $\{I_n\}$ be a 3-semisymmetric recursive information measure of multiplicative type M. Then, and only then, $\{I_n\}$ is given by*

$$I_n(P) = \begin{cases} M(p_1)L(p_1) + \sum_{i=2}^{n} M(p_i)[L(p_i) + c], & \text{if } M \text{ is a projection} \\ L(p_1) + (n-1)c, & \text{if } M \equiv 1 \\ bM(p_1) + c\sum_{i=2}^{n} M(p_i) - b, & \text{otherwise,} \end{cases} \tag{4.2.14}$$

for all $P \in \Gamma_n^o (n = 2, 3, \ldots)$, where $L : J \to \mathbb{R}$ satisfies (4.2.4), and $b, c \in \mathbb{R}$.

Proof: By Lemma 4.1.1, the map $\varphi : J \to \mathbb{R}$ defined by (4.2.1) satisfies (4.1.3). The possible explicit forms of such φ are given in Theorem 4.2.2. In each case ((4.2.6), (4.2.7), (4.2.8)), we see that, via (4.1.2), equation (4.2.14) holds for $n = 2$.

The validity of (4.2.12) for $n \geq 3$ now follows by induction, using recursivity (3.1.1). Conversely, (4.2.12) is clearly 3-semisymmetric and is recursive by construction.

This is the main result of Aczél and Ng (1983) which crowns the study of recursive and weakly symmetric probabilistic information measures on open domains. Evidently the recursive and 3-semisymmetric measures are 3-symmetric (compare Corollary 3.5.3) exactly when $c = 0$ in the case M is a projection, when $L \equiv 0$ in the case $M \equiv 1$, and when $b = c$ in the other case. So the results are very close (and perhaps expected) in each case except $M \equiv 1$. There the appearance of the logarithmic map L in the weakly symmetric situation was quite a surprise.

We close with a reminder that, although we have dealt with logarithmic maps L only on J in this section, such maps carry a natural extension to $\mathcal{P}$ (cf. Lemma 1.2.12).

4.3. Regular recursive measures

In this section we show how to specialize the general forms of 3-symmetric recursive measures to those which possess some "regularity" property such as measurability (in the sense of Lebesgue). This is a rather difficult problem in the case of measures which are recursive of type M where M is nonzero and additive as well as multiplicative, so that M is a projection.

As an illustration, taking $k = 1$ for the moment, the Shannon entropy $H_n(P) = -\sum_{i=1}^n p_i \log p_i$ is very regular, and it has the sum form $\sum_{i=1}^n f(p_i)$ with the very regular function $f(x) = -x \log x$. The catch is that it also has the sum form with the very "irregular" functions $f(x) = -x \log x + d(x)$ where $d : \mathbb{R} \to \mathbb{R}$ is any nontrivial real derivation. (Remember that d is nonmeasurable and that $\sum_{i=1}^n d(p_i) = 0$, $P \in \Gamma_n^0$.) Thus the function f is exactly of the form (3.4.24) occurring in Theorem 3.4.4, namely $f(x) = xL(x)$, where $L(x) = -\log x + d(x)/x$ and L is logarithmic. So a measure of information can be very regular and simultaneously be of the form $\sum_{i=1}^n M(p_i)L(p_i)$ with a very irregular logarithmic function L. Because of this phenomenon, we shall treat first the simpler case of recursivity of type M where M is not a projection.

Theorem 4.3.1. *Let $M : J \to \mathbb{R}$ be multiplicative but not a projection, and suppose a measure of information $\{I_n\}$ is 3-symmetric and recursive of type M. Then I_2 and I_3 are Lebesgue measurable, if and only if there exist a real constant c and k-tuple α such that*

$$I_n(p_1, \dots, p_n) = c \left[\sum_{i=1}^{n} p_i^{\alpha} - 1 \right], \quad P \in \Gamma_n^o, \tag{4.3.1}$$

($n = 2, 3, \dots$), where α is not a basic unit vector.

Proof: In one direction, it is obvious. For the converse, we may apply either Theorem 3.3.2 in case M is not additive, or Theorem 3.4.4 in case $M \equiv 0$, and obtain the representation

$$I_n(p_1, \dots, p_n) = c \left[\sum_{i=1}^{n} M(p_i) - 1 \right], \quad P \in \Gamma_n^o, \tag{4.3.2}$$

where M is not additive or $M \equiv 0$. If $M \equiv 0$, then we have (4.3.1) with $\alpha = (0, \dots, 0)$. Also, if $c = 0$, (4.3.2) gives (4.3.1) immediately. Now suppose M is not additive and $c \neq 0$.

Using (4.3.2), we calculate that

$$
\begin{aligned}
I_3(tq, &t(1-q), 1-t) - I_2(t, 1-t) \\
&= c\left[M(tq) + M(t(1-q)) - M(t)\right] \\
&= cM(t)\left[M(q) + M(1-q) - 1\right].
\end{aligned}
$$

By hypothesis, this is a Lebesgue measurable function of t. Since M is not a projection, there exists a value of q for which $M(q) + M(1-q) - 1 \neq 0$. Hence M is Lebesgue measurable, and therefore

$$M(t) = t^{\alpha}, \quad t \in J,$$

for some k-tuple α which is not a basic unit vector. With this, (4.3.2) becomes (4.3.1), and the proof is complete.

In the case M is a projection things are not so easy. Let us briefly explain the idea, which will be applied several times in this book. Assuming the measurability of I_2 it is sufficient to solve the fundamental equation of information (see Lemma 4.1.1). So the problem is to find the weakly regular solutions of a functional equation. Now the idea is to produce a differential equation from the original functional equation, then to solve the differential equation and thus to obtain the solutions of the functional equation. But since we have only assumed that our solutions are measurable we want to show that every measurable solution is in fact infinitely often differentiable. Because of the more complicated notion of differentiability in higher dimensions, we solve the fundamental equation first in the 1-dimensional case and then prove the result by induction on the dimension.

Because of the induction step it turns out that we have to consider in the 1-dimensional case a generalized fundamental equation of information with four unknown functions. Let us start with this result, where the main tools will be our general regularity theorem and a theorem on the differentiation of parametric integrals. For convenience we introduce the notation

$$D_2^1 = \{(\xi, \eta) \mid \xi, \eta, \xi + \eta \in]0, 1[\} .$$

Theorem 4.3.2. *Measurable functions $\varphi_1, \varphi_2, \varphi_3, \varphi_4 :]0, 1[\to \mathbb{R}$ satisfy*

$$\varphi_1(\xi) + (1 - \xi)\varphi_2 \left(\frac{\eta}{1 - \xi} \right) = \varphi_3(\eta) + (1 - \eta)\varphi_4 \left(\frac{\xi}{1 - \eta} \right), \qquad (4.3.3)$$

for all $(\xi, \eta) \in D_2^1$, if and only if they are of the form

$$
\begin{aligned}
\varphi_1(\xi) &= a_1 S(\xi) + (b_1 + a_3)\xi + a_4 - a_3 \\
\varphi_2(\xi) &= a_1 S(\xi) + a_2 \xi + a_3 \\
\varphi_3(\xi) &= a_1 S(\xi) + (a_2 + b_2)\xi + a_4 - b_2 \\
\varphi_4(\xi) &= a_1 S(\xi) + b_1 \xi + b_2
\end{aligned}
\qquad (4.3.4)
$$

where $S(\xi) = \xi \log \xi + (1 - \xi) \log(1 - \xi)$ and $a_i, b_j \in \mathbb{R}$ $(1 \le i \le 4, 1 \le j \le 2)$.

Proof. If φ_i, $1 \le i \le 4$, are given by (4.3.4), then the φ_i are measurable and an immediate calculation shows that (4.3.3) is satisfied. Now we prove the converse.

Step 1. We first prove the continuity of φ_1. Let us put in Theorem 1.3.2 $T =]0, 1[$, $D = D_2^1$, $s = r = 1$ and write (4.3.3) as

$$\varphi_1(\xi) = \sum_{i=2}^{4} h_i \left(\xi, \eta, \varphi_i(g_i(\xi, \eta)) \right)$$

with

$$g_2(\xi, \eta) = \frac{\eta}{1 - \xi}, \ g_3(\xi, \eta) = \eta, \ g_4(\xi, \eta) = \frac{\xi}{1 - \eta}$$
$$h_2(\xi, \eta, \zeta) = -(1 - \xi)\zeta, \ h_3(\xi, \eta, \zeta) = \zeta, \ h_4(\xi, \eta, \zeta) = (1 - \eta)\zeta.$$

Then the conditions (1), (2), (3) of Theorem 1.3.2 are satisfied. (In particular, for each $\xi \in]0, 1[$ and for all $(\xi, \eta) \in D_2^1$ we have $\frac{\partial}{\partial \eta} g_i(\xi, \eta) \neq 0$, $i = 2, 3, 4$.) Thus φ_1 is continuous in $]0, 1[$.

In the same manner the continuity of φ_3 can be proved. To prove the continuity of φ_2 and φ_4 we consider the bijective transformation $\tau : D_2^1 \to]0, 1[^2$ given by

$$\tau(\xi, \eta) = (u, t) = \left(\frac{\xi}{1 - \eta}, \frac{\eta}{1 - \xi} \right) \quad (\xi, \eta) \in D_2^1. \tag{4.3.5}$$

with its inverse

$$(\xi, \eta) = \tau^{-1}(u, t) = \left(\frac{u(1 - t)}{1 - ut}, \frac{t(1 - u)}{1 - ut} \right), \quad (u, t) \in]0, 1[^2. \tag{4.3.6}$$

Moreover, we have

$$1 - \xi = \frac{1 - u}{1 - ut}, \text{ and } 1 - \eta = \frac{1 - t}{1 - ut}. \tag{4.3.7}$$

Using (4.3.5)–(4.3.7) we rewrite (4.3.3) as

$$\varphi_2(t) = \frac{1 - ut}{1 - u} \left[\varphi_3 \left(\frac{t(1 - u)}{1 - ut} \right) - \varphi_1 \left(\frac{u(1 - t)}{1 - ut} \right) \right] + \frac{1 - t}{1 - u} \varphi_4(u) \tag{4.3.8}$$

for all $(u, t) \in]0, 1[^2$.

Now we apply Theorem 1.3.2 with $T =]0, 1[$, $D =]0, 1[^2$, $s = r = 1$,

$$G_1(t, u) = \frac{u(1 - t)}{1 - ut}, \ G_3(t, u) = \frac{t(1 - u)}{1 - ut}, \ G_4(t, u) = u,$$
$$H_1(t, u, v) = -\frac{1 - ut}{1 - u} \cdot v = -H_3(t, u, v), \ H_4(t, u, v) = \frac{1 - t}{1 - u} v.$$

Since again $\frac{\partial}{\partial u} G_i(t, u) \neq 0$ for all $(t, u) \in]0, 1[^2$, $i = 1, 3, 4$, we get the continuity of φ_2 on $]0, 1[$. Similarly we see that φ_4 is continuous on $]0, 1[$.

Step 2. We will show that $\varphi_1, \varphi_2, \varphi_3, \varphi_4$ are infinitely often differentiable in $]0,1[$. Let ξ be an arbitrary but fixed element of $]0,1[$. Choose $a, b \in]0,1[$ such that $0 < \xi < \xi + a < \xi + b < 1$. Then for all $\eta \in [a,b]$ we get

$$0 < a < \frac{a}{1-\xi} \le g_2(\xi, \eta) \le \frac{b}{1-\xi} < 1, \quad a \le g_3(\xi, \eta) \le b,$$

and

$$\frac{\xi}{1-a} \le g_4(\xi, \eta) \le \frac{\xi}{1-b} < 1.$$

Thus the image of $[a,b]$ under the map $\eta \mapsto g_i(\xi, \eta)$ is a closed interval of D_2^1 on which φ_i is integrable (since φ_i is continuous), $i = 2, 3, 4$. Integration of (4.3.3) with respect to η on $[a,b]$ and substitution of $g_i(\xi, \eta) = u$ (with the infinitely often differentiable solution $\eta = \gamma_i(\xi, u)$) yields (with $c = \int_a^b \varphi_3(\eta) d\eta$)

$$(b - a)\varphi_1(\xi) = \int_a^b \varphi_3(\eta) d\eta - \int_a^b (1 - \xi)\varphi_2 \left(\frac{\eta}{1 - \xi} \right) d\eta$$

$$+ \int_a^b (1 - \eta)\varphi_4 \left(\frac{\xi}{1 - \eta} \right) d\eta$$

$$= c - (1 - \xi) \int_{g_2(\xi,a)}^{g_2(\xi,b)} \varphi_2(u) \frac{\partial}{\partial u} \gamma_2(\xi, u) du$$

$$+ \int_{g_4(\xi,a)}^{g_4(\xi,b)} \xi u^{-1} \varphi_4(u) \frac{\partial}{\partial u} \gamma_4(\xi, u) du,$$

or

$$(b - a)\varphi_1(\xi) = c - (1 - \xi)^2 \int_{g_2(\xi,a)}^{g_2(\xi,b)} \varphi_2(u) du + \xi^2 \int_{g_4(\xi,a)}^{g_4(\xi,b)} u^{-3} \varphi_4(u) du. \quad (4.3.9)$$

Since φ_2 and φ_4 are continuous, the right hand side of (4.3.9) represents a continuously differentiable function at ξ and thus φ_1 is continuously differentiable (note that $b - a \ne 0$) at ξ. Since ξ was arbitrary, φ_1 is continuously differentiable in $]0, 1[$. In the same manner it can be proved that φ_3, φ_2 and φ_4 are continuously differentiable. (For instance, for fixed $t \in]0, 1[$ and for all $[c, d] \subset]0, 1[$, we have for all $u \in [c, d]$

$$\frac{c(1-t)}{1-ct} \leq G_1(t, u) \leq \frac{d(1-t)}{1-dt}, \quad c \leq G_4(t, u) \leq d,$$

$$\frac{t(1-d)}{1-dt} \leq G_3(t, u) \leq \frac{t(1-c)}{1-ct}.$$

Integration of (4.3.8) with respect to u and substitution of $G_i(t, u) = v$ shows that φ_2 is continuously differentiable in $]0, 1[$.) Now with the knowledge that φ_i is continuously differentiable (for $1 \leq i \leq 4$), we consider again (4.3.9). Using again the theorem concerning the differentiation of parametric integrals (see e.g. Dieudonné (1960)) we deduce that φ_1'' exists and is continuous on $]0, 1[$. In an analogous way $\varphi_2'', \varphi_3'', \varphi_4''$ are continuous on $]0, 1[$. Repeating this process and using the fact that g_i and γ_i $(i = 2, 4)$ are infinitely often differentiable we get the desired result and the proof of Step 2 is complete.

Step 3. Finally we prove that the φ_i's have the form (4.3.4). First we want to find a differential equation for φ_2 and φ_4. Because of the special form of functional equation (4.3.3) we can attain our goal by differentiating (4.3.3) with respect to ξ and the resulting equation with respect to η to obtain

$$\eta(1-\xi)^{-2} \varphi_2'' \left(\frac{\eta}{1-\xi} \right) = \xi(1-\eta)^{-2} \varphi_4'' \left(\frac{\xi}{1-\eta} \right)$$

or

$$(1-\eta) \frac{\eta}{1-\xi} \varphi_2'' \left(\frac{\eta}{1-\xi} \right) = (1-\xi) \frac{\xi}{1-\eta} \varphi_4'' \left(\frac{\xi}{1-\eta} \right). \tag{4.3.9}$$

for all $\xi, \eta, \xi + \eta \in]0, 1[$. Using the transformation τ of Step 1, this last equation goes over into

$$t(1-t) \varphi_2''(t) = s(1-s) \varphi_4''(s) = c \,(\text{say}), \quad s, t \in]0, 1[\tag{4.3.10}$$

where c is some real constant. The solution of this differential equation is (because of $\varphi_i''(t) = \frac{c}{t} + \frac{c}{1-t}$, $i = 1, 2$) given by

$$\varphi_2(t) = a_1 S(t) + a_2 t + a_3$$
$$\varphi_4(t) = a_1 S(t) + b_1 t + b_2$$

where $a_1, a_2, a_3, b_1, b_2 \in \mathbb{R}$. (Note that the coefficient of $S(t)$ in φ_2 and φ_4 is the same due to the same constant c in (4.3.10).) Substitution of φ_2 and φ_4 into (4.3.3) yields (after separation of the variables)

$$\varphi_1(\xi) - a_1\xi \log \xi - a_1(1-\xi)\log(1-\xi) - b_1\xi + a_3(1-\xi)$$
$$= \varphi_3(\eta) - a_1\eta \log \eta - a_1(1-\eta)\log(1-\eta) - a_2\eta + b_2(1-\eta)$$
$$= a_4 \text{ (say)},$$

which gives exactly (4.3.4).

Next we show by an induction argument how to get the k-dimensional measurable solutions of the fundamental equation of information using Theorem 4.3.2.

Theorem 4.3.3. *Let* $M : J \to \mathbb{R}$ *be a projection. Then the general solution* $\varphi : J \to \mathbb{R}$ *of (4.1.3) where φ is measurable in each variable, is given by*

$$\varphi(x) = M(x)(c + A \odot \log x) + M(1-x)(A \odot \log(1-x)), x \in J, \quad (4.3.11)$$

where $A \in \mathbb{R}^k$ *and* $c \in \mathbb{R}$.

Proof: It is an easy calculation to show that φ given by (4.3.11) satisfies (4.1.3). To prove the converse we use induction on the dimension k. If $k = 1$ then an application of Theorem 4.3.2 with the additional requirement $\varphi_1 = \varphi_2 = \varphi_3 = \varphi_4 = \varphi$ in (4.3.4) yields $\varphi(\xi) = a_1 S(\xi) + a_2\xi$, so that φ has the form (4.3.11). Now suppose $k \geq 2$ and let φ be a measurable solution of (4.1.3), where we may assume that $M(x) = M(\xi_1, \ldots, \xi_k) = \xi_1$. With the notation $x = (\xi, u)$ and $y = (\eta, v)$ (with $\xi = \eta_1$, $u = (\xi_2, \ldots, \xi_k)$, $\eta = \eta_1, v = (\eta_2, \ldots, \eta_k)$) (4.1.3) goes over into

$$\varphi(\xi, u) + (1-\xi)\varphi\left(\frac{\eta}{1-\xi}, \frac{v}{1-u}\right)$$
$$= \varphi(\eta, v) + (1-\eta)\varphi\left(\frac{\xi}{1-\eta}, \frac{u}{1-v}\right), \quad (4.3.12)$$

for all $\xi, \eta, \xi + \eta \in]0,1[$ and $u, v, u + v \in]0,1[^{k-1}$. Fixing u and v temporarily and defining

$$\left.\begin{array}{l} \varphi_1(\tau) := \varphi(\tau, u), \quad \varphi_2(\tau) := \varphi\left(\tau, \dfrac{v}{1-u}\right) \\[3mm] \varphi_3(\tau) := \varphi(\tau, v), \quad \varphi_4(\tau) := \varphi\left(\tau, \dfrac{u}{1-v}\right) \end{array}\right\} \quad (4.3.13)$$

for all $\tau \in]0, 1[$, equation (4.3.12) becomes (4.3.3). Since φ is measurable in each variable, the φ_i's are measurable and are given by (4.3.4). Comparing φ_1 and φ_3 in (4.3.4) with (4.3.13) and letting u, v vary again we get

$$\varphi(\xi, u) = a_1 S(\xi) + a_2(u)\xi + a_3(u), \quad \xi \in]0, 1[, \quad u \in]0, 1[^{k-1}. \qquad (4.3.14)$$

We substitute (4.3.14) into (4.1.3) and get, after cancellations,

$$\left(a_2(u) - a_3 \left(\frac{v}{1-u} \right) - a_2 \left(\frac{u}{1-v} \right) \right) \xi$$
$$+ a_3(u) + a_3 \left(\frac{v}{1-u} \right) - a_3(v) - a_3 \left(\frac{u}{1-v} \right)$$
$$= \left(a_2(v) - a_3 \left(\frac{u}{1-v} \right) - a_2 \left(\frac{v}{1-u} \right) \right) \eta.$$

Comparison of the coefficients of ξ and 1 yields

$$a_2(u) - a_3 \left(\frac{v}{1-u} \right) = a_2 \left(\frac{u}{1-v} \right) \qquad (4.3.15)$$

and

$$a_3(u) + a_3 \left(\frac{v}{1-u} \right) = a_3(v) + a_3 \left(\frac{u}{1-v} \right). \qquad (4.3.16)$$

Since (4.3.14) can be written as

$$\varphi(\xi, u) - a_1 S(\xi) = a_2(u)\xi + a_3(u), \qquad (4.3.17)$$

and since the left hand side in (4.3.17) is measurable in each variable we conclude that a_2 and a_3 are measurable in each variable, too. (Take different points $\xi_1, \xi_2 \in]0, 1[$. Substituting ξ_1 and ξ_2 into (4.3.17) and solving the obtained linear system for a_2 and a_3 we get

$$a_2(u) = (\xi_1 - \xi_2)^{-1} [\varphi(\xi_1, u) - a_1 S(\xi_1) - \varphi(\xi_2, u) + a_1 S(\xi_2)]$$
$$a_3(u) = (\xi_2 - \xi_1)^{-1} [\xi_2(\varphi(\xi_1, u) - a_1 S(\xi_1)) - \xi_1(\varphi(\xi_2, u) - a_1 S(\xi_2))]$$

which are measurable functions in each argument). Adding (4.3.15) and (4.3.16) and using the bijective transformation $\iota : D_2^{k-1} \to]0, 1[^{2(k-1)}$ given by

$$\iota(u, v) = (p, q) = \left(\frac{u}{1-v}, 1-v \right), (u, v) = \iota^{-1}(p, q) = (pq, 1-q)$$

we get the Pexider equation

$$(a_2 + a_3)(pq) = (a_2 + a_3)(p) + a_3(1 - q), \quad p, q \in]0, 1[^{k-1}.$$

Thus we get from Lemma 1.2.16 and Example 1.3.5(b)

$$(a_2 + a_3)(u) = A' \odot \log u + c, \ a_3(u) = A' \odot \log(1 - u) \qquad (4.3.18)$$

where $A' \in \mathbb{R}^{k-1}$ and $c \in \mathbb{R}$. Moreover (4.3.18) implies

$$a_2(u) = (a_2 + a_3)(u) - a_3(u) = A' \odot (\log u - \log(1 - u)) + c \qquad (4.3.19)$$

Thus (4.3.14) together with (4.3.18) and (4.3.19) leads to

$$
\begin{aligned}
\varphi(x) = \varphi(\xi, u) &= a_1 \big(\xi \log \xi + (1 - \xi) \log(1 - \xi)\big) \\
&\quad + \big[A' \odot (\log u - \log(1 - u)) + c\big]\xi + A' \odot \log(1 - u) \\
&= \xi(a_1 \log \xi + A' \odot \log u) + (1 - \xi)\big(a_1 \log(1 - \xi) \\
&\quad + A' \odot \log(1 - u)\big) + c\xi \\
&= M(x)(c + A \odot \log x) + M(1 - x)\big(A \odot \log(1 - x)\big)
\end{aligned}
$$

where $A = (a_1, A') \in \mathbb{R}^k$. This completes the proof.

We now prove the second main result in this section.

Theorem 4.3.4. *Let $M : J \to \mathbb{R}$ be a projection, and suppose a measure of information $\{I_n\}$ is 3-symmetric and recursive of type M. Then I_2 is Lebesgue measurable, if and only if there exist real constants $c_1, \ldots, c_k$ such that*

$$I_n(P) = \sum_{i=1}^{n} M(p_i) \sum_{j=1}^{k} c_j \log \pi_{ji} \qquad (4.3.20)$$

for all $P = (p_1, \ldots, p_n) \in \Gamma_n$, where $p_i = (\pi_{1i}, \ldots, \pi_{ki})$ for $i = 1, 2, \ldots, n$.

Proof: In one direction, it is a simple verification. For the converse, the function $\varphi : J \to \mathbb{R}$ defined by (4.1.2) is a symmetric information function of type M (see Lemma 4.1.1 and Remark 4.1.2) and has thus the form given in (4.3.11) with $c = 0$ (due to the symmetry condition $\varphi(x) = \varphi(1 - x)$). By recursivity we get the desired result

$$I_n(P) = \sum_{i=1}^{n} M(p_i)L(p_i)$$

where $L(p_i) = A \odot \log p_i$ with $A = (c_1, \ldots, c_k) \in \mathbb{R}^k$ and $\log p_i = (\log \pi_{1i}, \ldots, \log \pi_{ki})$.

Remark 4.3.5. The proof given here of Theorems 4.3.2-4.3.4 is due to the third author (unpublished) and makes use of ideas first developed in a paper of Tverberg (1958). For more results on generalizations of equation (4.3.3) we refer to Sander (1994), and Járai and Sander (1997).

Remark 4.3.6. We give now the announced brief outline of the original proof of Theorem 3.4.4 as proposed in Aczél (1980) and carried out in Aczél and Ng (1983). The idea is the same as in Theorem 4.3.3. Using induction on the dimension k we arrive at equation (4.3.3) which has been called the "generalized" (i.e. four function) fundamental equation of information (in one dimension). The problem of solving it (without measurability conditions) remained open for a rather long time until Maksa (1982) gave the general solution. First, he reduced it to the two (separate) problems of solving the systems

$$F(u+v, w) + G(u, v) + F(u+w, v) + G(u, w) = 0$$
$$F(tu, tv) = tF(u, v), \quad G(tu, tv) = tG(u, v)$$

and

$$H(u+v, w) + K(u, v) = H(u+w, v) + K(u, w)$$
$$H(tu, tv) = tH(u, v), \quad K(tu, tv) = tK(u, v)$$

for all $t, u, v, w > 0$. For these, it is necessary to apply, among other things, Theorem 3.4.1 twice (once for each system) and Theorem 2.2.4. Eventually one gets(for $1 \leq i \leq 4$)

$$\varphi_i(\xi) = \xi\ell(\xi) + (1-\xi)\ell(1-\xi) + a_i\xi + b_i, \; \xi > 0 \qquad (4.3.21)$$

where $\ell : \mathcal{P}_1 \to \mathbb{R}$ is logarithmic and the constants satisfy various conditions (compare with the measurable solutions φ_i presented in (4.3.4)). Then, Aczél and Ng used (4.3.21) in (4.3.13) to write φ in the form

$$\varphi(\xi, u) = \xi\ell(\xi) + (1-\xi)\ell(\xi) + a_1(u)\xi + a_2(u) \qquad (4.3.22)$$

for all $\xi > 0$, $u \in \mathcal{P}_{k-1}$. Substitution of (4.3.22) into the fundamental equation of information leads to (4.3.15) and (4.3.16), the general solutions of which can be determined as in the proof of Theorem 4.3.3. Finally, one arrives at

$$\varphi(x) = M(x)(L(x) + c) + M(1 - x)L(1 - x), \quad x \in J.$$

If φ is symmetric then $c = 0$. As in Theorem 4.3.4, one gets (3.4.44).

Remark 4.3.7. On the closed domain an improvement of Theorem 4.3.3 is known. If $k = 1$ and M is the identity function then Diderrich (1975, 1978, 1979, 1986) and Maksa (1980) showed that $\varphi(x) = cS(x)$, $x \in [0, 1]$, if $\varphi : [0, 1] \to \mathbb{R}$ is a symmetric solution of (4.1.3) (valid for $x, y \in [0, 1[$, $x + y \leq 1$) which is bounded on a set of positive Lebesgue measure. Maksa's proof for this result is simpler than the proof of Diderrich. He uses a result of deBruijn (1951) concerning the double difference property for bounded functions and a Steinhaus-type theorem.

Remark 4.3.8. We note also that a proof can be given for Theorem 4.3.4 which follows the same line of argument as the proof of Theorem 4.3.1 (first author, unpublished). Applying Theorem 3.4.4, we have the representation (3.4.44)

$$I_n(P) = I_n(p_1, ..., p_n) = \sum_{i=1}^{n} M(p_i)L(p_i).$$

Since M is a projection, we may assume for the moment that $M(p) = M(\pi_1, ..., \pi_k) = \pi_1$ and compute

$$\begin{aligned} I_3(p_1, p_2, p_3) &- I_2(p_1 + p_2, p_3) \\ &= \pi_{11}L(p_1) + \pi_{12}L(p_2) - (\pi_{11} + \pi_{12})L(p_1 + p_2), \end{aligned} \tag{4.3.23}$$

where $p_i = (\pi_{1i}, ..., \pi_{ki})$, for $i = 1, 2, 3$. If I_2 and I_3 are measurable, then the right hand side of (4.3.23) is measurable. Next, we write $L(p) = L(\pi_1, ...\pi_k) = \sum_{j=1}^{k} l_j(\pi_j)$ with each l_j logarithmic (cf. Lemma 1.2.12), and define

$$G_1(p_1, p_2) := \pi_{11}l_1(\pi_{11}) + \pi_{12}l_1(\pi_{12}) - (\pi_{11} + \pi_{12})l_1(\pi_{11} + \pi_{12}),$$

$$G_2(p_1, p_2) := \pi_{11} \sum_{j=2}^{k} l_j(\pi_{j1}) + \pi_{12} \sum_{j=2}^{k} l_j(\pi_{j2})$$

$$- (\pi_{11} + \pi_{12}) \sum_{j=2}^{k} l_j(\pi_{j1} + \pi_{j2}).$$

Now (4.3.23) can be written as

$$I_3(p_1, p_2, p_3) - I_2(p_1 + p_2, p_3) = G_1(p_1, p_2) + G_2(p_1, p_2),$$

so that $G_1 + G_2$ is λ^k-measurable in each variable. Simple calculations then show that both G_1 and G_2 are measurable. In fact, G_1 is really a function defined on D_2^1, and it can be extended in a straightforward way to a measurable function on $\mathbb{R}^2$. One can then apply a result of Laczkovich (1980) concerning the *double difference property* for measurable functions and conclude that

$$\pi l_1(\pi) = s(\pi) + a(\pi),$$

with s measurable and a additive ($\pi \in]0,1[$). Since l_1 is logarithmic, it is not too difficult to show now that

$$\pi l_1(\pi) = c_1 \pi \log \pi + [a(\pi) - \pi a(1)]. \tag{4.3.24}$$

Turning to G_2, since it is measurable in each variable it is easy to see that each l_j is measurable. Hence we get

$$l_j(\pi) = c_j \log \pi \qquad (2 \leq j \leq k).$$

Combining this with (4.3.24), we conclude that

$$I_n(P) = \sum_{i=1}^{n} M(p_i) L(p_i)$$

$$= \sum_{i=1}^{n} \pi_{1i} \sum_{j=1}^{k} l_j(\pi_{ji})$$

$$= \sum_{i=1}^{n} \pi_{1i} \sum_{j=1}^{k} c_j \log \pi_{ji}.$$

After recovering full generality by replacing π_{1i} by $M(p_i)$, this is (4.3.20).

Example 4.3.9. Let us point out that Theorem 4.3.1 and Theorem 4.3.4 contain many well-known characterization theorems as special cases. We mentioned already, at the end of section 3.3, that if α is not a basic unit vector (i.e. if M is not a projection), then the only information measures $\{I_n\}$ which are 3-symmetric and recursive of degree α are those of the form

$$I_n(p_1, \ldots, p_n) = c \left[\sum_{i=1}^{n} p_i^{\alpha} - 1 \right], \qquad P \in \Gamma_n^o, \tag{4.3.25}$$

for some constant c. Here the constant c serves only to determine the unit of information. If $k = 1$, then the normalization condition $I_2\left(\frac{1}{2},\frac{1}{2}\right) = 1$ leads just to the entropy of degree α (cf. (1.1.3)):

$$H_n^\alpha(p_1,\ldots,p_n) = (2^{1-\alpha} - 1)^{-1}\left(\sum_{i=1}^n p_i^\alpha - 1\right).$$

The justification for considering also negative values of α is, for example, given by the connection with the hyperbolic information

$$Y_n(P) = \left(\sum_{i=1}^n p_i^{-1} - 1\right) = 3H_n^{-1}(P).$$

If $k = 2$, then for $p_i = (\pi_{1i}, \pi_{2i})$ and $\alpha = (\alpha_1, \alpha_2)$, (4.3.25) becomes

$$I_n\begin{pmatrix}\pi_{11},\ldots,\pi_{1n}\\\pi_{21},\ldots,\pi_{2n}\end{pmatrix} = c\left[\sum_{i=1}^n \pi_{1i}^{\alpha_1}\pi_{2i}^{\alpha_2} - 1\right]. \qquad (4.3.26)$$

If we are looking for a measure of error (or inaccuracy) resulting from the choice of $(\pi_{21},\ldots,\pi_{2n})$ to estimate $(\pi_{11},\ldots,\pi_{1n})$, then a reasonable condition would be

$$I_2\begin{pmatrix}\frac{1}{2},\frac{1}{2}\\\frac{1}{2},\frac{1}{2}\end{pmatrix} = 0. \qquad (4.3.27)$$

Imposing this on (4.3.26), and requiring $c \neq 0$ (to avoid the trivial measure $I_n \equiv 0$ for all n) leads to $\alpha_2 = 1 - \alpha_1$. In this case, writing α now for α_1, we get for (4.3.26) a constant multiple of the error $\{E_n^\alpha\}$ of degree $\alpha(\neq 0,1)$:

$$I_n\begin{pmatrix}\pi_{11},\ldots,\pi_{1n}\\\pi_{21},\ldots,\pi_{2n}\end{pmatrix} = c\left[\sum_{i=1}^n \pi_{1i}^{\alpha}\pi_{2i}^{1-\alpha} - 1\right] = a\,E_n^\alpha(P)$$

(cf. (1.1.7)). Again, the value of a (or c) is determined by the definition of a unit of information (in this case, error).

In case α is a basic unit vector (say $\alpha = (0,\ldots,0,1,0,\ldots,0)$ with the "1" in the j-th coordinate position) then the only $\{I_n\}$ which are 3-symmetric and recursive of degree α are those of the form (3.4.44):

$$I_n(p_1,\ldots,p_n) = \sum_{i=1}^n \pi_{ji}\,L(p_i), \qquad P \in \Gamma_n^o,$$

where $p_i = (\pi_{1i}, \ldots, \pi_{ki})$ for $i = 1, \ldots, n$ (and $M(p_i) = \pi_{ji}$ for some fixed $j \in \{1, \ldots, k\}$). As we have seen, if I_2 is measurable, then we can choose an L with the same type of regularity, so that

$$L(p) = L(\pi_1, \ldots, \pi_k) = \sum_{m=1}^{k} c_m \log \pi_m, \quad p \in J,$$

in which case (3.4.44) takes the form

$$I_n(p_1, \ldots p_n) = \sum_{i=1}^{n} \pi_{ji} \sum_{m=1}^{k} c_m \log \pi_{mi}, \quad P \in \Gamma_n^o. \tag{4.3.28}$$

In particular, if $k = 1$ and we choose the usual normalization condition $I_2\left(\frac{1}{2}, \frac{1}{2}\right) = 1$, then we get exactly the Shannon entropy (cf. (1.1.1)):

$$H_n(p_1, \ldots, p_n) = -\sum_{i=1}^{n} p_i \log p_i.$$

If $k = 2$, then (4.3.28) takes the form

$$I_n \begin{pmatrix} \pi_{11}, \ldots, \pi_{1n} \\ \pi_{21}, \ldots, \pi_{2n} \end{pmatrix} = \sum_{i=1}^{n} \pi_{ji} (c_1 \log \pi_{1i} + c_2 \log \pi_{2i}),$$

where j is either 1 or 2. Without loss of generality, we take $j = 1$. Then the error condition (4.3.27) leads us to the conclusion $c_2 = -c_1$. Thus, we have a constant multiple of Kullback's error

$$I_n \begin{pmatrix} \pi_{11}, \ldots, \pi_{1n} \\ \pi_{21}, \ldots, \pi_{2n} \end{pmatrix} = c_1 E_n(P) = c_1 \sum_{i=1}^{n} \pi_{1i} \log_2 \frac{\pi_{1i}}{\pi_{2i}}$$

(cf. (1.1.6)). Moreover, this and all measures of the form (4.3.28), again taking $j = 1$, can be written as a linear combination of the Shannon entropy and Kerridge's inaccuracy (cf. (1.1.5))

$$K_n \begin{pmatrix} \pi_{11}, \ldots, \pi_{1n} \\ \pi_{21}, \ldots, \pi_{2n} \end{pmatrix} = -\sum_{i=1}^{n} \pi_{1i} \log_2 \pi_{2i},$$

namely

$$I_n(p_1, \ldots, p_n) = -c_1 H_n(\pi_{11}, \ldots, \pi_{1n}) - \sum_{m=2}^{k} c_m K_n \begin{pmatrix} \pi_{11}, \ldots, \pi_{1n} \\ \pi_{m1}, \ldots, \pi_{mn} \end{pmatrix}.$$

Thus we get, with appropriate additional conditions to determine the constants $c_1, \ldots, c_k$, characterization theorems for Shannon's entropy $\{H_n\}$, Kullback's error $\{E_n\}$ and Kerridge's inaccuracy $\{K_n\}$ among others.

Finally, let us mention that more general types of recursivity were introduced and studied in the first author's Ph.D. thesis, and the relevant results can be found in Ebanks (1976) and (1983a). In those cases it turned out that, under rather mild conditions, such generalized recursivities reduce to recursivity of multiplicative type. This gives added weight to the argument presented at the beginning of section 3.2.

4.4. Summary

Finally we give a brief summary of the main ideas and results of the Chapters 2, 3, and 4. In Chapter 2 we first determined all 4-symmetric and then all 4-semisymmetric information measures with the branching property (with the generating function G; see (2.1.1)). They are of the sum form

$$I_n(P) = \sum_{i=1}^{n} g(p_i), \quad P \in \Gamma_n^0, \tag{4.4.1}$$

respectively

$$I_n(P) = f(p_1) + \sum_{i=2}^{n} g(p_i) + h(p_n), \quad P \in \Gamma_n^0, \tag{4.4.2}$$

where f, g, h are arbitrary real-valued functions defined on J.

In the first case we arrived at the symmetric characteristic functional equation for branching measures of information

$$G(x, y) + G(x + y, z) = G(x, z) + G(x + z, y), \quad (x, y, z) \in D_3, \tag{4.4.3}$$

$$G(x, y) = G(y, x), \quad (x, y) \in D_2, \tag{4.4.4}$$

with the solution

$$G(x, y) = f(x) + f(y) - f(x + y), \quad (x, y) \in D_2, \tag{4.4.5}$$

for some $f : J \to \mathbb{R}$.

In the second case we were led only to equation (4.4.3) which has the solution

$$G(x, y) = f(x) + g(y) - f(x + y), \tag{4.4.6}$$

with $f, g : J \to \mathbf{R}$. Because of Lemma 2.1.2 the results of (4.4.1) and (4.4.2) are obtained from (4.4.5) and (4.4.6), respectively.

In Chapter 3 we specialized the generating function G to

$$G(x,y) = M(x + y)I_2 \left(\frac{x}{x + y}, \frac{y}{x + y} \right), \quad (x,y) \in D_2, \tag{4.4.7}$$

where $M : J \to \mathbf{R}$ is multiplicative. With this special function G we called $\{I_n\}$ a recursive information measure of multiplicative type M (recursive for short in this summary). A first key result was Lemma 3.2.2. If $\{I_n\}$ is 3-symmetric and recursive then $\{I_n\}$ is 4-symmetric and branching with generating function G, which satisfies in addition to (4.4.3) and (4.4.4) the M-homogeneity

$$G(tx, ty) = M(t)G(x, y) \tag{4.4.8}$$

whenever $t \in J, (x, y) \in D_2$. The interesting thing is that because of this one additional property (4.4.8) the function G can be determined very explicitly (see Corollary 3.5.1) as

$$G(x,y) = \begin{cases} M(x)L(x) + M(y)L(y) \\ \quad -M(x + y)L(x + y), & M \text{ projection} \\ c[M(x) + M(y) - M(x + y)], & \text{otherwise} \end{cases} \tag{4.4.9}$$

where $c \in \mathbf{R}$ and $L : \mathcal{P} \to \mathbf{R}$ is an arbitrary logarithmic map. Note that the function f in (4.4.5) is now $M \cdot L$ or $c \cdot M$ (up to an added additive function). To obtain all 3-symmetric recursive information measures we observed that by (4.4.7) we have (if $M \neq 0$)

$$I_2 \left(\frac{x}{x + y}, \frac{y}{x + y} \right) = [M(x + y)]^{-1}G(x, y), \quad (x, y) \in D_2, \tag{4.4.10}$$

which means that we have not only G but also I_2, and thus (by induction or by Lemma 2.1.2)

$$I_n(P) = \begin{cases} \displaystyle\sum_{i=1}^{n} M(p_i)L(p_i), & \text{if } M \text{ is a projection} \\ c \left(\displaystyle\sum_{i=1}^{n} M(p_i) - 1 \right), & \text{otherwise} . \end{cases} \tag{4.4.11}$$

(See Corollary 3.5.3.)

In Chapter 4 we examined the combination of symmetry and recursivity again, but from a different viewpoint. To obtain all 3-semisymmetric, recursive measures $\{I_n\}$ we could have taken a similar approach as in Chapter 3, deducing that $\{I_n\}$ is branching and 4-weak-symmetric, so that I_n has a special form of (4.4.2), namely

$$I_n(p) = k(p_1) + \sum_{i=1}^{n} g(p_i), \quad P \in \Gamma_n^0, \tag{4.4.12}$$

with generating function

$$G(x, y) = k(x) + g(y) - k(x + y), \quad (x, y) \in D_2, \tag{4.4.13}$$

for some functions $k, g : J \to \mathbb{R}$. Moreover G is M-homogeneous (4.4.8) but no longer symmetric (4.4.4). Instead of solving the system (4.4.13) and (4.4.8) we chose another way. Introducing the function $\varphi : J \to \mathbb{R}$ defined by

$$\varphi(p) := I_2(1 - p, p), \quad p \in J, \tag{4.4.14}$$

we found that φ satisfies the fundamental equation of information

$$\varphi(x) + M(1 - x)\varphi\left(\frac{y}{1 - x}\right) = \varphi(y) + M(1 - y)\varphi\left(\frac{x}{1 - y}\right) \tag{4.4.15}$$

for all $(x, y) \in D_2$. (If $\{I_n\}$ is 3-symmetric, then φ is also symmetric, that is

$$\varphi(x) = \varphi(1 - x).) \tag{4.4.16}$$

In order to determine I_n, it is sufficient to determine I_2, that is φ. Thus all information concerning I_n is concentrated in φ and in the functional equation (4.4.15). This is in analogy to Chapter 2 where all information concerning symmetric branching information measures I_n is given by the generating function G satisfying the equation (4.4.3) (and (4.4.4), in the case of full symmetry.)

Defining now $G' : D_2 \to \mathbb{R}$ by

$$G'(x, y) = \varphi(x) + M(1 - x)\varphi\left(\frac{y}{1 - x}\right) - \varphi(x + y), \quad (x, y) \in D_2 \tag{4.4.17}$$

where φ satisfies (4.4.14) (but not (4.4.16)), G' satisfies (4.4.3), (4.4.4) and (4.4.8). This means that G' is given by (4.4.9), and using (4.4.17) we get rather quickly the solution

$$
\varphi(x) = \begin{cases}
M(1-x)L(1-x) \\
\quad +M(x)[L(x)+c], & M \text{ projection} \\
L(1-x)+c, & M \equiv 1 \\
bM(1-x)+cM(x)-b, & \text{otherwise,}
\end{cases} \tag{4.4.18}
$$

which yields $\{I_n\}$ given by (4.2.14). We see that indeed $I_n(P)$ has the form (4.4.12) where k and g are given by (up to the addition of an additive function and a constant)

$$
\left.
\begin{aligned}
k(x) &= \begin{cases}
M(x)L(x), & \text{if } M \text{ is a projection,} \\
L(x), & \text{if } M = 1 \\
b[M(x)-1], & \text{otherwise,}
\end{cases} \\[2em]
g(x) &= \begin{cases}
M(x)[L(x)+c], & \text{if } M \text{ is a projection,} \\
c, & \text{if } M = 1 \\
cM(x), & \text{otherwise,}
\end{cases}
\end{aligned}
\right\} \tag{4.4.19}
$$

To determine the regular 3-symmetric measures of information $\{I_n\}$ which are recursive of type M, where M is multiplicative but not a projection, we see almost immediately that

$$
I_n(P) = c\left[\sum_{i=1}^n p_i^\alpha - 1\right], \quad P \in \Gamma_n^0 \tag{4.4.20}
$$

if I_2 and I_3 are Lebesgue measurable (here α is not a basic unit vector). Also we remark that if $\{I_n\}$ is 3-symmetric, α-recursive (where α is not a basic unit vector) then $\{I_n\}$ has the form (4.4.20) without any regularity condition.

If M is a projection then we see that, if I_2 is measurable, all 3-symmetric measures of information $\{I_n\}$ which are recursive of type M have the expected form

$$
I_n(P) = \sum_{i=1}^n M(p_i)(A \odot \log P_i) \tag{4.4.21}
$$

for some constant vector $A \in \mathbb{R}^k$, but some additional machinery is needed to prove this result. The advantage of all our considerations is a systematic treatment of measures with the branching property or property of recursivity, pointing out the connections between these types of measures and presenting simultaneously one general result from which all recent results in this area can be obtained rather easily.

CHAPTER 5

SUM FORM INFORMATION MEASURES AND ADDITIVITY PROPERTIES

5.1. Introduction

In Theorem 2.3.1, we saw that every 4-symmetric branching information measure $\{I_n\}$ has the sum form

$$I_n(P) = \sum_{i=1}^{n} F(p_i), \quad P \in \Gamma_n^o \quad (n = 2, 3, \ldots) \tag{5.1.1}$$

for some generating function $F : J \to \mathbb{R}$. In chapters 3 and 4, we found special measures of the form (5.1.1) resulting from strengthening the branching property to a recursivity property.

In the last four chapters of the book, we shall again find special measures of the form (5.1.1), called *sum form information measures*, but not by strengthening the branching property. Rather, we shall impose different kinds of conditions, called additivity (or generalized additivity) properties.

Before introducing the idea of additivity, let us note here that, while the general solutions of all problems posed up to this point in the book are known (and have been presented), the same is not true beyond this point. The reader will learn about some problems for which the general solutions are not known. On the other hand, the general *measurable* solutions of all problems will be given. Thus for the purposes of applications we still have a rather complete theory.

5.2. Additivity

As we did for branching and recursivity, we start with an intuitive motivation for (generalized) additivity properties. Suppose we want to know the outcomes of two "independent" (in the sense of probability theory) experiments A and B performed on the same sample space. That is, there are two independent characteristics to be identified. Let us say that experiment A may result in one of the outcomes $E_1, E_2, \ldots, E_\ell$, with corresponding probabilities $p_1, p_2, \ldots, p_\ell$, and that experiment B has outcomes $F_1, F_2, \ldots, F_m$, with corresponding probabilities $q_1, q_2, \ldots, q_m$. Here, independence means that the probability of realizing outcome E_i from experiment A and outcome F_j from experiment B is $p_i q_j$ ($i = 1, \ldots, \ell; j = 1, \ldots, m$). Thus, we could just as well perform the combined experiment C with outcomes $E_i \cap F_j$ and corresponding probabilities $p_i q_j$ ($i = 1, \ldots, \ell; j = 1, \ldots, m$). The basic assumption underlying the additivity property is that the information obtained from experiment C is the sum of the amounts of information received from experiments A and B. Now we make this idea precise.

Definition 5.2.1. For a fixed pair (ℓ, m) of integers greater than or equal to two, an information measure $\{I_n\}$ is said to be (ℓ, m)-*additive*, provided that

$$I_{\ell m}(P * Q) = I_\ell(P) + I_m(Q), \quad P \in \Gamma^o_\ell, Q \in \Gamma^o_m. \qquad (5.2.1)$$

Here $P * Q := (p_1 q_1, \ldots, p_1 q_m, \ldots, p_\ell q_1, \ldots, p_\ell q_m) \in \Gamma^o_{\ell m}$.

Therefore, an information measure $\{I_n\}$ which has the sum form (5.1.1) and is (ℓ, m)-additive is characterized by the functional equation

$$\sum_{i=1}^{\ell} \sum_{j=1}^{m} F(p_i q_j) = \sum_{i=1}^{\ell} F(p_i) + \sum_{j=1}^{m} F(q_j), \quad P \in \Gamma^o_\ell, Q \in \Gamma^o_m. \qquad (5.2.2)$$

If we can solve this equation for $F : J \to \mathbb{R}$, then (5.1.1) gives the explicit form of our information measure. We shall give all measurable solutions of (5.2.2) in Chapter 7. (The general solution of (5.2.2) is not known for any pair (ℓ, m).)

The objection may be raised that the (ℓ, m)-additivity imposes a certain "linearity" on the structure of $\{I_n\}$. (Cf. the case $M(p) = p$, $k = 1$, in Definition 3.1.1.) In fact, while the Shannon entropy satisfies (5.2.1), the entropies of degree $\alpha \neq 1$ do not. Thus, generalizations of (5.2.1) have been introduced to allow some additional flexibility in the choice of an information measure for certain applications. We propose two such generalizations in the next two sections.

5.3. Generalized additivities

Referring to our discussion at the beginning of the previous section, we may suppose (more generally) that the information obtained from experiment C is given by some polynomial function of the amounts of information received from experiments A and B.

Definition 5.3.1. An information measure $\{I_n\}$ is said to be *polynomially additive* if there exists a polynomial $g : \mathbb{R}^2 \to \mathbb{R}$ for which

$$I_{\ell m}(P * Q) = g[I_\ell(P), I_m(Q)], \quad P \in \Gamma_\ell^o, Q \in \Gamma_m^o, \tag{5.3.1}$$

holds for all $\ell, m \in \{2, 3, \ldots\}$.

Obviously, (5.2.1) is the special case $g(x, y) = x + y$ of (5.3.1). Surprisingly, however, (5.3.1) is not such an extreme generalization of (5.2.1) as it may appear at first glance. In order to demonstrate this, we prove first a lemma.

Lemma 5.3.2. *Suppose that $F : J \to \mathbb{R}$ satisfies*

$$\sum_{i=1}^{\ell} \sum_{j=1}^{m} F(p_i q_j) = C, \quad P \in \Gamma_\ell^o, Q \in \Gamma_m^o, \tag{5.3.2}$$

for some constant C and for $(\ell, m) \in \{(2, 2), (2, 3)\}$. Then, and only then, F is the restriction to J of an additive map on $\mathbb{R}^k$, with $F(1) = C$.

Proof: Clearly, additivity with $F(1) = C$ implies (5.3.2). For the converse, apply (5.3.2) twice, first for $P = (1 - p, p)$ and $Q = (1 - s - t, s + t)$, then for $P = (1 - p, p)$ and $Q = (1 - s - t, s, t)$. Comparing the results, we find that

$$F((1 - p)(s + t)) + F(p(s + t)) \\ = F((1 - p)s) + F(ps) + F((1 - p)t) + F(pt), \tag{5.3.3}$$

for all $p \in J$, $(s, t) \in D_2$. Taking $p = \frac{1}{2}$, we get

$$F\left(\frac{1}{2}(s + t)\right) = F\left(\frac{1}{2}s\right) + F\left(\frac{1}{2}t\right), \quad (s, t) \in D_2, \tag{5.3.4}$$

which means that F is additive on $\frac{1}{2}D_2$. Thus, restricting p to $]\frac{1}{2}, 1[^k$ in (5.3.3), we obtain by (5.3.4) that

$$F(p(s + t)) = F(ps) + F(pt), \quad p \in \left]\frac{1}{2}, 1\right]^k, \quad (s, t) \in D_2. \tag{5.3.5}$$

But any $(x, y) \in D_2$ can be written as (ps, pt) for $p \in]\frac{1}{2}, 1]^k$ and $(s, t) \in D_2$. (Let $\rho = max\{\frac{1}{2}, \xi_1 + \eta_1, \ldots, \xi_k + \eta_k\}$, $p = \left(\frac{1+\rho}{2}, \ldots, \frac{1+\rho}{2}\right)$, and $(s, t) = (p^{-1}x, p^{-1}y)$.) Hence (5.3.5) shows that F is additive on D_2. As we know, any such map is the restriction to J of an additive map on $\mathbb{R}^k$. Finally, (5.3.2) yields also $F(1) = C$, and the proof is complete.

We shall prove results more general than Lemma 5.3.2 later. For now, this is all we need to establish the following result (see Behara and Nath (1974), Ebanks (1984c)).

Theorem 5.3.3. *Suppose the information measure $\{I_n\}$ is (5.3.1) polynomially additive with the sum form (5.1.1), and that the range $\{I_n(\Gamma_n^o) \mid n = 2, 3, \ldots\}$ has infinite cardinality. Then, either*

(i) $g(x, y) = x + y + C$ for some constant C, so that (5.3.1) takes the form

$$I_{\ell m}(P * Q) = I_\ell(P) + I_m(Q) + C, \quad P \in \Gamma_\ell^o, Q \in \Gamma_m^o, \tag{5.3.6}$$

for all $\ell, m = 2, 3, \ldots$, or

(ii) $g(x, y) = Axy + B(x+y) + A^{-1}(B^2 - B)$ for some constants $A(\neq 0)$ and B, and (5.3.1) becomes

$$\begin{aligned} I_{\ell m}(P * Q) = B[I_\ell(P) + I_m(Q)] + A I_\ell(P) I_m(Q) \\ + A^{-1}(B^2 - B), \quad P \in \Gamma_\ell^o, Q \in \Gamma_m^o, \end{aligned} \tag{5.3.7}$$

for $\ell, m = 2, 3, \ldots$.

Proof: We show first that g must be a symmetric polynomial of degree at most one. Indeed, (5.3.1) and (5.1.1) yield

$$\begin{aligned} g[I_\ell(P), I_m(Q)] &= I_{\ell m}(P * Q) \\ &= \sum_{i=1}^{\ell} \sum_{j=1}^{m} F(p_i q_j) \\ &= I_{m\ell}(Q * P) \\ &= g[I_m(Q), I_\ell(P)] \end{aligned}$$

and also

$$\begin{aligned} g\{g[I_\ell(P), I_m(Q)], I_n(R)\} \\ = g\{I_{\ell m}(P * Q), I_n(R)\} \\ = I_{\ell m n}(P * Q * R) \\ = g\{I_\ell(P), I_{mn}(Q * R)\} \\ = g\{I_\ell(P), g[I_m(Q), I_n(R)]\}. \end{aligned}$$

Thus g is symmetric $g(x,y) = g(y,x)$ and associative $g(g(x,y),z) = g(x,g(y,z))$, for all $x,y,z \in \{I_n(\Gamma_n) \mid n = 2,3,\ldots\}$. Let $g(x,y)$ be of degree a in x and degree b in y $(a,b \in \mathbb{N})$. Under the assumption of infinite cardinality of the range of $\{I_n\}$, the symmetry of g implies $a = b$, and the associativity implies $a^2 = a$. Hence $a = b = 0$ or $a = b = 1$.

Next, we show that in fact $a = b = 1$ (i.e. g is of degree one). Suppose that $a = b = 0$. Then $g(x,y) = C$ for some constant C, and (5.3.1) with (5.1.1) gives exactly equation (5.3.2) for F. So Lemma 5.3.2 shows that F is (the restriction to J of) an additive map with $F(1) = C$. But then (5.1.1) reduces to $I_n(P) = C$ for all $P \in \Gamma_n^o$ $(n = 2,3,\ldots)$, contradicting the infinite cardinality of the range of $\{I_n\}$. Therefore $a = b = 1$.

Because of symmetry, g has the form

$$g(x,y) = Axy + B(x+y) + C, \tag{5.3.8}$$

for some constants A, B, C. Moreover, the associativity of g implies $g(g(x,x),y) = g(x,g(x,y))$. Using (5.3.8) we get

$$(AC+B)y + 2B^2x = B^2y + (AC + B + B^2)x,$$

that is

$$AC = B^2 - B. \tag{5.3.9}$$

If $A = 0$, then (5.3.9) gives $B \in \{0,1\}$. Again, to avoid constancy of $\{I_n\}$, we must have $B = 1$. In this case, (5.3.8) and (5.3.1) yield (5.3.6). If $A \neq 0$, then (5.3.9) gives $C = A^{-1}(B^2 - B)$, and (5.3.8) and (5.3.1) yield (5.3.7). This completes the proof of the theorem.

Let us consider now the consequences of Theorem 5.3.3. It suggests that we restrict our attention to generalized additivities of the form (5.3.6) and (5.3.7). But (5.3.6) is essentially not more general than (5.2.1). Indeed, if $\{I_n'\}$ satisfies (5.3.6), then the information measure $\{I_n\}$ defined by

$$I_n(P) := I_n'(P) + C, \quad P \in \Gamma_n^o \ (n = 2,3,\ldots)$$

satisfies (5.2.1). Therefore $\{I_n'\}$ differs from an additive measure only by an additive constant (the same for all n).

Similarly, if $\{I_n''\}$ satisfies (5.3.7), then the measure $\{I_n\}$ defined by

$$I_n(P) := \lambda\left[AI_n''(P) + (B-1)\right], \quad P \in \Gamma_n^o \ (n = 2,3,\ldots), \tag{5.3.10}$$

satisfies

$$I_{\ell m}(P * Q) = I_\ell(P) + I_m(Q) + \lambda I_\ell(P) I_m(Q), \qquad (5.3.11)$$

for all $P \in \Gamma_\ell^o$, $Q \in \Gamma_m^o$. That brings us to the following.

Definition 5.3.4. For a fixed pair (ℓ, m) of integers greater than or equal to two, and for a fixed real parameter λ, an information measure $\{I_n\}$ is said to be (ℓ, m)-*additive of type* λ if it satisfies equation (5.3.11) for all $P \in \Gamma_\ell^o$, $Q \in \Gamma_m^o$.

Thus we see that a measure $\{I_n''\}$ satisfying (5.3.7) differs from a measure which is (5.3.11) additive of type λ by the linear transformation (5.3.10).

Hence we may focus our attention on (5.2.1) additive measures and on measures which are (5.3.11) additive of type λ. This we do in the next three chapters. Observe also that (5.2.1) is the special case $\lambda = 0$ of (5.3.11). On the other hand, methods of solution for $\lambda \neq 0$ differ somewhat from those used for $\lambda = 0$. Moreover, general solutions of most problems involving $\lambda \neq 0$ are known, whereas general solutions in the case $\lambda = 0$ are not known. Nonetheless, we present the general measurable solutions of all problems ($\lambda \neq 0$ and $\lambda = 0$).

We would also like to point out here that, for any real α, the entropy of degree α satisfies (5.3.11) with $\lambda = 2^{1-\alpha} - 1$. (So $\lambda = 0$ corresponds to $\alpha = 1$.) For this reason, (5.3.11) with λ replaced by $2^{1-\alpha} - 1$ is also known as *additivity of degree* α.

An information measure $\{I_n\}$ with the sum form (5.1.1) which is (ℓ, m) -additive of type λ is characterized by the functional equation

$$\sum_{i=1}^{\ell} \sum_{j=1}^{m} F(p_i q_j) = \sum_{i=1}^{\ell} F(p_i) + \sum_{j=1}^{m} F(q_j) + \lambda \sum_{i=1}^{\ell} F(p_i) \sum_{j=1}^{m} F(q_j), \quad (5.3.12)$$

valid for all $P \in \Gamma_\ell^o$ and $Q \in \Gamma_m^o$. Again, note that for $\lambda = 0$ this reduces to (5.2.2), the measurable solutions of which will be given in Chapter 7. Chapter 8 contains the results in the case $\lambda \neq 0$. There, we present not only the measurable solutions, but also the general solutions in case either ℓ or m is at least 3.

For (5.3.12), including the special case (5.2.2), proofs are usually somewhat more difficult when either ℓ or m is equal to 2 than when both ℓ and m are at least 3. The subcase $\ell = m = 2$ is the most difficult of all, and it sometimes permits a curiously larger class of solutions.

5.4. Weighted additivity

Referring once more to our discussion at the beginning of section 2, we may suppose (again, more generally than (5.2.1)) that the information obtained from experiment C is a weighted sum of the amounts of information received from experiments A and B. More precisely, we assume that this relationship takes the form

$$I_{\ell m}(P * Q) = \omega_1(P)I_m(Q) + \omega_2(Q)I_\ell(P) \qquad (5.4.1)$$

for some functions ("weights") ω_1 and ω_2.

Now let us discuss what form these weights ω_1 and ω_2 should take. Since the operation $*$ is associative, we calculate that

$$
\begin{aligned}
\omega_1(P * Q)&I_n(R) + \omega_2(R)\left[\omega_1(P)I_m(Q) + \omega_2(Q)I_\ell(P)\right] \\
&= \omega_1(P * Q)I_n(R) + \omega_2(R)I_{\ell m}(P * Q) \\
&= I_{\ell mn}(P * Q * R) \\
&= \omega_1(P)I_{mn}(Q * R) + \omega_2(Q * R)I_\ell(P) \\
&= \omega_1(P)\left[\omega_1(Q)I_n(R) + \omega_2(R)I_m(Q)\right] + \omega_2(Q * R)I_\ell(P).
\end{aligned}
$$

Comparing the two extremes of this line of equations, and assuming a sufficient "fullness" of the range of $\{I_n\}$ (as in the previous section), we find that

$$\omega_1(P * Q) = \omega_1(P)\omega_1(Q), \qquad \omega_2(Q * R) = \omega_2(Q)\omega_2(R). \qquad (5.4.2)$$

Furthermore, since we are assuming that $\{I_n\}$ has the sum form (5.1.1), it can be seen in (5.4.1) that ω_1 and ω_2 also have the sum form, provided there exist distributions $P^* \in \Gamma_\ell^0$, $Q^* \in \Gamma_m^0$ such that $I_\ell(P^*) \neq 0$, respectively $I_m(Q^*) \neq 0$. For example, if $a := I_m(Q^*) \neq 0$ and if F is the generating function of $\{I_n\}$ then we get

$$\sum_{i=1}^{\ell}\sum_{j=1}^{m} F(p_i q_j^*) - \omega_2(Q^*)\sum_{i=1}^{\ell} F(p_i) = a \cdot \omega_1(P).$$

Putting $F^*(p) = \sum_{j=1}^{m} F(pq_j^*)$ and $b := \omega_2(Q^*)$ we obtain

$$\omega_1(P) = \sum_{i=1}^{\ell}(F^*(p_i) - bF(p_i))a^{-1} =: \sum_{i=1}^{\ell} f_1(p_i).$$

Inserting

$$\omega_s(P) = \sum_{i=1}^{n} f_s(p_i), \quad P \in \Gamma_n^o \quad (n = 2, 3, \dots; s = 1, 2), \tag{5.4.3}$$

into (5.4.1) and (5.4.2), we find

$$\sum_{i=1}^{\ell}\sum_{j=1}^{m} F(p_i q_j) = \sum_{i=1}^{\ell} f_1(p_i) \sum_{j=1}^{m} F(q_j) + \sum_{i=1}^{\ell} F(p_i) \sum_{j=1}^{m} f_2(q_j), \tag{5.4.4}$$

where f_s must satisfy $(s = 1, 2)$

$$\sum_{i=1}^{\ell} f_s(p_i) \sum_{j=1}^{m} f_s(q_j) = \sum_{i=1}^{\ell}\sum_{j=1}^{m} f_s(p_i q_j). \tag{5.4.5}$$

Let us rewrite this last equation as

$$\sum_{i=1}^{\ell}\sum_{j=1}^{m} [f_s(p_i q_j) - f_s(p_i) f_s(q_j)] = 0,$$

$s = 1, 2$. If we ignore for one moment the double sum, then it seems that f_s is multiplicative. On the other hand, any additive function $B : J \to \mathbb{R}$ clearly satisfies (5.4.5) if $B(1)^2 = B(1)$, that is $B(1) \in \{0, 1\}$. It will later turn out that indeed the general solution of (5.4.5) is given by the sum of a multiplicative and an additive function (so that the additive term corresponds to the "double sum"; compare Lemma 5.3.2.). Thus we shall see that the general solution of (5.4.5) has one of the forms (depending on whether $f_s = 0$ or $f_s \neq 0$):

(i) $f_s(p) = B_s(p)$, where B_s additive and $B_s(1) \in \{0, 1\}$; or

(ii) $f_s(p) = M_s(p) + A_s(p)$, where M_s multiplicative and A_s additive with $A_s(1) = 0$.

In case (i), (5.4.3) yields $\omega_s \equiv 0$ or $\omega_s \equiv 1$; in case (ii), $\omega_s(P) = \sum_{i=1}^{n} M_s(p_i)$ for all $P \in \Gamma_n^o$ $(n = 2, 3, \dots)$. Thus in either case we have

$$\omega_s(P) = \sum_{i=1}^{n} M_s(p_i), \quad P \in \Gamma_n^o \quad (n = 2, 3, \dots; s = 1, 2),$$

where M_s is multiplicative. (In case (i), take $M_s \equiv 0$, respectively $M_s(p) = p$.) Inserting this into (5.4.1), we arrive at the equation in the following definition.

Definition 5.4.1. For a fixed pair (ℓ, m) of integers greater than or equal to 2, we say that $\{I_n\}$ is *weighted (ℓ, m)-additive of type (M_1, M_2)* if it satisfies

$$I_{\ell m}(P * Q) = \sum_{i=1}^{\ell} M_1(p_i) I_m(Q) + \sum_{j=1}^{m} M_2(q_j) I_\ell(P), \qquad (5.4.6)$$

for all $P \in \Gamma_\ell^o$, $Q \in \Gamma_m^o$, where $M_1, M_2 : J \to \mathbb{R}$ are multiplicative functions.

Thus the fundamental sum form equation of weighted (ℓ, m)-additivity is

$$\sum_{i=1}^{\ell} \sum_{j=1}^{m} F(p_i q_j) = \sum_{i=1}^{\ell} \sum_{j=1}^{m} [M_1(p_i) F(q_j) + M_2(q_j) F(p_i)], \qquad (5.4.7)$$

valid for all $P \in \Gamma_\ell^o$, $Q \in \Gamma_m^o$. Again, this is a generalization of (5.2.2), to which (5.4.7) reduces in the case M_1 and M_2 are projections. In the case M_1 and M_2 are not both projections, we give the measurable solutions of (5.4.7) in Chapter 9. The general solutions are given also, provided that ℓ and m are greater than 2. Again, $\ell = m = 2$ is exceptional.

Finally, we point out that the entropy of degree (α, β) fulfills (5.4.6) when $M_1(q) = q^\beta$ and $M_2(p) = p^\alpha$.

Remark 5.4.2. All measurable solutions of equation (5.4.4) (without (5.4.5)) were determined in the 1-dimensional case on the closed domain in Sander (1990). The main result is that essentially $f_1(p) = p^\alpha$ and $f_2(p) = p^\beta$ for some $\alpha, \beta \in \mathbb{R}$. This supports the above considerations and justifies why only two special forms of "nonadditive" measures of information were investigated in the literature, namely the additivities of type λ and of type (M_1, M_2).

Remark 5.4.3. Weighted (ℓ, m)-additive information measures of type (M_1, M_2) are also known as (ℓ, m)-additive sum form information measures of multiplicative type (M_1, M_2). In Chapter 9, we will use this latter terminology.

CHAPTER 6

BASIC SUM FORM FUNCTIONAL EQUATIONS

6.1. Introduction

In the last chapter, we discussed the problems of determining explicit forms for information measures of sum form which satisfy also certain additivity or generalized additivity properties. We saw that these problems boiled down to solving various sum form functional equations, namely (5.2.2), (5.3.12) and (5.4.7).

To derive the measurable solutions of these equations, we proceed as follows. In sections 2 and 3 of this chapter, we present the general measurable solution of the functional equation

$$f_1(pq) + f_2(p(1-q)) + f_3((1-p)q) + f_4((1-p)(1-q)) = g(p) + h(q), \quad (6.1.1)$$

for all $p, q \in J$, where $f_i, g, h : J \to \mathbb{R}$ $(i = 1, \ldots, 4)$. This functional equation is fundamental in our investigation, for reasons which we can indicate now. If $f_1 = f_2 = f_3 = f_4 = F$ and $g(p) = h(p) = F(p) + F(1 - p)$, then (6.1.1) becomes exactly (5.2.2) for $\ell = m = 2$. Furthermore, equation (5.2.2) for $\ell, m \geq 2$ can always be reduced to an equation of the form (6.1.1).

In section 4, we present further results which are basic to the study of sum form functional equations. We shall use these results to find general solutions of (5.3.12) and (5.4.7). First, we give the general solution of

$$\sum_{i=1}^{\ell} f(p_i) = C, \qquad P \in \Gamma_\ell^o,$$

for some constant C, fixed integer $\ell \geq 3$, and $f : J \to \mathbb{R}$. We have already seen some indication of the importance of such an equation in Lemma 5.3.2 and its application to Theorem 5.3.3. Then we will obtain the general solutions of the functional equations

$$\sum_{i=1}^{2} \{f(pq_i) - f(p)f(q_i)\} = 0, \qquad (6.1.2)$$

117

$$\sum_{i=1}^{2} \{f(pq_i) - f(p)M(q_i)\} = 0, \tag{6.1.3}$$

$$\sum_{i=1}^{2} \{f(pq_i) - M(p)f(q_i) - M(q_i)f(p)\} = 0, \tag{6.1.4}$$

for all $p \in J$, $Q \in \Gamma_2^o$, where $M : J \to \mathbb{R}$ is multiplicative. These equations are instrumental in solving (5.3.12) and (5.4.7), but may also be interesting in their own right because of their relationship to basic Cauchy functional equations.

Finally, in section 5 we present some results concerning the linear independence of additive and multiplicative functions. After this, we will be ready to solve equations (5.2.2), (5.3.12) and (5.4.7) in chapters 7, 8 and 9, respectively.

6.2. Measurable solutions of (6.1.1) in the 1-dimensional case

As already pointed out, one of our main goals is to find the measurable solutions of (5.2.2) if $\ell \geq 2$, $m \geq 2$. The strategy is similar to what was used in the proof of Theorem 4.3.3. We shall solve (5.2.2) by induction on the dimension k. Because of the inductive step it is necessary to solve in the 1-dimensional case the more complicated equation (6.1.1) (instead of (5.2.2)). This will be done by proving that all measurable solutions of (6.1.1) are infinitely often differentiable and satisfy a certain differential equation, the solutions of which lead to the solutions of (6.1.1). That will be done in this section.

By a nontrivial induction we will obtain in the next section not only the measurable solutions of (5.2.2) but also the measurable solutions of (6.1.1) in the k-dimensional case. We need the k-dimensional measurable solutions of (6.1.1) since equation (5.2.2) in the case $\ell \geq 2$, $m \geq 3$ or $\ell \geq 3$, $m \geq 2$ can be transformed into an equation of the form (6.1.1). As a first step we prove the following result which concerns a generalization of functional equation (6.1.1). It is due to Losonczi (1993) and is a slight generalization of a result of Kannappan and Ng (1985).

Theorem 6.2.1. *Let* $f_i, g_j, h_j :]0, 1[\to \mathbb{R}$ *$(i = 1, 2, 3, 4; j = 1, 2)$ satisfy the functional equation*

$$\begin{aligned} f_1(pq) + f_2(p(1-q)) &+ f_3((1-p)q) + f_4((1-p)(1-q)) \\ &= g_1(p)h_1(q) + g_2(p)h_2(q) \end{aligned} \tag{6.2.1}$$

for all $p, q \in]0, 1[$. If f_1, f_2, f_3, f_4 are measurable on $]0, 1[$, and if g_1, g_2 and h_1, h_2 are linearly independent, then f_i, g_j, h_j are infinitely often differentiable in $]0, 1[$.

Proof. Let us first give an outline of the steps needed in the proof.

Step 1. If f_1, f_2, f_3, f_4 are measurable (respectively continuous, n-times differentiable) then g_1, g_2, h_1, h_2 are measurable (respectively continuous, n-times differentiable), too.

Step 2. The functions $f_1, f_2, f_3, f_4, g_1, g_2, h_1, h_2$ are continuous on $]0, 1[$.

Step 3. The functions g_1, g_2, h_1, h_2 are continuously differentiable on $]0, 1[$.

Step 4. The functions f_1, f_2, f_3, f_4 are continuously differentiable on $]0, 1[$.

Step 5. The functions f_i, g_j, h_j are infinitely often differentiable on $]0, 1[$.

Verification of Step 1. The linear independence of g_1, g_2 implies that there exist points $p_1, p_2 \in]0, 1[$ such that

$$a := \begin{vmatrix} g_1(p_1) & g_2(p_1) \\ g_1(p_2) & g_2(p_2) \end{vmatrix} \neq 0 \tag{6.2.2}$$

(see Aczél (1966), p. 201). Denoting the left-hand side of equation (6.2.1) by $L(p, q)$ and substituting into (6.2.1) first $p = p_1$, then $p = p_2$, we get

$$L(p_1, q) = g_1(p_1)h_1(q) + g_2(p_1)h_2(q), \tag{6.2.3}$$

$$L(p_2, q) = g_1(p_2)h_1(q) + g_2(p_2)h_2(q) \tag{6.2.4}$$

for $q \in]0, 1[$. The solution of this linear system for h_1, h_2 yields (by (6.2.2))

$$h_1(q) = a^{-1}\big[g_2(p_2)L(p_1, q) - g_2(p_1)L(p_2, q)\big] \tag{6.2.5}$$

and

$$h_2(q) = a^{-1}\big[g_1(p_1)L(p_2, q) - g_1(p_2)L(p_1, q)\big] \tag{6.2.6}$$

Thus h_j $(j = 1, 2)$ is measurable (respectively continuous, n-times differentiable) since h_1 and h_2 are linear combinations of the measurable (respectively continuous, n-times differentiable) functions $L(p_i, \cdot)$, $i = 1, 2$. In the same manner the measurability of g_1, g_2 can be proved, starting with the linear independence of h_1, h_2.

Verifications of Step 2. Because of Step 1 we know that all functions in (6.2.1) are measurable. Now we want to apply Remark 1.3.3. With $x = pq$, $y = q$, equation (6.2.1) becomes

$$f_1(x) = -\sum_{i=2}^{4} f_i(G_i(x,y)) + g_1(G_5(x,y))h_1(G_6(x,y))$$
$$+ g_2(G_7(x,y))h_2(G_8(x,y)) \tag{6.2.7}$$

for $0 < x < y < 1$, where

$$G_2(x,y) = \frac{x}{y} - x, \ \ G_3(x,y) = y - x, \ \ G_4(x,y) = 1 + x - y - \frac{x}{y}, \tag{6.2.8}$$

$$G_5(x,y) = G_7(x,y) = \frac{x}{y}, \ \ G_6(x,y) = G_8(x,y) = y. \tag{6.2.9}$$

Putting $f_5 := g_1$, $f_6 := h_1$, $f_7 := g_2$, $f_8 := h_2$, (6.2.7) goes over into

$$f_1(x) = h(x,y,\, f_2(G_2(x,y)),\ldots,\, f_8(G_8(x,y))) \tag{6.2.10}$$

where

$$h(x,y,z_2,\ldots,z_8) = -z_2 - z_3 - z_4 + z_5 \cdot z_6 + z_7 \cdot z_8 \tag{6.2.11}$$

is a continuous function. Defining $s = r = 1$, $Y_2 = \ldots = Y_8 =]0,1[$, $T =]0,1[$, $D = \{(x,y) \subset \mathbb{R}^2 \mid 0 < x < y < 1\}$, we see that assumptions (ii) and (iii) of Theorem 1.3.2 are satisfied. Indeed, for every $x \in T =]0,1[$ an immediate calculation shows that

$$\frac{\partial}{\partial y}G_i(x,y) \neq 0, \ \ \ \text{if } y \neq \sqrt{x} \tag{6.2.12}$$

$(i = 2,\ldots,8)$. (Note that $\frac{\partial}{\partial y}G_4(x,y) = \frac{x}{y^2} - 1 \neq 0$ if $y \neq \sqrt{x}$). Thus by Remark 1.3.3, f_1 is continuous. In exactly the same manner the continuity of $f_2,\ldots,f_8$ follow and Step 2 is proven.

Verification of Step 3. By Step 2 we know that f_i ($1 \le i \le 4$) and g_j, h_j ($j = 1,2$) are continuous. Let us define $K_j, L_j :]0,1[\to \mathbb{R}$ ($j = 1,2$) by

$$K_j(x) := \int_{\frac{1}{2}}^{x} g_j(u)du, \ \ \ \ \ L_j(x) := \int_{\frac{1}{2}}^{x} h_j(u)du.$$

Then K_1, K_2 and L_1, L_2 are linearly independent, too. For example, from $a_1 K_1(x) + b_1 K_2(x) = 0$ for some constants a_1, b_1 we get by differentiation $a_1 g_1(x) + b_1 g_2(x) = 0$ which implies $a_1 = b_1 = 0$ since g_1, g_2 are linearly independent. Now we proceed as in Step 1. There exist points $p_1, p_2, q_1, q_2 \in {]0,1[}$ such that

$$
b := \begin{vmatrix} K_1(p_1) & K_2(p_1) \\ K_1(p_2) & K_2(p_2) \end{vmatrix} \neq 0, \quad c := \begin{vmatrix} L_1(q_1) & L_2(q_1) \\ L_1(q_2) & L_2(q_2) \end{vmatrix} \neq 0.
$$

Integration of (6.2.1) with respect to p from $\frac{1}{2}$ to p_i $(i = 1, 2)$ yields (substituting $pq = u$ and $G_i(p, q) = u$, $i = 2, 3, 4$ (see Step 2))

$$
\frac{1}{q} \int_{\frac{q}{2}}^{p_i q} f_1(u)\,du + \frac{1}{1-q} \int_{\frac{1-q}{2}}^{(1-q)p_i} f_2(u)\,du
$$

$$
- \frac{1}{q} \int_{\frac{q}{2}}^{(1-p_i)q} f_3(u)\,du - \frac{1}{1-q} \int_{\frac{1-q}{2}}^{(1-p_i)(1-q)} f_4(u)\,du \tag{6.2.13}
$$

$$
= K_1(p_i)h_1(q) + K_2(p_i)h_2(q).
$$

Denoting the left-hand side in (6.2.13) by $S(p_i, q)$ we have a linear system for the unknowns $h_1(q)$ and $h_2(q)$ (compare with equations (6.2.3) and (6.2.4)). As in Step 1 we find that $h_i(q)$ is a linear combination of $S(p_1, q)$ and $S(p_2, q)$ (see equations (6.2.5) and (6.2.6)). Thus $h_i(q)$ is continuously differentiable (using the theorem on differentiation of parametric integrals; see Dieudonné (1960), p.172), since $S(p_i, q)$, $i = 1, 2$ is continuously differentiable. In a completely similar way we deduce that g_1 and g_2 are continuously differentiable, too.

Verification of Step 4. By Step 2 and Step 3 we know that f_i $(i = 1, 2, 3, 4)$ is continuous and that g_1, g_2, h_1, h_2 are continuously differentiable on ${]0,1[}$. We will prove that f_1 is continuously differentiable. (Similarly, it can be shown that $f_2, f_3, f_4 \in C^1({]0,1[})$.) Let $[\alpha, \beta]$ be an arbitrary subinterval of ${]0,1[}$ and choose the interval $[\gamma, \delta]$ such that $\sqrt{\beta} < \gamma < \delta < 1$. Thus

$$
[\alpha, \beta] \times [\gamma, \delta] \subset D = \{(x, y) \subset \mathbb{R}^2 \mid 0 < x < y < 1\} \tag{6.2.14}
$$

(see Step 2). Let us now integrate equation (6.2.7) with respect to y on $[\gamma, \delta]$. Substituting $G_i(x, y) = u$ $(i = 2, 3, 4)$ and recalling (6.2.8) and (6.2.10) we arrive (as in equation (4.3.9)) at

$$(\delta - \gamma) f_1(x) = -\sum_{i=2}^{4} \int_{G_i(x,\gamma)}^{G_i(x,\delta)} f_i(u) \frac{\partial}{\partial u} \gamma_i(x, u) du$$

$$+ \int_{\gamma}^{\delta} g_1\left(\frac{x}{y}\right) h_1(y) dy + \int_{\gamma}^{\delta} g_2\left(\frac{x}{y}\right) h_2(y) dy,$$

(6.2.15)

where $y = \gamma_i(x, u)$ $(i = 2, 3, 4)$ is the infinitely often differentiable solution of $G_i(x, y) = u$. (Note that $G_i(x, y) = u$ can indeed be solved for y whenever $x \in [\alpha, \beta]$, because of our condition $\sqrt{\beta} < \gamma$ and (6.2.12).) In the first sum on the right-hand side of (6.2.15) the functions f_2, f_3, f_4 are continuous and $G_2, G_3, G_4, \gamma_2, \gamma_3, \gamma_4$ are infinitely often differentiable. Thus the theorem on differentiation of parametric integrals (see Dieudonné (1960), p. 172) assures that this first sum is continuously differentiable in $[\alpha, \beta]$. But the last two summands on the right-hand side of (6.2.15) are in $C^1([\alpha, \beta])$, too, since $g_1, h_1, g_2, h_2 \in C^1([\alpha, \beta])$. Thus $f_1 \in C^1([\alpha, \beta])$. Since $[\alpha, \beta]$ was arbitrary, $f_1 \in C^1(]0, 1[)$.

Verification of Step 5. By repeated applications of the reasoning in Steps 2-4, we get that f_i, g_j, h_j $(i = 1, 2, 3, 4; j = 1, 2)$ are infinitely often differentiable. This completes the proof of Theorem 6.2.1.

Remark 6.2.2. (a) Theorem 6.2.1 implies that $f_1, f_2, f_3, f_4, g, h \in C^\infty(]0, 1[)$ if these functions satisfy (6.1.1) and if f_1, f_2, f_3, f_4 are measurable and g, h are not constant. Indeed, put $g_1 = g$, $g_2 \equiv 1$ and $h_1 \equiv 1$, $h_2 = h$ into (6.2.1) and apply Theorem 6.2.1.

(b) If either g or h (or both) on the right-hand side in (6.1.1) is constant, then from the proof of Theorem 6.2.1 it is again clear that $f_i \in C^\infty(]0, 1[)$ whenever f_1, f_2, f_3, f_4 are measurable. (The verifications of the steps become simpler in this case.)

(c) We have proved Theorem 6.2.1 for the equation (6.2.1) rather than equation (6.1.1) because we need the more general equation (6.2.1) later in the study of the $(2,2)$-case of equations (5.3.12) and (5.4.7).

(d) The following result will be used later and generalizes slightly the argument given in the verification of Step 1.

Let $f : \mathbb{R}^k \times \mathbb{R} \to \mathbb{R}$ be measurable in each variable and let f have the representation

$$f(p,r) = \sum_{i=1}^{n} g_i(p)h_i(r) \qquad (6.2.16)$$

for functions $g_i : \mathbb{R}^k \to \mathbb{R}$ and $h_i : \mathbb{R} \to \mathbb{R}$, $i = 1,\ldots,n$. If $h_1,\ldots h_n$ are linearly independent then $g_1,\ldots,g_n$ are measurable. (And symmetrically, if $g_1,\ldots,g_n$ are linearly independent then $h_1,\ldots,h_n$ are measurable.)

Proof. By the linear independence of $h_1,\ldots,h_n$ there exist points $r_1,\ldots,r_n$ such that $\det (h_i(r_j))_{i,j=1}^{n} \neq 0$. Substituting $r = r_j$ $(j = 1,\ldots,n)$ in (6.2.16) and solving the resulting system for $g_1,\ldots,g_n$ we get

$$g_i(p) = \sum_{j=1}^{n} a_{ij} f(p,r_j)$$

for some constants $a_{ij} \in \mathbb{R}$, which means that g_i is measurable $(i = 1,\ldots,n)$.

(e) Finally, note that actually we have proved a little more than the statement in Theorem 6.2.1. Indeed, the following result is valid:

Let $f_i, g_j, h_j :]0,1[\to K$ $(i = 1,2,3,4;\ j = 1,2,\ldots,N;\ K$ is $\mathbb{R}$ or $\mathbb{C})$ satisfy the functional equation

$$f_1(pq) + f_2(p(1-q)) + f_3((1-p)q) + f_4((1-p)(1-q))$$
$$= \sum_{i=1}^{N} g_i(p)h_i(q) \qquad (6.2.17)$$

for all $p,q \in]0,1[$. If f_1,f_2,f_3,f_4 are measurable on $]0,1[$ and if $h_1,\ldots,h_N$ and $g_1,\ldots,g_N$ are linearly independent, then f_i,g_j,h_j are infinitely often differentiable in $]0,1[$.

The proof uses obvious modifications of Steps 1-5 of Theorem 6.2.1, and takes into account also the above remark 6.2.2.(d).

In the rest of this section we determine explicitly the measurable solutions of (6.1.1). We start with some auxiliary results.

Lemma 6.2.3. A measurable function $F :]0,1[\to \mathbb{R}$ satisfies

$$F(pq) + F(p(1-q)) + F((1-p)q) + F((1-p)(1-q)) = 0 \qquad (6.2.18)$$

for all $p, q \in]0, 1[$, if and only if there exists a constant a for which

$$F(p) = 4ap - a, \qquad p \in]0, 1[. \tag{6.2.19}$$

Proof: Putting $f_1 = f_2 = f_3 = f_4 = F$ and $g_1 = g_2 = h_1 = h_2 = 0$ in (6.2.1), we get from Remark 6.2.2.(b) that F is infinitely often differentiable. Let us show that

$$F''(p) = 0, \qquad p \in]0, 1[. \tag{6.2.20}$$

Putting $q = \frac{1}{2}$ into (6.2.18) we get

$$F\left(\frac{p}{2}\right) + F\left(\frac{1-p}{2}\right) = 0, \qquad p \in]0, 1[$$

and thus

$$F''\left(\frac{p}{2}\right) + F''\left(\frac{1-p}{2}\right) = 0, \qquad p \in]0, 1[. \tag{6.2.21}$$

If we now differentiate the equation (6.2.18) twice with respect respect to q then we arrive at

$$\begin{aligned} p^2 F''(pq) = &- p^2 F''(p(1-q)) - (1-p)^2 F''((1-p)q) \\ &- (1-p)^2 F''((1-p)(1-q)) \end{aligned} \tag{6.2.22}$$

for all $p, q \in]0, 1[$. Substituting here $q = \frac{1}{2}$ we get

$$p^2 F''\left(\frac{p}{2}\right) = -(1-p)^2 F''\left(\frac{1-p}{2}\right), \qquad p \in]0, 1[.$$

Thus (6.2.21) implies

$$F''\left(\frac{p}{2}\right)(2p - 1) = 0, \qquad p \in]0, 1[,$$

and hence

$$F''\left(\frac{p}{2}\right) = 0$$

for all $p \in]0, 1[$, $p \neq \frac{1}{2}$. But (6.2.21) with $p = \frac{1}{2}$ yields $F''(\frac{1}{4}) = 0$ so that (6.2.20) is valid for all $p \in]0, \frac{1}{2}[$. Now consider an arbitrary but fixed $t \in [\frac{1}{2}, 1[$. Then there exist $p, q \in]0, 1[$ such that $t = pq$. Because $p > \frac{1}{2}$ and $q > \frac{1}{2}$ we have $(1-p)(1-q), p(1-q), q(1-p) \in]0, \frac{1}{2}[$ so that (6.2.22) implies $F''(t) = F''(pq) = 0$. Thus equation (6.2.20) holds, and integration of it yields

$$F(p) = bp - a, \qquad p \in]0, 1[,$$

for some constants a, b. Substitution of this function F into (6.2.18) yields $-4a + b = 0$, which leads to (6.2.19). Obviously (6.2.19) satisfies (6.2.18) and thus the lemma is proven.

Lemma 6.2.4. *A measurable function* $F :]0, 1[\to \mathbb{R}$ *satisfies*

$$F(pq) - F(p(1-q)) - F((1-p)q) + F((1-p)(1-q)) = 0 \qquad (6.2.23)$$

for all $p, q \in]0, 1[$, *if and only if there exist constants* a, b, c *such that*

$$F(p) = a(p^2 - p) + b \log p + c, \qquad p \in]0, 1[. \qquad (6.2.24)$$

Proof: It is clear that F given by (6.2.24) satisfies (6.2.23). To prove the converse, first observe that F is differentiable by Remark 6.2.2.(b), then differentiate (6.2.23) with respect to p and the resulting equation with respect to q. With the notation

$$F_1(p) = F'(p) + pF''(p) = [pF'(p)]', \qquad p \in]0, 1[, \qquad (6.2.25)$$

we find that F_1 satisfies equation (6.2.18). Thus Lemma 6.2.3 and (6.2.25) yield

$$F_1(p) = 4ap - a = [pF'(p)]', \qquad p \in]0, 1[,$$

for some constant $a \in \mathbb{R}$. Hence there exist constants b, c such that

$$pF'(p) = 2ap^2 - ap + b,$$

$$F(p) = ap^2 - ap + b \log p + c.$$

Lemma 6.2.5. *The measurable functions* $F, G :]0, 1[\to \mathbb{R}$ *satisfy*

$$F(pq) + G(p(1-q)) + G((1-p)q) + F((1-p)(1-q)) = 0 \qquad (6.2.26)$$

for all $p, q \in]0, 1[$, *if and only if there exist constants* a, b, c, d *such that*

$$F(p) = bp^2 + (2a - b)p - c \log p - d - a, \qquad (6.2.27)$$

$$G(p) = -bp^2 + (2a + b)p + c \log p + d. \qquad (6.2.28)$$

Proof: It is easy to check that (6.2.27) and (6.2.28) imply (6.2.26). Conversely, replacing p by $1 - p$ in (6.2.26) and adding this equation to (6.2.26), we obtain

$$(F + G)(pq) + (F + G)(p(1 - q))$$
$$+ (F + G)((1 - p)q) + (F + G)((1 - p)(1 - q)) = 0$$

for all $p, q \in]0, 1[$. Thus Lemma 6.2.3 shows that

$$F(p) + G(p) = 4ap - a, \qquad p \in]0, 1[$$

for some constant a. Defining

$$H(p) := G(p) - 2ap = 2ap - a - F(p), \qquad (6.2.29)$$

we deduce from (6.2.26) that

$$H(pq) - H(p(1 - q)) - H((1 - p)q) + H((1 - p)(1 - q)) = 0.$$

By Lemma 6.2.4, we get

$$H(p) = -b\,(p^2 - p) + c\,\log p + d$$

for some constants b, c, d, which together with (6.2.29) leads to the desired forms for F and G.

Lemma 6.2.6. *The only functions $F, G :]0, 1[\to \mathbb{R}$ satisfying*

$$F(pq) + G(p(1 - q)) - G((1 - p)q) - F((1 - p)(1 - q)) = 0 \qquad (6.2.30)$$

are given by

$$F(p) = a \qquad and \qquad G(p) = b \qquad (6.2.31)$$

for some constants a, b.

Proof: From

$$F(pq) - F((1 - p)(1 - q)) = G((1 - p)q) - G(p(1 - q))$$

and the symmetry of the left side and the antisymmetry of the right side in p and q we obtain

$$F(pq) - F((1-p)(1-q)) = 0 = G((1-p)q) - G(p(1-q)). \qquad (6.2.32)$$

Putting $u = pq$ in (6.2.32) we get

$$F(u) = F\left(1 - p - \frac{u}{p} + u\right), \qquad p \in]u, 1[$$

for each $u \in]0, 1[$. This is equivalent to

$$F(u) = F(t), \qquad t \in]0, (1 - \sqrt{u})^2[$$

for all $u \in]0, 1[$, which says that for every $u \in]0, 1[$ the function F is constant on the interval $]0, (1 - \sqrt{u})^2[$. Thus F is constant on $]0, 1[$. Similarly G is also constant.

Lemma 6.2.7. *The measurable functions $f_i :]0, 1[\to \mathbb{R}$ $(i = 1, 2, 3, 4)$ satisfy*

$$f_1(pq) + f_2(p(1-q)) + f_3((1-p)q) + f_4((1-p)(1-q)) = 0 \qquad (6.2.33)$$

for all $p, q \in]0, 1[$, if and only if there exist constants $a, b, c, d_1, d_2, d_3, d_4$ such that

$$f_i(p) = bp^2 + (a-b)p - c\log p + d_i, \quad i = 1, 4, \qquad (6.2.34)$$

$$f_i(p) = -bp^2 + (a+b)p + c\log p + d_i, \quad i = 2, 3, \qquad (6.2.35)$$

with $a + d_1 + d_2 + d_3 + d_4 = 0$.

Proof: Replace p by $1-p$ and q by $1-q$ in (6.2.33), then add and subtract the resulting equation to (6.2.33) to obtain

$$
\begin{aligned}
(f_1 + f_4)(pq) + (f_2 + f_3)(p(1-q)) + (f_2 + f_3)((1-p)q) \\
+ (f_1 + f_4)((1-p)(1-q)) = 0,
\end{aligned}
\qquad (6.2.36)
$$

respectively

$$
\begin{aligned}
(f_1 - f_4)(pq) + (f_2 - f_3)(p(1-q)) - (f_2 - f_3)((1-p)q) \\
- (f_1 - f_4)((1-p)(1-q)) = 0.
\end{aligned}
\qquad (6.2.37)
$$

By applying Lemmas 6.2.5 and 6.2.6 to these last two equations we get on
the one hand

$$(f_1 + f_4)(p) = 2bp^2 + (2a - 2b)p - 2c\log p + e_1$$
$$(f_2 + f_3)(p) = -2bp^2 + (2a + 2b)p + 2c\log p + e_2,$$

with $a + e_1 + e_2 = 0$, and on the other hand

$$(f_1 - f_4)(p) = e_3 \quad \text{and} \quad (f_2 - f_3)(p) = e_4.$$

By adding and subtracting the expressions for $f_1 + f_4$ and $f_1 - f_4$ and for
$f_2 + f_3$ and $f_2 - f_3$, we arrive at (6.2.34) and (6.2.35), respectively, where
$d_1 = (e_1 + e_3)/2$, $d_2 = (e_2 + e_4)/2$, $d_3 = (e_2 - e_4)/2$ and $d_4 = (e_1 - e_3)/2$.
Moreover, $e_1 + e_2 = d_1 + d_2 + d_3 + d_4$. Thus the lemma is established.

Theorem 6.2.8. *The measurable functions* $f_i, g, h :]0,1[\to \mathbb{R}$ *(i = 1, 2, 3, 4)*
satisfy (6.1.1)

$$f_1(pq) + f_2(p(1-q)) + f_3((1-p)q) + f_4((1-p)(1-q)) = g(p) + h(q)$$

for all $p, q \in]0, 1[$, *if and only if there exist constants* $A, B, C, C_i, D_j, E_i \in \mathbb{R}$
(1 ≤ i ≤ 4, 1 ≤ j ≤ 6) such that

$$f_i(p) = 4Ap^3 + (B - 9A)p^2 + Cp\log p \\ + C_ip + E_i\log p + D_i, \quad i = 1, 4, \tag{6.2.38}$$

$$f_i(p) = 4Ap^3 - (B + 9A)p^2 + Cp\log p \\ + C_ip + E_i\log p + D_i, \quad i = 2, 3, \tag{6.2.39}$$

$$g(p) = -6Ap^2 + (6A - 2B + C_2 - C_4)p \\ + C\left[p\log p + (1-p)\log(1-p)\right] \\ + (E_1 + E_2)\log p + (E_3 + E_4)\log(1-p) + D_5, \tag{6.2.40}$$

$$h(p) = -6Ap^2 + (6A - 2B + C_3 - C_4)p \\ + C\left[p\log p + (1-p)\log(1-p)\right] \\ + (E_1 + E_3)\log p + (E_2 + E_4)\log(1-p) + D_6 \tag{6.2.41}$$

with

$$4B + C_1 - C_2 - C_3 + C_4 = 0$$

and

$$B - 5A + C_4 + D_1 + D_2 + D_3 + D_4 = D_5 + D_6.$$

Proof: By Remark 6.2.2.(b) we know that all f_i's are infinitely often differentiable. Differentiate (6.1.10) with respect to p and the resultant with respect to q to arrive at

$$F_1(pq) - F_2(p(1-q)) - F_3((1-p)q) + F_4((1-p)(1-q)) = 0,$$

where

$$F_i(p) := f_i'(p) + p\, f_i''(p) = [pf_i'(p)]', \qquad i = 1, 2, 3, 4. \qquad (6.2.42)$$

Because of Lemma 6.2.7 we obtain

$$F_i(p) = bp^2 + (a - b)p - c\log p + d_i, \quad i = 1, 4, \qquad (6.2.43)$$

$$F_i(p) = bp^2 - (a + b)p - c\log p - d_i, \quad i = 2, 3, \qquad (6.2.44)$$

for some constants a, b, c, d_i with $a + d_1 + d_2 + d_3 + d_4 = 0$. Using (6.2.42) we get from (6.2.43) and (6.2.44)

$$f_i(p) = \frac{b}{9}p^3 + \frac{a-b}{4}p^2 - c[p\log p - 2p] + d_i p + E_i \log p + D_i, \quad i = 1, 4,$$

$$f_i(p) = \frac{b}{9}p^3 - \frac{a+b}{4}p^2 - c[p\log p - 2p] - d_i p + E_i \log p + D_i, \quad i = 2, 3,$$

where D_i, E_i are constants. Putting $A = b/36$, $B = a/4$, $C = -c$, $C_i = 2c + d_i$ if $i = 1, 4$, and $C_i = 2c - d_i$ if $i = 2, 3$, the last two equations go over into (6.2.38) and (6.2.39). Moreover we get $a + d_1 + d_2 + d_3 + d_4 = 4B + C_1 - C_2 - C_3 + C_4 = 0$. Substituting (6.2.38) and (6.2.39) into (6.1.1), we get (after some simplifications)

$$\begin{aligned}
g(p) + h(q) = &- 6Ap^2 - 6Aq^2 \\
&+ (6A - 2B + C_2 - C_4)p + (6A - 2B + C_3 - C_4)q \\
&+ C[p\log p + (1 - p)\log(1 - p)] \\
&+ C[q\log q + (1 - q)\log(1 - q)] \\
&+ (E_1 + E_2)\log p + (E_3 + E_4)\log(1 - p) \\
&+ (E_1 + E_3)\log q + (E_2 + E_4)\log(1 - q) \\
&+ B - 5A + C_4 + D_1 + D_2 + D_3 + D_4.
\end{aligned}$$

Hence we get the desired forms of g and h (since the variables p and q can be separated). Since the converse is a straightforward verification, the theorem is proven.

6.3. Measurable solutions of (6.1.1) in higher dimensions

The aim of this section is to generalize Theorem 6.2.8 to higher dimensions, presenting the general measurable solution of (6.1.1) for functions $f_i, g, h :]0, 1[^k \to \mathbb{R}$. For $k = 2$, this result is due to Kannappan and Ng (1985), and in the general k-dimensional case the result is contained in Abou-Zaid's unpublished thesis. By introducing appropriate notation and operations we simplify the presentation of Abou-Zaid's proof. Indeed, the proof is now very similar to the one given in the 2-dimensional case.

Definition 6.3.1. Let $a = (a_1, a_2, ..., a_k) \in \mathbb{R}^k$, let $C = (C_{ij})$ be a $k \times k$ matrix, let $\log p = (\log \pi_1, \log \pi_2, ..., \log \pi_k) \in \mathbb{R}^k$, $p^r = (\pi_1^r, \pi_2^r, ..., \pi_k^r) \in \mathbb{R}^k$, $r \in \mathbb{N}$, and $\mathbf{1} = (1, 1, ..., 1) \in \mathbb{R}^k$. Then we define

$$
\left.
\begin{aligned}
&(i) \quad a \odot p^r = \sum_{i=1}^{k} a_i \, \pi_i^r, \\[2mm]
&(ii) \quad a \odot \mathbf{1} = \mathbf{1} \odot a = \sum_{i=1}^{k} a_i, \\[2mm]
&(iii) \quad p \odot C \odot \log p = \sum_{i=1}^{k} \sum_{j=1}^{k} \pi_i C_{ij} \log \pi_j, \\[2mm]
&(iv) \quad \mathbf{1} \odot C \odot \log p = \sum_{i=1}^{k} \sum_{j=1}^{k} C_{ij} \log \pi_j.
\end{aligned}
\right\}
\qquad (6.3.1)
$$

Thus the operation in (i), (ii) is still the usual scalar product (or inner product) in $\mathbb{R}^k$. Products (iii) and (iv), which contain a "sandwich" matrix C, are simply the usual matrix multiplication. Now we are able to present Abou-Zaid's (1984) result which, because of the definition (6.3.1), looks completely analogous to Theorem 6.2.8.

Theorem 6.3.2. *Let $f_\ell, g, h : J \to \mathbb{R}$, $1 \le \ell \le 4$, be measurable functions in each variable. Then these functions satisfy the functional equation (6.1.1) for all $p, q \in J$, if and only if there exist a $k \times k$ matrix $C = (C_{ij})$, vectors*

$a, b, c_\ell, e_\ell \in \mathbb{R}^k$ $(\ell = 1, 2, 3, 4)$ *and constants* $d_\ell \in \mathbb{R}$ $(1 \le \ell \le 6)$ *such that*

$$f_\ell(p) = 4a \odot p^3 + (b - 9a) \odot p^2 + p \odot C \odot \log p$$
$$+ c_\ell \odot p + e_\ell \odot \log p + d_\ell, \qquad \ell = 1, 4, \tag{6.3.2}$$

$$f_\ell(p) = 4a \odot p^3 - (b + 9a) \odot p^2 + p \odot C \odot \log p$$
$$+ c_\ell \odot p + e_\ell \odot \log p + d_\ell, \qquad \ell = 2, 3, \tag{6.3.3}$$

$$g(p) = -6a \odot p^2 + (6a - 2b + c_2 - c_4) \odot p$$
$$+ p \odot C \odot \log p + (1 - p) \odot C \odot \log(1 - p) \tag{6.3.4}$$
$$+ (e_1 + e_2) \odot \log p + (e_3 + e_4) \odot \log(1 - p) + d_5,$$

$$h(p) = -6a \odot p^2 + (6a - 2b + c_3 - c_4) \odot p$$
$$+ p \odot C \odot \log p + (1 - p) \odot C \odot \log(1 - p) \tag{6.3.5}$$
$$+ (e_1 + e_3) \odot \log p + (e_2 + e_4) \odot \log(1 - p) + d_6$$

for all $p \in J$, *where*

$$4b + c_1 - c_2 - c_3 + c_4 = 0 \tag{6.3.6}$$

$$b \odot 1 - 5a \odot 1 + c_4 \odot 1 + d_1 + d_2 + d_3 + d_4 = d_5 + d_6. \tag{6.3.7}$$

Proof: If f_ℓ, g and h are given by (6.3.2) - (6.3.5) with the conditions (6.3.6) and (6.3.7), then they satisfy the functional equation (6.1.1) (which can be verified by using the notation of definition (6.3.1) and a tedious but straightforward calculation). To prove the converse, we use the induction principle. If $k = 1$ then (6.3.2) - (6.3.7) reduce to the result in Theorem 6.2.8. We now suppose that the result is valid for all positive integers up to $k - 1$, $k \ge 2$. To prove that (6.3.2) - (6.3.7) are also valid in the k-dimensional case we rewrite the functional equation (6.1.1) in the form

$$f_1(pq, rs) + f_2(p(1 - q), r(1 - s)) + f_3((1 - p)q), (1 - r)s)$$
$$+ f_4((1 - p)(1 - q), (1 - r)(1 - s)) \tag{6.3.8}$$
$$= g(p, r) + h(q, s),$$

where $p, q \in]0, 1[^{k-1}$ and $r, s \in]0, 1[$.

Thus for each fixed $r, s \in]0, 1[$, the equation (6.3.8) is just (6.1.1) for all $p, q \in]0, 1[^{k-1}$, which by the induction hypothesis leads to

$$f_1(p, rs) = 4A \odot p^3 + (B - 9A) \odot p^2 + p \odot C' \odot \log p$$
$$+ C_1 \odot p + E_1 \odot \log p + D_1 \tag{6.3.9}$$

$$f_2(p, r(1-s)) = 4A \odot p^3 - (B+9A) \odot p^2 + p \odot C' \odot \log p$$
$$+ C_2 \odot p + E_2 \odot \log p + D_2 \tag{6.3.10}$$

$$f_3(p, (1-r)s) = 4A \odot p^3 - (B+9A) \odot p^2 + p \odot C' \odot \log p$$
$$+ C_3 \odot p + E_3 \odot \log p + D_3 \tag{6.3.11}$$

$$f_4(p, (1-r)(1-s)) = 4A \odot p^3 + (B-9A) \odot p^2 + p \odot C' \odot \log p$$
$$+ C_4 \odot p + E_4 \odot \log p + D_4 \tag{6.3.12}$$

$$g(p, r) = -6A \odot p^2 + (6A - 2B + C_2 - C_4) \odot p$$
$$+ p \odot C' \odot \log p + (1-p) \odot C' \odot \log(1-p) \tag{6.3.13}$$
$$+ (E_1 + E_2) \odot \log p + (E_3 + E_4) \odot \log(1-p) + D_5$$

$$h(p, s) = -6A \odot p^2 + (6A - 2B + C_3 - C_4) \odot p$$
$$+ p \odot C' \odot \log p + (1-p) \odot C' \odot \log(1-p) \tag{6.3.14}$$
$$+ (E_1 + E_3) \odot \log p + (E_2 + E_4) \odot \log(1-p) + D_6$$

for all $p \in]0, 1[^{k-1}$ and $r, s \in]0, 1[$ with

$$4B + C_1 - C_2 - C_3 + C_4 = 0, \tag{6.3.15}$$

$$B \odot 1 - 5A \odot 1 + C_4 \odot 1 + D_1 + D_2 + D_3 + D_4 = D_5 + D_6, \tag{6.3.16}$$

where $C' = (C_{ij})$ is a $(k-1) \times (k-1)$ matrix, and where $A, B, C_\ell, E_\ell \in \mathbb{R}^{k-1}$, $\ell = 1, 2, 3, 4$, and $D_\ell \in \mathbb{R}$, $1 \leq \ell \leq 6$. Letting r, s vary again we see that all components of A, B, C_ℓ, E_ℓ, all C_{ij} and D_ℓ are functions of r and s. Thus it is enough to determine the explicit forms of all these functions, which we now proceed to do.

From the forms of g and h, we see that

$$A(r, s) = A(r) = A(s)$$

for every $r, s \in]0, 1[$. Thus $A(r, s)$ is a constant, say

$$A(r, s) = A \in \mathbb{R}^{k-1}.$$

Also from the forms of f_1, f_4 and A we find that $B(r, s) - 9A$ is a function of rs and is also a function of $(1-r)(1-s)$. This implies that $B(r, s)$ is constant, say

$$B(r, s) = B \in \mathbb{R}^{k-1}.$$

(See the argument in the proof of Lemma 6.2.6.)

As with A, it happens also that all $C_{ij}(r,s)$, $1 \leq i,j \leq k-1$ are constant. Thus $C' = (C_{ij}) = (C_{ij}(r,s))$ is a constant $(k-1) \times (k-1)$ matrix, where $C_{ij} \in \mathbb{R}$.

From the measurability of the functions f_1, f_2, f_3, f_4 we observe by applying Remark 6.2.2.(d) componentwise since $p^3, p^2, p\log p, p, \log p, 1$ are linearly independent in $]0,1[)$ that $C_1(r,s), C_2(r,s), C_3(r,s), C_4(r,s)$ are measurable functions of rs, $r(1-s)$, $(1-r)s$ and $(1-r)(1-s)$ respectively. That is,

$$\left.\begin{array}{l} C_1(r,s) = C_1(rs), \\ C_2(r,s) = C_2(r(1-s)), \\ C_3(r,s) = C_3((1-r)s), \\ C_4(r,s) = C_4((1-r)(1-s)), \end{array}\right\} \quad (6.3.17)$$

where C_1, C_2, C_3 and C_4 are measurable. Thus from the form of g and (6.3.15) we get, respectively,

$$6A - 2B + C_2(r(1-s)) - C_4((1-r)(1-s)) = \text{a function of } r \text{ only}, \quad (6.3.18)$$

$$4B + C_1(rs) - C_2(r(1-s)) - C_3((1-r)s) + C_4((1-r)(1-s)) = 0. \quad (6.3.19)$$

Therefore the $k-1$ components of each C_ℓ, $1 \leq \ell \leq 4$ in (6.3.19) satisfy the conditions of Lemma 6.2.7. If we write these solutions again in vector form and substitute them into condition (6.3.18), then we arrive at

$$C_\ell(r) = E \log r + K_\ell, \quad 1 \leq \ell \leq 4, \quad (6.3.20)$$

where $E = (\bar{e}_1, \bar{e}_2, ..., \bar{e}_{k-1})$, and $B, K_\ell \in \mathbb{R}^{k-1}$ satisfy

$$4B + K_1 - K_2 - K_3 + K_4 = 0. \quad (6.3.21)$$

By considering the terms involving $\log p$ alone in the forms of f_1, f_2, f_3, f_4, by a linear independence argument again follows that all $k-1$ components of E_ℓ, $1 \leq \ell \leq 4$ are measurable functions (because of Remark 6.2.2.(d)) which can be written in the same form as (6.3.17). Using these forms (6.3.13) and (6.3.14) by considering the terms containing $\log p$ and $\log(1-p)$ in ℓ, we get

$$E_1(rs) + E_2(r(1-s)) = \text{a function of } r,$$

$$E_3((1-r)s) + E_4((1-r)(1-s)) = \text{a function of } r,$$

$$E_1(rs) + E_3((1-r)s) = \text{a function of } s,$$

$$E_2(r(1-s)) + E_4((1-r)(1-s)) = \text{a function of } s.$$

Each equation is a Pexider equation, with measurable solutions (written again in vector form) given by

$$E_\ell(r) = E'r + L_\ell, \qquad 1 \le \ell \le 4, \tag{6.3.22}$$

where $E' = (e'_1, ..., e'_{k-1})$, $L_\ell \in \mathbb{R}^{k-1}$.

As above we find by considering the terms independent of p in the forms of f_1, f_2, f_3, f_4, g, h that D_1, D_2, D_3, D_4 can be written in the form (6.3.17) whereas

$$D_5(r,s) = D_5(r), \qquad D_6(r,s) = D_6(s),$$

where all D_ℓ, $1 \le \ell \le 6$ are measurable functions. Using this fact together with (6.3.17) and (6.3.20), the equation (6.3.16) leads to

$$
\begin{aligned}
D_1(rs) &+ D_2(r(1-s)) + D_3((1-r)s) + D_4((1-r)(1-s)) \\
&= \{D_5(r) - (E\log(1-r)) \odot 1\} \\
&+ \{D_6(s) - (E\log(1-s)) \odot 1\} + (5A - B - K_4) \odot 1\}.
\end{aligned} \tag{6.3.23}
$$

Since the D_ℓ's are real-valued functions, by applying Theorem 6.2.8 to equation (6.3.23) we obtain

$$
\begin{aligned}
D_\ell(r) =& 4A'r^3 + (B' - 9A')r^2 \\
&+ C''r\log r + K'_\ell r + L'_\ell \log r + d_\ell, \quad \ell = 1,4,
\end{aligned} \tag{6.3.24}
$$

$$
\begin{aligned}
D_l(r) =& 4A'r^3 - (B' + 9A')r^2 \\
&+ C''r\log r + K'_\ell r + L'_\ell \log r + d_\ell, \quad \ell = 2,3,
\end{aligned} \tag{6.3.25}
$$

$$
\begin{aligned}
D_5(r) =& -6A'r^2 + (6A' - 2B' + K'_2 - K'_4)r \\
&+ C''[r\log r + (1-r)\log(1-r)] \\
&+ (L'_1 + L'_2)\log r + (L'_3 + L'_4)\log(1-r)\} \\
&+ (E\log(1-r)) \odot 1 + d_5,
\end{aligned} \tag{6.3.26}
$$

$$
\begin{aligned}
D_6(r) =& -6A'r^2 + (6A' - 2B' + K'_3 - K'_4)r \\
&+ C''[r\log r + (1-r)\log(1-r)] \\
&+ (L'_1 + L'_3)\log r + (L'_2 + L'_4)\log(1-r)\} \\
&+ (E\log(1-r)) \odot 1 + d_6,
\end{aligned} \tag{6.3.27}
$$

where $A', B', C''', K'_\ell, L'_\ell, d_\ell$ are real constants satisfying

$$4B' + K'_1 - K'_2 - K'_3 + K'_4 = 0, \tag{6.3.28}$$

$$B' - 5A' + K'_4 + d_1 + d_2 + d_3 + d_4$$
$$= d_5 + d_6 + (5A - B - K_4) \odot 1. \tag{6.3.29}$$

Thus we have obtained the explicit forms of A, B, C_ℓ, E_ℓ ($\ell = 1, 2, 3, 4$), D_ℓ ($1 \le \ell \le 6$), and C_{ij} ($1 \le i, j \le k - 1$). Substitution of these functions into (6.3.9) and (6.3.13) leads to

$$
\begin{aligned}
f_1(p, r) = & 4A \odot p^3 + (B - 9A) \odot p^2 + p \odot C' \odot \log p \\
& + (E \log r + K_1) \odot p + (E'r + L_1) \odot \log p + 4A'r^3 \\
& + (B' - 9A')r^2 + C''r \log r + K'_1 r + L'_1 \log r + d_1
\end{aligned} \tag{6.3.30}
$$

$$
\begin{aligned}
g(p, r) = & -6A \odot p^2 \\
& + (6A - 2B + E \log r - E \log(1 - r) + K_2 - K_4) \odot p \\
& + p \odot C' \odot \log p + (1 - p) \odot C' \odot \log(1 - p) \\
& + (E'r + L_1 + L_2) \odot \log p \\
& + (E'(1 - r) + L_3 + L_4) \odot \log(1 - p) \\
& - 6A'r^2 + (6A' - 2B' + K'_2 - K'_4)r \\
& + C''r \log r + C''(1 - r) \log(1 - r) \\
& + (L'_1 + L'_2) \log r + (L'_3 + L'_4) \log(1 - r) \\
& + (E \log(1 - r)) \odot 1 + d_5
\end{aligned} \tag{6.3.31}
$$

for all $p \in]0, 1[^{k-1}$, $r \in]0, 1[$. Thus (6.3.30) and (6.3.31) go over into (6.3.2) and (6.3.4) by defining

$$P = (p, r) \in]0, 1[^k, \quad a = (A, A') \in \mathbb{R}^k, \quad b = (B, B') \in \mathbb{R}^k,$$

$$c_\ell = (K_\ell, K'_\ell) \in \mathbb{R}^k, \quad e_\ell = (L_\ell, L'_\ell) \in \mathbb{R}^k, \quad 1 \le \ell \le 4,$$

$$
C = \begin{pmatrix} & & \vdots & \bar{e}_1 \\ & C' & \vdots & \vdots \\ \cdots & \cdots & \vdots & \bar{e}_{k-1} \\ e'_1 & \cdots & e'_{k-1} & C'' \end{pmatrix}. \tag{6.3.32}
$$

In the same manner we can establish the forms of f_2, f_3, f_4 and h. Moreover (6.3.6) and (6.3.7) are valid because of (6.3.21), (6.3.28), (6.3.29) and (6.3.32). This completes the proof.

We also note here the following consequence which we will need later.

Corollary 6.3.3. *A measurable function* $f :]0,1[^k \to \mathbb{R}$ *satisfies the functional equation*

$$f(pq) + f(p(1-q)) + f((1-p)q) + f((1-p)(1-q))$$
$$= f(p) + f(1-p) + f(q) + f(1-q) \tag{6.3.33}$$

for all $p, q \in]0,1[^k$, *if and only if there exists a* $k \times k$ *matrix* $C = (C_{ij})$, *vectors* $a, c \in \mathbb{R}^k$ *and a constant* $d \in \mathbb{R}$ *such that*

$$f(p) = 4a \odot p^3 - 9a \odot p^2 + p \odot C \odot \log p + c \odot p + d \tag{6.3.34}$$

with

$$-5a \odot 1 + c \odot 1 = 0 \tag{6.3.35}$$

Proof: Let us first put $f_1 = f_2 = f_3 = f_4 = f$ in (6.1.1). Then Theorem 6.3.2 implies

$$b = 0, \quad c_1 = c_2 = c_3 = c_4 =: c,$$
$$e_1 = e_2 = e_3 = e_4 =: e, \quad d_1 = d_2 = d_3 = d_4 =: d. \tag{6.3.36}$$

Moreover (6.3.7) implies

$$(-5a + c) \odot 1 + 4d = d_5 + d_6. \tag{6.3.37}$$

Because of

$$f(p) = 4a \odot p^3 - 9a \odot p^2 + p \odot C \odot \log p + c \odot p + e \odot \log p + d \tag{6.3.38}$$

and

$$f(p) + f(1-p) = -6a \odot p^2 + 6a \odot p + (-5a + c) \odot 1$$
$$+ p \odot C \odot \log p + (1-p) \odot C \odot \log(1-p) \tag{6.3.39}$$
$$+ e \odot (\log p + \log(1-p)) + 2d,$$

the condition $g(p) = f(p) + f(1-p) = h(p)$ implies $d_5 = d_6 = 2d$ and $e = 0$, so that (6.3.37) and (6.3.38) go over into (6.3.34) and (6.3.35).

The following result generalizes Lemma 6.2.3.

Corollary 6.3.4. *A measurable function* $f : J \to \mathbb{R}$ *satisfies*

$$f(pq) + f(p(1-q)) + f((1-p)q) + f((1-p)(1-q)) = 0, \quad p, q \in J, \tag{6.3.40}$$

if and only if there exists a vector $A \in \mathbb{R}^k$ such that

$$f(p) = A \odot (4p - 1), \qquad p \in J. \tag{6.3.41}$$

Proof: We proceed in the same manner as the proof of Corollary 6.3.3 and arrive at the equations (6.3.36) to (6.3.38). Now we put $g = h = 0$ so that instead of (6.3.39) we get $d_5 = d_6 = 0$, $a = e = 0$ and $C = (C_{ij}) =$ zero matrix. Thus (6.3.37) and (6.3.38) yield

$$f(p) = c \odot p + d, \qquad c \odot 1 + 4d = 0.$$

Putting $A = \frac{1}{4}c$, we have (6.3.41).

6.4. Some generalized Cauchy equations

As pointed out in the beginning of this chapter we now present the general solutions of some functional equations which are related to the basic Cauchy functional equations. We deal first with the functional equation

$$\sum_{i=1}^{n} f(p_i) = c, \qquad P \in \Gamma_n^o, \tag{6.4.1}$$

where c is a constant. Consider with the following heuristic argument. Let $n = 3$ and $k = 1$. Putting $p_1 = x$, $p_2 = y$ and $g(x) = c - f(1-x)$ into (6.4.1) we get the Pexider equation

$$g(x + y) = f(x) + f(y),$$

so that we expect that the solution of (6.4.1) is an additive function plus a constant. The following theorem establishes this result.

Theorem 6.4.1. *Let $n \geq 3$ be a fixed integer and let $c \in \mathbb{R}$ be a constant. The function $f : J \to \mathbb{R}$ satisfies the functional equation (6.4.1), if and only if there is an additive function $A : \mathbb{R}^k \to \mathbb{R}$ and a constant $b \in \mathbb{R}$ such that*

$$f(p) = A(p) + b, \qquad p \in J, \tag{6.4.2}$$

and

$$A(1) + nb = c. \tag{6.4.3}$$

Proof: We choose $u \in]0, 1[$ such that $(n-3)u < 1$, and let $a = 1 - (n-3)u$. (Note that a is dependent upon u.) For arbitrary $x, y, x + y \in]0, a[^k$ we put

$$p_1 = x, \quad p_2 = y, \quad p_3 = \mathbf{a} - x - y, \quad p_4 = p_5 = \cdots = p_n = \mathbf{u}$$

into (6.4.1) and get

$$f(x) + f(y) + f(\mathbf{a} - x - y) + (n-3)f(\mathbf{u}) = c. \tag{6.4.4}$$

(Remember that $p = (p_1, p_2, ..., p_n) \in \Gamma_n^o$ is a $k \times n$ matrix with columns $p_i = (p_{1i}, ..., p_{ki})^T$, $1 \leq i \leq n$; moreover $\mathbf{u}, \mathbf{a}$ stand for constant vectors of $\mathbb{R}^k$ having the same components u, a respectively.)

Let us define $d = c - (n-3)f(\mathbf{u})$. Then we obtain from (6.4.4)

$$\begin{aligned}
f(x + y) + f(z) &= d - f((\mathbf{a} - x - y) - z) \\
&= d - f((\mathbf{a} - x) - y - z) \\
&= f(x) + f(y + z)
\end{aligned}$$

for all $x, y, z \in]0, a[^k$ with $x + y + z \in]0, a[^k$. That is,

$$f(x + y) - f(x) = f(z + y) - f(y) =: g(y). \tag{6.4.5}$$

Because of

$$f(x) + g(y) = f(x + y) = f(y + x) = f(y) + g(x),$$

we get from Lemma 1.2.1

$$f(x) - g(x) = f(y) - g(y) =: b,$$

or $g(x) = f(x) - b$. Using this form of g and (6.4.5) we see that g is additive in $\Delta = \{(x, y) \mid x, y, x + y \in]0, a[^k\}$. Since g has a unique additive extension to $\mathbb{R}^k$ (see Lemma 1.2.4 and Lemma 1.2.5), we have

$$f(p) = A(p) + b, \qquad p \in]0, a[^k \tag{6.4.6}$$

where b is a real constant and $A : \mathbb{R}^k \to \mathbb{R}$ is additive. If we repeat the above consideration for $u' \in]0, 1[$ with $(n-3)u' < 1$ and $a' := 1 - (n-3)u'$, then we get the representation

$$f(p) = A'(p) + b', \qquad p \in]0, a'[^k, \tag{6.4.7}$$

for some additive function $A' : \mathbb{R}^k \to \mathbb{R}$ and some constant b'. Thus (6.4.6) and (6.4.7) imply

$$A(p) + b = A'(p) + b', \qquad p \in]0, a''[^k,$$

where $a'' = \min\{a, a'\}$. But then $A = A'$ and $b = b'$. Therefore the representation (6.4.6) is valid for all $a \in]0, 1[$ and we obtain (6.4.2):

$$f(p) = A(p) + b, \qquad p \in \bigcup_{a \in]0,1[}]0, a[^k = J.$$

Substituting (6.4.2) into (6.4.1) we arrive at (6.4.3). Since the converse is obviously satisfied, the theorem is proven.

Next we derive the general solutions of the three functional equations (cf. (6.1.2)-(6.1.4))

$$f(pq) + f(p(1-q)) = f(p)\{f(q) + f(1-q)\}, \qquad p, q \in J, \qquad (6.4.9)$$

$$f(pq) + f(p(1-q)) = f(p)\{M(q) + M(1-q)\}, \qquad p, q \in J, \qquad (6.4.10)$$

$$\begin{aligned} f(pq) + f(p(1-q)) &= f(p)\{M(q) + M(1-q)\} \\ &+ M(p)\{f(q) + f(1-q)\}, \quad p, q \in J, \end{aligned} \qquad (6.4.11)$$

where $f, M : J \to \mathbb{R}$ and M is multiplicative.

First we present the solutions of (6.4.9) and (6.4.10). The solution of equation (6.4.11) will be obtained later from the solution of (6.4.10). In the 1-dimensional case, equations (6.4.9) and (6.4.10) have been solved by Maksa (1987) and by Kannappan and Sahoo (1993), respectively. (Maksa's solution is given on the closed interval $[0, 1]$ but works also on the open interval.) The functional equation (6.4.10) was solved in the k-dimensional case by Sahoo and Sander (1989), but the method presented in their paper does not work for the k-dimensional case of equation (6.4.9). The following proof is an adaptation of Maksa's proof in (1987) and Kannappan and Sahoo's proof in (1993).

Let us start with the following notations.

Definition 6.4.2. Let $a = (a_1, ..., a_k)$, $b = (b_1, ..., b_k)$ in $\mathbb{R}^k$ and let $f, M : J \to \mathbb{R}$.

(1) $R(a, b) := \{x = (x_1, ..., x_n) \mid a_i < x_i < b_i, 1 \leq i \leq n\}$

(2) $\Omega_f := \{x \in J \mid f(tx) = f(t)f(x), \; \forall t \in J\}$

(3) $\Omega_{f,M} := \{x \in J \mid f(tx) = f(t)M(x), \; \forall t \in J\}$.

The sets Ω_f and $\Omega_{f,M}$ play central roles in the determination of the general solutions of the functional equations (6.4.9) and (6.4.10), respectively, as is seen in the next three lemmas.

Lemma 6.4.3. (i) *If $f : J \to \mathbb{R}$ satisfies (6.4.9) and if $f(x_o) = 0$ for some $x_o \in \Omega_f$, then $f \equiv 0$ and $\Omega_f = J$.*

(ii) *Let $f, M : J \to \mathbb{R}$ satisfy (6.4.10) where M is multiplicative. If $M(x_o) = 0$ for some $x_o \in J$, then $M \equiv 0$, $f \equiv 0$ and $\Omega_{f,M} = J$.*

Proof: (i) If $f(x_o) = 0$ for $x_o \in \Omega_f$, then the definition of Ω_f shows that $f \equiv 0$ on $R(0, x_o)$. Note that (6.4.9) with $q = \frac{1}{2}$ yields

$$f\left(\frac{1}{2}p\right) = f(p) f\left(\frac{1}{2}\right) \qquad (6.4.12)$$

for all $p \in J$, that is $\frac{1}{2} \in \Omega_f$. We show next that $f(\frac{1}{2}) = 0$. From (6.4.12), by induction we get $f(\frac{1}{2^k}) = f(\frac{1}{2})^k$ for $k = 1, 2, \cdots$. But for sufficiently large k we have $\frac{1}{2^k} \in R(0, x_o)$, hence $f(\frac{1}{2^k}) = 0$ for such k and we arrive at $f(\frac{1}{2}) = 0$. Because of (6.4.12), we deduce that $f \equiv 0$ on $R(0, \frac{1}{2})$.

Next, let $q \in R(\frac{1}{2}, 1)$, so that $1 - q \in R(0, \frac{1}{2})$. Then $f(1 - q) = 0 = f(p(1-q))$ for all $p \in J$. So (6.4.9) shows that $q \in \Omega_f$. Moreover, it follows that $f(q^k) = f(q)^k$ $(k = 1, 2, 3, ...)$. Since $q^k \in R(0, \frac{1}{2})$ for sufficiently large k, we get $f(q) = 0$. Thus $f \equiv 0$ on $R(0, \frac{1}{2}) \cup R(\frac{1}{2}, 1)$.

Finally, consider an arbitrary $y \in J$. Letting $p = \frac{1}{2}(1 + y)$ and $q = \frac{2y}{1+y}$ in (6.4.9), we find that

$$f(y) + f\left(\frac{1-y}{2}\right) = f\left(\frac{1+y}{2}\right) \left\{ f\left(\frac{2y}{1+y}\right) + f\left(\frac{1-y}{1+y}\right) \right\}.$$

But $\frac{1}{2}(1 - y) \in R(0, \frac{1}{2})$ and $\frac{1}{2}(1 + y) \in R(\frac{1}{2}, 1)$, hence $f(\frac{1}{2}(1 - y)) = f(\frac{1}{2}(1 + y)) = 0$, which leads to $f(y) = 0$. Therefore $f \equiv 0$ on J.

(ii) If $M(x_o) = 0$ for some $x_o \in J$, then the multiplicativity of M forces M to be the zero map on J. Now (6.4.10) with $M = 0$ goes over into $f(pq) + f(p(1 - q)) = 0$ for $p, q \in J$. Putting $q = \frac{1}{2}$, respectively $p = \frac{1}{2}$ into this last equation we get $f(\frac{p}{2}) = 0$ and $f(\frac{1-q}{2}) = -f(\frac{q}{2}) = 0$ for all $p, q \in J$. Thus $f \equiv 0$ on $R(0, \frac{1}{2}) \cup R(\frac{1}{2}, 1)$. As in part (i) we get that $f \equiv 0$ on J, and this completes the proof of Lemma 6.4.3.

Next, we list some key properties of the sets Ω_f and $\Omega_{f,M}$.

Lemma 6.4.4. (i) *If* $f : J \to \mathbb{R}$ *satisfies (6.4.9), then* Ω_f *has the following properties:*

(a) $\frac{1}{2} \in \Omega_f$;

(b) *if* $x \in \Omega_f$, *then* $1 - x \in \Omega_f$;

(c) *if* $x, y \in \Omega_f$, *then* $xy \in \Omega_f$;

(d) *if* $x, y \in \Omega_f$ *and* $\frac{x}{y} \in J$, *then* $\frac{x}{y} \in \Omega_f$;

(e) *if* $x, y \in \Omega_f$ *and* $x - y \in J$, *then* $x - y \in \Omega_f$;

(f) *if* $x, y \in \Omega_f$ *and* $x + y \in J$, *then* $x + y \in \Omega_f$.

Thus, Ω_f *is nonempty and closed under the four basic algebraic operations whenever these operations lead to a result in* J.

(ii) *If* $f, M : J \to \mathbb{R}$ *satisfy (6.4.10) and* M *is multiplicative, then* $\Omega_{f,M}$ *has the same properties (a)–(f) above.*

Proof: (ii) If $M = 0$, then $f = 0$ by Lemma 6.4.3.(ii). In this case, $\Omega_{f,M} = J$ and all parts of the lemma obviously hold. Henceforth, we assume that $M \neq 0$, thus M is nowhere zero.

Part (a) follows from (6.4.10) by putting $q = \frac{1}{2}$. Part (b) follows immediately from (6.4.10). For (c), observe that $x, y \in \Omega_{f,M}$ implies

$$f(txy) = f(tx)\, M(y) = f(t)\, M(x)\, M(y) = f(t)\, M(xy) \qquad (6.4.13)$$

for all $t \in J$. To get (d), use the fact that for any $t \in J$, $x, y \in \Omega_{f,M}$,

$$f\left(\frac{tx}{y}\right) M(y) = f(tx) = f(t)\, M(x) = f(t)\, M\left(\frac{x}{y}\right) M(y), \qquad (6.4.14)$$

and $M(y) \neq 0$ by hypothesis. For (e), combine (b), (c) and (d) with the facts that $\frac{y}{x} \in J$ and $x - y = x(1 - \frac{y}{x})$. Finally, if $x, y \in \Omega_{f,M}$ and $x + y \in J$, then $1 - x$ and $(1 - x) - y$ are in $\Omega_{f,M}$, by (b) and then (e), respectively. Thus $1 - (x + y) \in \Omega_{f,M}$ and hence $x + y \in \Omega_{f,M}$ by (b) again. This establishes (f) and completes the proof of part (ii).

To prove (i) we see that $f \equiv 0$ implies $\Omega_f = J$ so that (a)-(f) are clearly valid. Otherwise, by Lemma 6.4.3 we may assume that f is nowhere zero on Ω_f. Now the proof of the statements (a)-(f) is completely analogous to the proof of part (ii), replacing equation (6.4.10), M and $\Omega_{f,M}$ by equation (6.4.9), f and Ω_f, respectively, and using in (6.4.13) and (6.4.14) the assumption $y \in \Omega_f$ (that is $f(y) \neq 0$) instead of the fact that M is multiplicative and nowhere zero. This completes the proof.

Lemma 6.4.5. (i) *Suppose $f : J \to \mathbb{R}$ satisfies (6.4.9). If Ω_f contains a nonempty open set, then $\Omega_f = J$ and f is a multiplicative function on J.*
(ii) *Suppose $f, M : J \to \mathbb{R}$ satisfy (6.4.10) where M is multiplicative. If the set $\Omega_{f,M}$ contains a nonempty open set, then $\Omega_{f,M} = J$ and $f = cM$ for some constant c.*

Proof: (i) Suppose Ω_f contains a nonempty open set $G \subset J$. Then $G - G = \{x - y \mid x, y \in G\}$ is an open ball centered at the origin in $\mathbb{R}^k$. By part (e) of Lemma 6.4.4, the set $A = (G - G) \cap J$ is contained in Ω_f. Now, given any point x in J, there exists a positive integer k such that $\frac{1}{k}x \in A \subset \Omega_f$. By part (f) of Lemma 6.4.4 (applied inductively), $x = k(\frac{1}{k}x) \in \Omega_f$. Hence $\Omega_f = J$, which means that f is multiplicative on the set J.

(ii) Using the same procedure as in the proof of (i), it can be shown that $\Omega_{f,M} = J$, that is,

$$f(tx) = f(t) M(x), \qquad t, x \in J. \tag{6.4.13}$$

If $M \equiv 0$, then $f \equiv 0$ and we are done. Otherwise, using the fact that $f(tx) = f(xt)$, we get from (6.4.13)

$$f(t) M(x) = f(x) M(t), \qquad t, x \in J.$$

Putting $x = x_o$ (*constant*) and using the fact that M is nowhere zero, we arrive at $f(t) = c M(t)$ for all $t \in J$, and the proof is finished.

Now we are ready to give the general solutions of the equations (6.4.9) and (6.4.10). The result in both cases is that the general solution takes on one of two forms depending on whether Ω_f or $\Omega_{f,M}$ contains a nonempty open set or not.

Theorem 6.4.6. (i) *The map $f : J \to \mathbb{R}$ satisfies (6.4.9), if and only if either f is multiplicative or f is the restriction to J of an additive map on $\mathbb{R}^k$.*
(ii) *Let M be multiplicative. Then $f, M : J \to \mathbb{R}$ satisfy equation (6.4.10), if and only if either $f = cM$ for some constant c or f and M are restrictions to J of an additive map and a projection on $\mathbb{R}^k$, respectively.*

Proof: The forms of f given in (i) and (ii), obviously satisfy (6.4.9) and (6.4.10), respectively. To prove the converse we first prove (ii). If $\Omega_{f,M}$ contains a nonempty open set, then Lemma 6.4.5 shows that $f = cM$ for some constant c. Let us suppose now that $\Omega_{f,M}$ does not contain any nonempty open set in $\mathbb{R}^k$.

By Lemma 6.4.3.(ii) and Lemma 6.4.5, M is nowhere zero. Also, the map $\phi : J^2 \to \mathbb{R}$ defined by

$$\phi(t, x) = f(tx) - f(t) M(x), \qquad t, x \in J, \qquad (6.4.14)$$

is not identically zero. By (6.4.10), we have

$$\phi(t, x) + \phi(t, 1 - x) = 0, \qquad t, x \in J, \qquad (6.4.15)$$

and in particular

$$\phi\left(t, \frac{1}{2}\right) = 0, \qquad t \in J. \qquad (6.4.16)$$

For any $(x, y) \in D_2$, $t \in J$, put $q = \frac{x}{x+y}$ and $p = t(x + y)$ or $p = x + y$ into (6.4.10) to get, respectively,

$$f(tx) + f(ty) = f(t(x + y)) \left\{ M\left(\frac{x}{x + y}\right) + M\left(\frac{y}{x + y}\right) \right\}, \qquad (6.4.17)$$

$$f(x) + f(y) = f(x + y) \left\{ M\left(\frac{x}{x + y}\right) + M\left(\frac{y}{x + y}\right) \right\}. \qquad (6.4.18)$$

Using (6.4.18) and the multiplicativity of M we calculate

$$\begin{aligned}
M(x &+ y)[\phi(t, x) + \phi(t, y)] \\
&= M(x + y)[f(tx) - M(x)f(t) + f(ty) - M(y)f(t)] \\
&= M(x + y)[f(tx) + f(ty)] - M(x + y)[M(x) + M(y)]f(t) \\
&= M(x + y)f(tx + ty) \left[M\left(\frac{x}{x + y}\right) + M\left(\frac{y}{x + y}\right) \right] \\
&\quad - M(x + y)[M(x) + M(y)]f(t) \\
&= f(tx + ty)[M(x) + M(y)] - M(x + y)[M(x) + M(y)]f(t) \\
&= [M(x) + M(y)][f(tx + ty) - M(x + y)f(t)] \\
&= [M(x) + M(y)]\phi(t, x + y).
\end{aligned}$$

That is, for all $t \in J$, $(x, y) \in D_2$, we have

$$M(x + y) \{\phi(t, x) + \phi(t, y)\} = \phi(t, x + y) \{M(x) + M(y)\}. \qquad (6.4.19)$$

Replacing y by $1 - x - y$ in (6.4.19), and using (6.4.15), we have also

$$M(1 - y)\{\phi(t, x) - \phi(t, x + y)\} = -\phi(t, y)\{M(x) + M(1 - x - y)\}. \quad (6.4.20)$$

Next multiply both sides of (6.4.20) by $M(x) + M(y)$ and then use (6.4.19) to eliminate the $\phi(t, x + y)$ term. This leads to

$$\begin{aligned}
\phi(t, x)\, M(1 - y)\, &\{M(x + y) - M(x) - M(y)\} \\
&= \phi(t, y)\, \{[M(x) + M(y)]\, [M(x) + M(1 - x - y)] \quad (6.4.21) \\
&\quad - M(x + y)\, M(1 - y)\},
\end{aligned}$$

for all $t \in J$, $(x, y) \in D_2$.

Let $y \in \Omega_{f,M}$ and choose $x \in J \setminus \Omega_{f,M}$ with $(x, y) \in D_2$. By hypothesis, there exists a $t_o \in J$ for which $\phi(t_o, x) \neq 0$ and $\phi(t_o, y) = 0$. Since $M(1-y) \neq 0$, (6.4.21) yields (for $t = t_o$)

$$M(x + y) = M(x) + M(y), \quad (6.4.22)$$

for $x \in J \setminus \Omega_{f,M}$ and $y \in \Omega_{f,M}$ with $(x, y) \in D_2$.

Now let $x, y \in \Omega_{f,M}$ and $x + y \in J$. Then there exists $x_o \in R(0, 1 - x - y)$ such that $x + x_o \in J \setminus \Omega_{f,M}$, since otherwise $\Omega_{f,M}$ would contain the open set $R(x, 1 - y)$. By Lemma 6.4.4 (f), we see that $x_o \in J \setminus \Omega_{f,M}$ and that $x + y \in \Omega_{f,M}$, hence $x + x_o + y \in J \setminus \Omega_{f,M}$ by Lemma 6.4.4 (e). Thus, by (6.4.22),

$$\begin{aligned}
M(x + y) &= M(x + y + x_o) - M(x_o) \\
&= M(x + x_o) + M(y) - M(x_o) \\
&= M(x) + M(y).
\end{aligned}$$

What we have now is that (6.4.22) is valid for all $(x, y) \in D_2$ with $y \in \Omega_{f,M}$. In particular,

$$M\left(x + \frac{1}{2}\right) = M(x) + M\left(\frac{1}{2}\right), \qquad x \in R\left(0, \frac{1}{2}\right),$$

since $\frac{1}{2} \in \Omega_{f,M}$ by Lemma 6.4.4 (a). Putting $y = \frac{1}{2}$ in (6.4.20) and using (6.4.16) and the fact that $M(\frac{1}{2}) \neq 0$, we have $\phi(t, x) = \phi\left(t, x + \frac{1}{2}\right)$, which means that

$$f\left(tx + \frac{t}{2}\right) = f(tx) + f(t)\left\{M\left(x + \frac{1}{2}\right) - M(x)\right\}$$
$$= f(tx) + f(t)\,M\left(\frac{1}{2}\right)$$
$$= f(tx) + f\left(\frac{t}{2}\right)$$

for all $t \in J$ and $x \in R(0, \frac{1}{2})$. Therefore

$$f(u + v) = f(u) + f(v), \qquad u, v \in R\left(0, \frac{1}{2}\right). \qquad (6.4.23)$$

Now for any $(u, v) \in D_2$, we have $\frac{u}{2}, \frac{v}{2} \in R(0, \frac{1}{2})$, so that (6.4.14), (6.4.16) and (6.4.23) imply

$$M\left(\frac{1}{2}\right) f(u + v) = f\left(\frac{u}{2} + \frac{v}{2}\right)$$
$$= f\left(\frac{u}{2}\right) + f\left(\frac{v}{2}\right)$$
$$= M\left(\frac{1}{2}\right)\{f(u) + f(v)\}.$$

Hence

$$f(u + v) = f(u) + f(v), \qquad (u, v) \in D_2, \qquad (6.4.24)$$

from which it follows that f is the restriction to J of an additive function on $\mathbb{R}^k$.

Moreover, f is nowhere zero, since otherwise $\Omega_{f,M} = J$ contrary to our assumption. So (6.4.18) now yields

$$M\left(\frac{x}{x + y}\right) + M\left(\frac{y}{x + y}\right) = 1, \qquad (x, y) \in D_2.$$

Multiplying by $M(x + y)$, we have

$$M(x) + M(y) = M(x + y), \qquad (x, y) \in D_2.$$

Hence M, too, is the restriction to J of an additive function on $\mathbb{R}^k$. Finally, since $M \neq 0$, M must be the restriction to J of a projection. This completes the proof of part (ii).

Now we assume that $f : J \to \mathbb{R}$ satisfies (6.4.9). If Ω_f contains a nonempty open set then Lemma 6.4.5 proves one part of statement (i). If Ω_f does not contain any nonempty open set, then f is not a multiplicative function on J, for otherwise $\Omega_f = J$. In particular, f is not the zero map, hence f never takes the value zero on Ω_f, by Lemma 6.4.3.(i). Another consequence is that the map $\phi : J^2 \to \mathbb{R}$ defined by

$$\phi(t, x) := f(tx) - f(t) f(x), \qquad t, x \in J$$

is not identically zero. By (6.4.9), we have again (6.4.15) and (6.4.16). Proceeding as in the proof of part (ii), we replace all occurrences of M by f, and we use where necessary the definition of Ω_f in place of the multiplicativity of M and the definition of $\Omega_{f,M}$. In so doing, we arrive finally at (6.4.24) again, and we are done.

Now we are ready to give the solution of the remaining equation (6.4.11). This can be done by transforming (6.4.11) into an equation of type (6.4.10).

Theorem 6.4.7. *Let M be multiplicative. Then $f, M : J \to \mathbb{R}$ satisfy (6.4.11), if and only if*

$$f(p) = M(p)L(p) + A(p), \tag{6.4.25}$$

where $L : J \to \mathbb{R}$ is a logarithmic function, and $A : J \to \mathbb{R}$ is additive. Here $A(1) = 0$ if M is additive and $M \neq 0$, otherwise $A \equiv 0$.

Proof: The "if" part is a simple verification. For the converse we observe first that the case $M = 0$ leads to

$$f(pq) + f(p(1 - q)) = 0$$

which results, as in the proof of Lemma 6.4.3, in $f = 0$. Thus (6.4.25) is satisfied and in the following we may assume that M is nowhere zero.

For every $t \in J$ we define

$$F_t(p) := f(pt) - M(t)f(p) - f(t)M(p), \qquad p \in J. \tag{6.4.26}$$

Thus the equation (6.4.11) can be rewritten as

$$F_t(p) = -F_t(1 - p) \quad p \in J. \tag{6.4.27}$$

We now show that F_t satisfies (6.4.10). Replacing p by pt in (6.4.11) we get

$$
\begin{aligned}
f(qpt) + f((1-q)pt) = {} & f(pt)\{M(q) + M(1-q)\} \\
& + M(p)M(t)\{f(q) + f(1-q)\}.
\end{aligned}
\tag{6.4.28}
$$

Using the definition of F_t, (6.4.11) and (6.4.28), we get the desired result:

$$
\begin{aligned}
& F_t(pq) + F_t(p(1-q)) \\
={} & \{f(qpt) + f((1-q)pt)\} - M(t)\{f(pq) + f(p(1-q))\} \\
& - f(t)M(p)\{M(q) + M(1-q)\} \\
={} & f(pt)\{M(q) + M(1-q)\} + M(p)M(t)\{f(q) + f(1-q)\} \\
& - M(t)f(p)\{M(q) + M(1-q)\} \\
& - M(p)M(t)\{f(q) + f(1-q)\} - f(t)M(p)\{M(q) + M(1-q)\} \\
={} & \{f(pt) - M(t)f(p) - f(t)M(p)\}\{M(q) + M(1-q)\} \\
={} & F_t(p)\{M(q) + M(1-q)\}.
\end{aligned}
$$

By Theorem 6.4.6.(ii) we have the following two cases:
Either

$$
F_t(p) = c(t)\,M(p), \qquad t, p \in J,
\tag{6.4.29}
$$

or

$$
F_t(p) = A(p, t) \qquad \text{with } M \text{ additive and nonzero},
\tag{6.4.30}
$$

where $A : \mathbb{R}^k \times J \to \mathbb{R}$ is additive in its first variable and $c : J \to \mathbb{R}$ is an arbitrary function.

Treating the case (6.4.29) first, from (6.4.26) we obtain

$$
\begin{aligned}
F_t(p) &= c(t)\,M(p) \\
&= f(pt) - M(t)\,f(p) - f(t)\,M(p), \quad t, p \in J.
\end{aligned}
\tag{6.4.31}
$$

Because of the symmetry in p and t on the right side in (6.4.31) we have

$$
c(t)\,M(p) = c(p)\,M(t), \qquad t, p \in J.
\tag{6.4.32}
$$

Since M is nowhere zero, (6.4.32) with $p = \frac{1}{2}$ goes over into

$$
c(t) = a\,M(t), \qquad t \in J
$$

where $a = c(\frac{1}{2})/M(\frac{1}{2})$. Thus $F_t(p) = aM(t)M(p)$ and (6.4.31) becomes

$$f(pt) - aM(pt) = M(t)f(p) + f(t)M(p)$$

or, using the multiplicativity of M,

$$\frac{f(pt) - aM(pt)}{M(pt)} = \frac{f(p)}{M(p)} + \frac{f(t)}{M(t)}, \qquad p, t \in J. \qquad (6.4.33)$$

But (6.4.33) is a Pexider equation with the solutions

$$f(p) - a\, M(p) = \{L(p) + 2b\}M(p) \qquad (6.4.34)$$

and

$$f(p) = \{L(p) + b\}M(p), \qquad (6.4.35)$$

where $L : J \to \mathbb{R}$ is a logarithmic function and b is a constant. But the representation of f in (6.4.34) and (6.4.35) gives $b = -a$. Finally, the substitution of f given by (6.4.35) into (6.4.11) leads to $bM(p)(M(q) + M(1-q)) = 0$. Putting here $p = q = \frac{1}{2}$ we get $b = 0$ so that f has the form (6.4.25).

Now we turn to the case (6.4.30) and define $G : D_2 \to \mathbb{R}$ by

$$G(x, y) := f(x) + f(y) - f(x+y) \qquad (6.4.36)$$

for all $(x, y) \in D_2$. Using the fact that

$$A(x, t) + A(y, t) - A(x+y, t) = 0, \qquad (6.4.37)$$

we deduce from (6.4.30), (6.4.36) and (6.4.37)

$$0 = G(xt, yt) - M(t)G(x, y) - f(t)\{M(x) + M(y) - M(x+y)\}.$$

That is (since M is additive, too)

$$G(xt, yt) = M(t)G(x, y) \qquad (6.4.38)$$

for all $(x, y) \in D_2$ and $t \in J$. Since G defined by (6.4.36) satisfies (3.2.3)-(3.2.5), Remark 3.5.4 implies that

$$f(x) = M(x)L(x) + A(x), \qquad x \in J \qquad (6.4.39)$$

for some additive function $A : \mathbb{R}^k \to \mathbb{R}$ and for some logarithmic function $L : \mathcal{P} \to \mathbb{R}$. Substitution of (6.4.39) into (6.4.11), using $M \neq 0$ and $M(1) = 1$, yields

$$A(p)[1 - M(q) - M(1 - q)] = A(1)M(p).$$

Thus, if M is additive and nonzero we get $A(1) = 0$. Hence f has the form (6.4.25), and this completes the proof of the theorem.

Finally, let us present the following result which will be used rather often.

Theorem 6.4.8. *Let $M : J \to \mathbb{R}$ be a multiplicative function satisfying*

$$M(p) + M(1 - p) = a, \qquad p \in J, \tag{6.4.40}$$

for some constant a. Then either M is additive and $a = 1$, or $M = a = 0$, or $M = 1$ and $a = 2$.

Proof: Equation (6.4.40) yields, using the multiplicativity of M,

$$M(pq) + M(p(1 - q)) = M(p)\{M(q) + M(1 - q)\} = a\, M(p). \tag{6.4.41}$$

If $a = 1$, then M is additive. If $a = 0$, then Lemma 6.4.3.(ii) implies $M = 0$. (Note that with $p = q = \frac{1}{2}$ we get $M(\frac{1}{4}) = 0$ from (6.4.41).) Finally, if $a \neq 0, 1$, then (6.4.41) is a Pexider equation with the solution

$$M(p) = B(p) + b, \qquad a\, M(p) = B(p) + 2b, \qquad p \in J, \tag{6.4.42}$$

where B is additive and b is an arbitrary constant. But (6.4.42) yields $M(p) = \frac{b}{a-1}$ since $a \neq 1$. The only constant multiplicative maps are $M = 0$ and $M = 1$. But $M = 0$ is impossible since $a \neq 0$, therefore $M = 1$ and $a = 2$. This completes the proof.

6.5. Some linear independence results on additive and multiplicative functions

Later we need some results which give sufficient conditions for constant, additive and multiplicative functions to be linearly independent.

Lemma 6.5.1. *Let $A : J \to \mathbb{R}$ be additive and let $M_1, M_2 : J \to \mathbb{R}$ be multiplicative functions. If*

$$M_1(p) = a\, M_2(p) + A(p) + b, \qquad p \in J, \tag{6.5.1}$$

for some constants a and b, then we have

$$a\left\{M_1(q) - M_2(q)\right\}\left\{M_2(p + r) - M_2(p) - M_2(r)\right\} = b\left\{M_1(q) - 1\right\} \quad (6.5.2)$$

for all $p, q, r \in J$ *with* $p + r \in J$.

Proof: From (6.5.1) we get

$$A(pq)$$
$$= M_1(pq) - a\, M_2(pq) - b$$
$$= M_1(p)M_1(q) - a\, M_2(p)M_2(q) - b$$
$$= \{a\, M_2(p) + A(p) + b\}\{a\, M_2(q) + A(q) + b\} - a\, M_2(p)M_2(q) - b,$$

that is,

$$\begin{aligned}
A(pq) = \;& a\,(a-1)\, M_2(p)M_2(q) + a\, M_2(p)\{A(q) + b\}\\
& + a\, M_2(q)\{A(p) + b\} + A(p)A(q)\\
& + b\,\{A(p) + A(q)\} + b^2 - b.
\end{aligned} \quad (6.5.3)$$

Now consider $p, q, r \in J$ with $p + r \in J$. From the additivity of A we have

$$A((p+r)q) = A(pq) + A(rq). \quad (6.5.4)$$

Making use of (6.5.3) on both sides of equation (6.5.4) we arrive at

$$a\left\{M_2(p+r) - M_2(p) - M_2(r)\right\}\{aM_2(q) + A(q) + b - M_2(q)\}$$

$$= b\{aM_2(q) + A(q) + b - 1\}. \quad (6.5.5)$$

Using (6.5.1) we get (6.5.2) from (6.5.5) and the proof of the lemma is complete.

Lemma 6.5.2. *Let* $A : J \to \mathbb{R}$ *be a nonzero additive function and let* $M_1, M_2 : J \to \mathbb{R}$ *be multiplicative functions; moreover, suppose* M_1 *is neither additive nor constant. Then the following statements are valid:*
(i) $\{1, M_1, A\}$ *is a linearly independent set.*
(ii) *If* $M_1 \neq M_2$, *then* $\{1, M_1 - M_2, A\}$ *is linearly independent.*

Proof: To prove the first statement let us suppose that there are constants $a, b, c \in \mathbb{R}$ such that

$$a\, M_1(p) + b\, A(p) + c = 0. \quad (6.5.6)$$

If $a \neq 0$ then (6.5.6) implies

$$M_1(p) = -\frac{b}{a} A(p) - \frac{c}{a}, \qquad (6.5.7)$$

from which it follows that $b \neq 0$. (Otherwise we get from (6.5.7) the contradiction that M_1 is a constant.) Since $-\frac{b}{a} A(p)$ is an additive function, we obtain from (6.5.7) (by using Lemma 6.5.1)

$$\frac{c}{a} \{M_1(q) - 1\} = 0, \qquad q \in J. \qquad (6.5.8)$$

If $c = 0$ then we get by (6.5.7) that M_1 is additive, which contradicts the hypothesis. If $c \neq 0$ then (6.5.8) shows that M_1 is a constant which is also a contradiction. Thus $a = 0$ in (6.5.6). Now the assumption that $b \neq 0$ in (6.5.6) yields $A(p) = -\frac{c}{b}$. Since the only constant additive function is the zero function, this is again a contradiction. Hence b must be zero and thus also $c = 0$. This proves that $\{1, M_1, A\}$ is linearly independent.

To prove the second statement of the lemma, let us assume that there are constants $a, b, c \in \mathbb{R}$ such that

$$a \{M_1(p) - M_2(p)\} + b \, A(p) + c = 0. \qquad (6.5.9)$$

Let us first consider the case $a = 0$. If $b = 0$ then obviously $c = 0$ and we are done. If $b \neq 0$ then $A(p) = -\frac{c}{b}$. As in part (i) we get the contradiction that A is zero. Now we will show that the case $a \neq 0$ also leads to a contradiction. In this case we rewrite (6.5.9) as

$$M_1(p) = M_2(p) - \frac{b}{a} A(p) - \frac{c}{a}. \qquad (6.5.10)$$

If M_2 is additive, then $A'(p) := M_2(p) - \frac{b}{a} A(p)$ is also additive. Moreover, $A' \neq 0$ since otherwise (6.5.10) shows that M_1 is constant. But now the representation

$$M_1(p) = A'(p) - \frac{c}{a}$$

contradicts the fact that $\{1, M_1, A'\}$ is linearly independent by part (i) of this lemma. Thus M_2 is not additive. But now we deduce from (6.5.9) and Lemma 6.5.1 that

$$\{M_1(q) - M_2(q)\}\{M_2(p+r) - M_2(p) - M_2(r)\} = -\frac{c}{a} \{M_1(q) - 1\}. \quad (6.5.11)$$

By hypothesis, there exists $q^* \in J$ such that $d = M_1(q^*) - M_2(q^*) \neq 0$. Substituting $q = q^*$ into (6.5.11) we get

$$M_2(p + r) - M_2(p) - M_2(r) = -\frac{c}{ad}\{M_1(q^*) - 1\} = a_1,$$

for some constant a_1, or

$$\{M_2(p + r) + a_1\} = \{M_2(p) + a_1\} + \{M_2(r) + a_1\}.$$

This implies that $M_2(p) + a_1$ is additive, that is

$$M_2(p) + a_1 = A_1(p) \qquad (6.5.12)$$

for some additive function A_1. If $A_1 = 0$ then M_2 is constant, and from the representation (6.5.10) we get a contradiction to part (i) of this lemma. On the other hand, if A_1 is nonzero, then M_2 is nonconstant. But now (6.5.12) again contradicts statement (i) of this lemma. Thus the case $a \neq 0$ cannot occur and the lemma is established.

Lemma 6.5.3. *Let $M_1, M_2 : J \to \mathbb{R}$ be multiplicative functions with $M_1 \neq M_2$.*
(i) If M_1 is not a projection, or if M_2 is not a projection, then for every $m \in \mathbb{N}$, $m \geq 3$, there is a $P' \in \Gamma_m^o$ such that

$$\sum_{j=1}^{m} \{M_1(p'_j) - M_2(p'_j)\} \neq 0. \qquad (6.5.13)$$

(ii) If either M_1 or M_2 is nonconstant and not additive, then for every $m \in \mathbb{N}, m \geq 3$, and for every $d \in \mathbb{R}$, there exists $P' \in \Gamma_m^o$ for which

$$\sum_{j=1}^{m} \{M_1(p'_j) - M_2(p'_j)\} \neq d. \qquad (6.5.14)$$

Proof: (i) Without loss of generality, we may assume that M_1 is not a projection. Suppose, contrary to (6.5.13), that

$$\sum_{j=1}^{m} \{M_1(p_j) - M_2(p_j)\} = 0, \qquad P \in \Gamma_m^o. \qquad (6.5.15)$$

Since $m \geq 3$, Theorem 6.4.1 yields

$$M_1(p) := M_2(p) + A(p) + b, \tag{6.5.16}$$

for some additive map $A : \mathbb{R}^k \to \mathbb{R}$ and constant b satisfying

$$A(1) + mb = 0. \tag{6.5.17}$$

By Lemma 6.5.2, part (ii), equation (6.5.16) implies that either $A = 0$ or M_1 is constant. (M_1 is not projection by hypothesis, hence M_1 is not additive.) In the first case ($A = 0$), (6.5.17) yields $b = 0$ and then (6.5.16) gives $M_1 = M_2$, a contradiction. Thus $A \neq 0$ and M_1 must be constant, say

$$M_1 = c \in \{0, 1\}. \tag{6.5.18}$$

Now, by part (i) of Lemma 6.5.2, (6.5.16) and (6.5.18) require (since $A \neq 0$) that M_2 is either additive or a constant. But M_2 cannot be constant, for then (6.5.16) with (6.5.18) would force A to be constant, and the only constant additive function is the zero function. Therefore M_2 is a nonconstant additive map. That is,

$$M_2 \text{ is a projection.} \tag{6.5.19}$$

Now, substituting (6.5.18) and (6.5.19) into (6.5.15), we get $0 = mc - 1 \in \{-1, m-1\}$, a contradiction. This completes the proof of part (i).

Part (ii) can be proved in a similar fashion. We assume, without loss of generality that M_1 is nonconstant and not additive. Let us suppose, contrary to (6.5.14), that

$$\sum_{j=1}^{m} \{M_1(p_j) - M_2(p_j)\} = d, \qquad P \in \Gamma_m^o. \tag{6.5.20}$$

By part (i) we may assume that $d \neq 0$. Since $m \geq 3$, then again Theorem 6.4.1 leads to (6.5.16), with $A : \mathbb{R}^k \to \mathbb{R}$ additive and constant b such that

$$A(1) + mb = d. \tag{6.5.21}$$

Again we apply Lemma 6.5.2 to (6.5.16) and conclude this time that $A = 0$ holds. Now (6.5.16) simplifies to

$$M_1(p) - M_2(p) = b, \qquad p \in J.$$

But then $M_1 - M_2$ is constant, which is again contrary to Lemma 6.5.2. Thus (6.5.14) is established, and the proof is complete.

In the next lemma we use the fact that an additive function $A : J \to \mathbb{R}$ can be additively extended to $\mathbb{R}^k$.

Lemma 6.5.4. *Let $A : J \to \mathbb{R}$ be additive and let $M_1, M_2 : J \to \mathbb{R}$ be multiplicative functions with $M_1 \neq M_2$. Moreover, suppose M_1 is not a projection (or M_2 is not a projection). If $F : J \to \mathbb{R}$, defined by*

$$F(p) = A(p) + c\left\{M_1(p) - M_2(p)\right\} + d, \qquad p \in J, \qquad (6.5.22)$$

$(c, d \in \mathbb{R})$ satisfies (5.4.7) for $\ell \geq 3$, $m \geq 2$ or for $\ell \geq 2$, $m \geq 3$ then

$$A(1) = 0 \quad \text{or} \quad d = 0 \quad \text{or} \quad A(1) + d = 0. \qquad (6.5.23)$$

Moreover, if at least one of M_1 or M_2 is nonconstant and not additive, then

$$A(1) = d = 0. \qquad (6.5.24)$$

Proof: Without loss of generality we may assume that M_1 is not a projection and that $\ell \geq 3$, $m \geq 2$. (The other three cases can be handled in the same manner.) Substitution of the function F given by (6.5.22) into (5.4.7) yields

$$A(1) + \ell m d = \{A(1) + md\} \sum_{i=1}^{\ell} M_1(p_i) + \{A(1) + \ell d\} \sum_{j=1}^{m} M_2(q_j) \quad (6.5.25)$$

for all $P \in \Gamma_\ell^o$, $Q \in \Gamma_m^o$. If $M_1 = 1$ or $M_1 = 0$, then (6.5.25) goes over into

$$A(1)(1 - \ell) = \{A(1) + \ell d\} \sum_{j=1}^{m} M_2(q_j) \qquad (6.5.26)$$

or

$$A(1) + \ell m d = \{A(1) + \ell d\} \sum_{j=1}^{m} M_2(q_j), \qquad (6.5.27)$$

respectively. If in (6.5.26) $M_2 = 0$ or M_2 is a projection then we get $A(1) = 0$, respectively $d + A(1) = 0$. If in (6.5.26) M_2 is nonconstant and not additive, and if we assume that $A(1) + \ell d \neq 0$, then $\sum_{j=1}^{m} M_2(q_j) = a$ for some constant a. Such a representation is impossible if $m \geq 3$, by Theorem 6.4.1 and Lemma 6.5.2; it is also impossible if $m = 2$, by Theorem 6.4.8. Thus $A(1) + \ell d = 0$ and (6.5.26) implies $A(1) = 0$ and thus $d = 0$, too.

If in (6.5.27) $M_2 = 1$ or M_2 is additive then we get $A(1) = 0$, respectively $d = 0$. On the other hand, if in (6.5.27) M_2 is nonconstant and not additive, then we get again $A(1) + \ell d = 0$ which implies $A(1) + \ell m d = 0$. Thus $A(1) = d = 0$.

Finally, suppose M_1 is nonconstant and not additive. Assume that $A(1) + md \neq 0$ in (6.5.25). Then fixing Q in (6.5.25) we have $\sum_{i=1}^{\ell} M_1(p_i) = c$ for some $c \in \mathbb{R}$ which leads to contradictions. Thus $A(1) + md = 0$. Now we consider some subcases. If $M_2 = 1$ or $M_2 = 0$ then (6.5.25) yields $A(1) = 0$, respectively $A(1) + \ell m d = 0$, so that in both cases we get $A(1) = d = 0$. If M_2 is additive in (6.5.25) then we get $(m-1)d = 0$ which implies $A(1) = d = 0$. Finally let M_2 be nonconstant and not additive in (6.5.25). Then we get $A(1) + \ell d = 0$ (since otherwise we would have $\sum_{j=1}^{m} M_2(q_j)$ constant, which is impossible). Thus $A(1) + \ell m d = 0$ and hence $A(1) = d = 0$. Thus (6.5.23) holds in every case.

Moreover, a reexamination of the proof reveals that (6.5.24) holds whenever at least one of M_1 or M_2 is nonconstant and not additive. This completes the proof of the lemma.

CHAPTER 7

ADDITIVE SUM FORM INFORMATION MEASURES

7.1. Introduction

The main objective in this chapter is to determine all additive information measures with the sum property (5.1.1), where the generating function $F : J \to \mathbb{R}$ is measurable. That is, we present the measurable solutions of the functional equation

$$\sum_{i=1}^{\ell}\sum_{j=1}^{m} F(p_i q_j) = \sum_{i=1}^{\ell} F(p_i) + \sum_{j=1}^{m} F(q_j), \quad P \in \Gamma_\ell^o, Q \in \Gamma_m^o \qquad (5.2.2)$$

for a fixed pair (ℓ, m) of integers, $\ell, m \geq 2$. We remark that it is an open problem to find the general solution of (5.2.2). As an indication of the difficulty of solving this problem, consider a map $F :]0, 1[\to \mathbb{R}$ of the form

$$F(x) = \frac{d^2(x)}{x}, \qquad x \in]0, 1[,$$

where d is a nonzero real derivation. Recalling that such d is additive and satisfies $d(xy) = xd(y) + yd(x)$, we see that also

$$d(1) = 0, \qquad \sum_{i=1}^{n} d(p_i) = 0$$

if $k = 1$ and $P \in \Gamma_n^o$. Using these properties of d, we find that

$$\sum_{i=1}^{\ell}\sum_{j=1}^{m} F(p_i q_j) = \sum_{i=1}^{\ell}\sum_{j=1}^{m} \left[\frac{p_i^2 d^2(q_j)}{p_i q_j} + \frac{q_j^2 d^2(p_i)}{p_i q_j} + 2d(p_i)d(q_j) \right]$$

$$= \sum_{j=1}^{m} \frac{d^2(q_j)}{q_j} + \sum_{i=1}^{\ell} \frac{d^2(p_i)}{p_i} + 2d \left(\sum_{i=1}^{\ell} p_i \right) d \left(\sum_{j=1}^{m} q_j \right)$$

$$= \sum_{i=1}^{\ell} F(q_j) + \sum_{i=1}^{\ell} F(p_i).$$

Hence F satisfies (5.2.2), but it is not measurable. (In fact, F is even nonnegative.)

7.2. Solution of (5.2.2) in the case $m \geq 3$ or $\ell \geq 3$

We now present our main result, which can be found in Sahoo (1995a); the case $k = 2$ was settled in two papers by Kannappan and Sahoo (1985b) and by Ebanks (1985). Here the presentation of these results makes use of the notation introduced in Definition 6.3.1.

Theorem 7.2.1. *Let $F : J \to \mathbb{R}$ be measurable in each variable. Then F satisfies the functional equation (5.2.2) for a fixed pair of integers $\ell \geq 2$, $m \geq 3$ or $\ell \geq 3$, $m \geq 2$, if and only if there exists a $k \times k$ matrix $C = (C_{ij})$, a vector $a \in \mathbb{R}^k$ and a constant $b \in \mathbb{R}$ such that*

$$F(p) = p \odot C \odot \log p + a \odot p + b, \qquad p \in J, \tag{7.2.1}$$

with

$$a \odot 1 = (lm - \ell - m)b. \tag{7.2.2}$$

Proof: First, we treat the case $\ell \geq 3$ and $m \geq 3$. Choose a pair of constants $u, v \in\,]0, 1[$ such that $(\ell - 2)u < 1$ and $(m - 2)v < 1$; moreover we put

$$s = 1 - (\ell - 2)u, \qquad t = 1 - (m - 2)v$$

$$\underline{u} = (u, ..., u)^T, \underline{v} = (v, ..., v)^T, \underline{s} = (s, ..., s)^T, \underline{t} = (t, ..., t)^T \in J.$$

(Remember that a^T is the transpose of the row vector $a \in \mathbb{R}^k$.) Now let $p = (p_1, ..., p_k)^T \in\,]0, s[^k$, $q = (q_1, ..., q_k)^T \in\,]0, t[^k$ and substitute

$$P = (p, \underline{s} - p, \underline{u}, ..., \underline{u}) = \begin{pmatrix} p_1 & s - p_1 & u & \cdots & u \\ p_2 & s - p_2 & u & \cdots & u \\ \vdots & \vdots & \vdots & \ddots & \vdots \\ p_k & s - p_k & u & \cdots & u \end{pmatrix} \in \Gamma_\ell^o$$

and

$$Q = (q, \underline{t} - q, \underline{v}, ..., \underline{v}) = \begin{pmatrix} q_1 & t - q_1 & v & \cdots & v \\ q_2 & t - q_2 & v & \cdots & v \\ \vdots & \vdots & \vdots & \ddots & \vdots \\ q_k & t - q_k & v & \cdots & v \end{pmatrix} \in \Gamma_m^o$$

into (5.2.2) to get

$$
\begin{aligned}
F(pq) &+ F(p(\underline{t} - q)) + F((\underline{s} - p)q) + F((\underline{s} - p)(\underline{t} - q)) \\
&+ (m - 2)F(p\underline{v}) + (m - 2)F((\underline{s} - p)\underline{v}) + (\ell - 2)F(q\underline{u}) \\
&+ (\ell - 2)F((\underline{t} - q)\underline{u}) + (\ell - 2)(m - 2)F(\underline{u}\underline{v}) \\
&= F(p) + F(\underline{s} - p) + (\ell - 2)F(\underline{u}) \\
&\quad + F(q) + F(\underline{t} - q) + (m - 2)F(\underline{v})
\end{aligned}
\tag{7.2.3}
$$

for all $p \in]0, s[^k$ and $q \in]0, t[^k$. Substituting $x = \frac{p}{\underline{s}}$ and $y = \frac{q}{\underline{t}}$ into (7.2.3) and defining

$$f(x) := F(\underline{s}\,\underline{t}\,x), \tag{7.2.4}$$

$$\begin{aligned}
g(x) := &F(\underline{s}\,x) + F(\underline{s}(1-x)) + (\ell - 2)F(\underline{u}) \\
&- (m-2)F(\underline{s}\,\underline{v}\,x) - (m-2)F(\underline{s}\,\underline{v}\,(1-x)) \\
&- (\ell - 2)(m-2)F(\underline{u}\,\underline{v})
\end{aligned}$$

$$\begin{aligned}
h(x) := &F(\underline{t}\,x) + F(\underline{t}(1-x)) + (m-2)F(\underline{v}) \\
&- (\ell - 2)F(\underline{t}\,\underline{u}\,x) - (\ell - 2)F(\underline{t}\,\underline{u}\,(1-x)),
\end{aligned}$$

then equation (7.2.3) goes over into

$$f(xy) + f(x(1-y)) + f((1-x)y) + f((1-x)(1-y)) = g(x) + h(y) \quad (7.2.5)$$

for all $x, y \in J$. The measurable solutions of (7.2.5) are given in Theorem 6.3.2. At this point, we need only the form of f. From Theorem 6.3.2 and (7.2.4) we obtain

$$F(p) = p \odot C \odot \log p + a_1 \odot p^3 + a_2 \odot p^2 + a_3 \odot p + a_4 \odot \log p + b, \quad (7.2.6)$$

for all $p \in J$, where $C = (C_{ij})$ is a $k \times k$ matrix, a_1, a_2, a_3, a_4 are vectors in $\mathbb{R}^k$, and b is a real constant. Note that (7.2.6) holds indeed for all $p \in J$, since $\underline{s}, \underline{t}$ can be chosen arbitrarily near $\mathbf{1} = (1, ..., 1)$ (and since the matrix, the vectors and the constants are the same for each $\underline{s}, \underline{t} \in J$ due to the linear independence of the terms $p_i^3, p_i^2, p_i, p_i \log p_j, 1$ $(i, j = 1, 2, ..., k)$ occurring in (7.2.6). (See also the argument given in the proof of Theorem (6.4.1).)

Now we substitute the form of F given by (7.2.6) into (5.2.2) and obtain

$$\sum_{i=1}^{\ell} \sum_{j=1}^{m} \{p_i q_j \odot C \odot \log p_i q_j + a_1 \odot p_i^3 q_j^3 + a_2 \odot p_i^2 q_j^2$$

$$+ a_3 \odot p_i q_j + a_4 \odot \log p_i q_j\} + m\ell b$$

$$= \sum_{i=1}^{\ell} \{p_i \odot C \odot \log p_i + a_1 \odot p_i^3 + a_2 \odot p_i^2$$

$$+ a_3 \odot p_i + a_4 \odot \log p_i\} + \ell b$$

$$+ \sum_{j=1}^{m} \{q_j \odot C \odot \log q_j + a_1 \odot q_j^3 + a_2 \odot q_j^2$$

$$+ a_3 \odot q_j + a_4 \odot \log q_j\} + mb,$$

for all $P = (p_1, p_2, ..., p_l) \in \Gamma_\ell^o$ and $Q = (q_1, q_2, ..., q_m) \in \Gamma_m^o$. Using Definition 6.3.1, this equation reduces to

$$\sum_{i=1}^{\ell} \sum_{j=1}^{m} \{a_1 \odot p_i^3 q_j^3 + a_2 \odot p_i^2 q_j^2 + a_4 \odot \log p_i q_j\} + a_3 \odot 1 + mlb$$

$$= \sum_{i=1}^{\ell} \{a_1 \odot p_i^3 + a_2 \odot p_i^2 + a_4 \odot \log p_i\} + a_3 \odot 1 + \ell b$$

$$+ \sum_{j=1}^{m} \{a_1 \odot q_j^3 + a_2 \odot q_j^2 + a_4 \odot \log q_j\} + a_3 \odot 1 + mb.$$

(For instance, if $a_3 = (\alpha_{31}, ..., \alpha_{3r}, ..., \alpha_{3k})$, $p_i = (\pi_{1i}, ..., \pi_{ri}, ..., \pi_{ki})^T$, $q_j = (\xi_{1j}, ..., \xi_{rj}, ..., \xi_{kj})^T$, then $p_i q_j = (\pi_{1i}\xi_{1j}, ..., \pi_{ri}\xi_{rj}, ..., \pi_{ki}\xi_{kj})^T$ so that $\sum_{i=1}^{\ell} \sum_{j=1}^{m} a_3 \odot p_i q_j = \sum_{r=1}^{k} \sum_{i=1}^{l} \sum_{j=1}^{m} \alpha_{3r} \pi_{ri}\xi_{rj} = \sum_{r=1}^{k} \alpha_{3r} = a_3 \odot 1$; moreover $\sum_{r=1}^{k} \sum_{s=1}^{k} \sum_{i=1}^{\ell} \sum_{j=1}^{m} \pi_{ri}\xi_{rj} C_{rs} \log \pi_{si}\xi_{sj} = \sum_{r=1}^{k} \sum_{s=1}^{k} \sum_{i=1}^{\ell} \pi_{ri} C_{rs} \log \pi_{si} + \sum_{r=1}^{k} \sum_{s=1}^{k} \sum_{j=1}^{m} \xi_{rj} C_{rs} \log \xi_{sj}$.) Now comparing coefficients in (7.2.7) we see that

$$a_1 = a_2 = a_4 = 0, \qquad a_3 \odot 1 = (\ell m - \ell - m)b,$$

so that F, given by (7.2.6) has the form (7.2.1) with (7.2.2). Since functions of this form satisfy equation (5.2.2) too, the proof is complete if $\ell \geq 3$ and $m \geq 3$.

Without loss of generality, we consider now the remaining case $\ell = 2$ and $m \geq 3$. The proof is analogous to the proof above with a slight modification. Choose $v, \underline{v}, t, \underline{t}$ as above. Moreover put $s = 1$ and $\underline{s} = 1 = (1, ..., 1) \in \mathbb{R}^k$. Now we proceed as above, but substituting $P = (p, 1 - p) \in \Gamma_2^o$ (instead of $(p, \underline{s} - p, \underline{u}, ..., \underline{u})$). Again we arrive at (7.2.3) - (7.2.4), where now $\underline{s} = 1$ and $\ell = 2$ (so that the terms with the factor $(\ell - 2)$ disappear). Since equation (7.2.5) remains unchanged we arrive at the same result as above, and this completes the proof of the theorem.

We are now ready to prove the main result of this section. For the following remember that $P \in \Gamma_n^o$ can be written as $P = (p_1, p_2, ..., p_n)$ or as $P = (P_1, P_2, ..., P_k)^T$, $P_j = (\pi_{j1}, ..., \pi_{jn}) \in \Gamma_n^o$, $1 \leq j \leq k$ (see (1.1.10)-(1.1.12)). Instead of the column vector we agree on writing a row vector.

Theorem 7.2.2. Let $I_n : \Gamma_n^o \to \mathbb{R}$ have the sum form

$$I_n(P) = \sum_{i=1}^{n} F(p_i), \qquad P \in \Gamma_n^o \qquad (5.1.1)$$

for some measurable function $F : J \to \mathbb{R}$. Then $\{I_n\}$ is (ℓ, m)-additive for fixed $\ell \geq 2$, $m \geq 3$ or $\ell \geq 3$, $m \geq 2$, if and only if I_n is given by

$$I_n(P) = I_n(P_1, P_2, ..., P_k)$$

$$= \sum_{i=1}^{k} c_i H_n(P_i) + \sum_{i=1}^{k} \sum_{\substack{j=1 \\ j \neq i}}^{k} C_{ij} K_n(P_i, P_j) \qquad (7.2.8)$$

$$+ (\ell m - \ell - m)b + nb,$$

where b, c_i, C_{ij} are constants $(i, j = 1, 2, ..., k; j \neq i)$.

Proof: Since the suppositions of Theorem 7.2.1 are satisfied we know that F has the form (7.2.1) and satisfies (7.2.2) for some $k \times k$ matrix $C = (C_{rs})$, a vector $a \in \mathbb{R}^k$, and a constant $b \in \mathbb{R}$. Substituting this form of F into (5.1.1) we obtain (using (7.2.2))

$$I_n(P) = \sum_{i=1}^{n} p_i \odot C \odot \log p_i + \sum_{i=1}^{n} a \odot p_i + nb$$

$$= \sum_{r=1}^{k} \sum_{s=1}^{k} \sum_{i=1}^{n} \pi_{ri} C_{rs} \log \pi_{si} + a \odot 1 + nb$$

$$= -\sum_{r=1}^{k} \sum_{s=1}^{k} C_{rs} K_n(P_r, P_s) + (\ell m - \ell - m)b + nb,$$

which has the form given by (7.2.8) if we define $c'_r = -C_{rr}$, $C'_{rs} = -C_{rs}$ $(r, s = 1, 2, ..., k; r \neq s)$ and using the fact that $K_n(P, P) = H_n(P)$. Since $H_n^0(P_j) = n - 1$, $1 \leq j \leq k$, we can rewrite (7.2.8) as

$$I_n(P) = \sum_{i=1}^{k} c_i H_n(P_i) + bk^{-1} \sum_{i=1}^{k} H_n^0(P_i)$$

$$+ \sum_{i=1}^{k} \sum_{\substack{j=1 \\ j \neq i}}^{k} C_{ij} K_n(P_i, P_j) + (\ell m - \ell - m + 1)b. \qquad (7.2.9)$$

Thus I_n is a linear combination of Kerridge's inaccuracies, entropies of degree one and zero, and a constant, dependent upon ℓ and m. This completes the proof.

Now, for $k = 1$, (7.2.9) yields

$$I_n(P) = cH_n^1(P) + bH_n^0(P) + a$$

where $a = (\ell m - \ell - m + 1)b$. If in Theorem 7.2.2, $\{I_n\}$ is both (ℓ, m)-additive and (ℓ', m')-additive such that

$$\ell m - \ell - m \neq \ell'm' - \ell' - m',$$

then $b = 0$ (which is the case if $\{I_n\}$ is (ℓ, m)-additive for all $\ell \geq 2$, $m \geq 3$).

Note that if $k \geq 2$ and if $b = 0$ then $\{I_n\}$ is a linear combination of inaccuracies and Shannon entropies. Moreover, Theorem 7.2.2 says that the information measure $H_n : \Gamma_n^o \to \mathbb{R}$, defined by

$$H_n(P) = H_n(p_1, p_2, ..., p_k) = \sum_{i=1}^{n} p_i \odot C \odot \log p_i, \qquad (7.2.10)$$

$$1 \odot C \odot 1 = -1,$$

(where $C = (C_{rs})$ is a $k \times k$ matrix) can be considered as a natural generalization of the Shannon entropy. Indeed the conditions

$$I_n(P) = \sum_{i=1}^{n} F(p_i), \qquad F(p) = p \odot C \odot \log p$$

and

$$I_2 \left(\frac{1}{2}, \frac{1}{2} \right) = 1$$

lead to (7.2.10). Note that the condition $1 \odot C \odot 1 = -1$ means

$$\sum_{r=1}^{k} \sum_{s=1}^{k} C_{rs} = -1,$$

so that in the case $k = 1$ we get the "usual" Shannon entropy.

7.3. Solution of (5.2.2) in the case $\ell = 2 = m$

We now determine all measurable sum form information measures which are (2,2)-additive. Surprisingly, it will turn out that these information measures have a more general form than (7.2.8). As in section 2 we present first the form of the generating function F.

Theorem 7.3.1. *Let $F : J \to \mathbb{R}$ be measurable in each variable. Then F satisfies the functional equation (5.2.2) for $\ell = 2 = m$, if and only if there* *exists a $k \times k$ matrix $C = (C_{ij})$, vectors $a, d \in \mathbb{R}^k$ and a constant $b \in \mathbb{R}$ such that*

$$F(p) = 4d \odot p^3 - 9d \odot p^2 + p \odot C \odot \log p + a \odot p + b \quad (7.3.1)$$

with

$$a \odot 1 = 5d \odot 1 \quad (7.3.2)$$

Proof: This result is nothing else but Corollary 6.3.3 since equation (5.2.2) for $\ell = 2 = m$ is exactly equation (6.3.33).

Theorem 7.3.2. *Let $I_n : \Gamma_n^o \to \mathbb{R}$ have the sum form (5.1.1) for some measurable function $F : J \to \mathbb{R}$. Then $\{I_n\}$ is (2,2)-additive, if and only if I_n is given by*

$$I_n(P) = I_n(P_1, P_2, ..., P_k)$$
$$= 2 \sum_{i=1}^{k} d_i H_n^3(P_i) - 3 \sum_{i=1}^{k} d_i H_n^2(P_i) + \sum_{i=1}^{k} c_i H_n^1(P_i)$$
$$+ bk^{-1} \sum_{i=1}^{k} H_n^0(P_i) + \sum_{i=1}^{k} \sum_{\substack{j=1 \\ j \neq i}}^{k} C_{ij} K_n(P_i, P_j) + nb \quad (7.3.3)$$

where b, c_i, d_i, C_{ij} are constants $(i, j = 1, 2, ..., k; i \neq j)$.

Proof: By Theorem 7.3.1, F has the form (7.3.1) with (7.3.2) for some $k \times k$ matrix $C = (C_{ij})$, vectors $a, d = (d_1, ..., d_k) \in \mathbb{R}^k$ and a constant $b \in \mathbb{R}$. Substituting this form of F into (5.1.1) we get

$$I_n(P) = 4 \sum_{i=1}^{n} d \odot p_i^3 - 9 \sum_{i=1}^{n} d \odot p_i^2 + a \odot 1 + \sum_{i=1}^{n} p_i \odot C \odot \log p_i + nb. \quad (7.3.4)$$

As in the proof of Theorem 7.2.2 the last two summands in (7.3.4) yield the last four summands in (7.3.3). Thus we have only to show that the first three summands in (7.3.4) coincide with the first two summands in (7.3.3). But this is obvious because of the following calculation (where we make use of

(7.3.2)):

$$4 \sum_{i=1}^{n} d \odot p_i^3 - 9 \sum_{i=1}^{n} d \odot p_i^2 + a \odot 1$$

$$= 4 \sum_{r=1}^{k} d_r \left(\sum_{i=1}^{n} p_{ri}^3 - 1 + 1 \right) - 9 \sum_{r=1}^{k} d_r \left(\sum_{i=1}^{n} p_{ri}^2 - 1 + 1 \right) + 5 d \odot 1$$

$$= -3 \sum_{r=1}^{k} d_r H_n^3(P_r) + 4 \sum_{r=1}^{k} d_r + \frac{9}{2} \sum_{r=1}^{k} d_r H_n^2(P_r) - 9 \sum_{r=1}^{k} d_r + 5 \sum_{r=1}^{k} d_r$$

$$= 2 \sum_{r=1}^{k} d'_r H_n^3(P_r) - 3 \sum_{r=1}^{k} d'_r H_n^2(P_r),$$

where $d'_r = -\frac{3}{2} d_r$ $(r = 1, 2, ..., k)$. Since I_n given by (7.3.3) is (2,2)-additive, the proof is complete.

Thus we have the result that the (2,2)-additive and measurable information measures are essentially linear combinations of entropies of degree 3, 2, 1 and 0 and inaccuracies.

From Theorem 7.2.2 and Theorem 7.3.2 we easily obtain the following result.

Corollary 7.3.3. *Let $I_n : \Gamma_n^o \to \mathbb{R}$ have the sum form (5.1.1) for some measurable function $F : J \to \mathbb{R}$. Then $\{I_n\}$ is (ℓ, m)-additive for all $\ell \geq 2$, $m \geq 2$, if and only if*

$$I_n(P) = I_n(P_1, ..., P_k)$$
$$= \sum_{i=1}^{k} c_i H_n(P_i) + \sum_{i=1}^{k} \sum_{\substack{j=1 \\ j \neq i}}^{k} C_{ij} K_n(P_i, P_j), \tag{7.3.5}$$

where c_i, C_{ij} are constants $(i, j = 1, 2, ..., k; i \neq j)$.

Remark 7.3.4. Comparing Theorem 7.3.4 with Theorem 7.2.1 it is obvious that the case $\ell = m = 2$ in equation (5.2.2) is the most difficult one. But from a mathematical point of view it is rather interesting, since "unexpected solutions" in the form of a polynomial of degree 3 occur. The first result in this direction was given by Daróczy and Járai (1979) for $k = 1$. Results for $k = 2$ and for general k were proved by Kannappan and Ng (1979, 1980, 1985) and Abou-Zaid (1984), respectively. Less is known about the general solution of equation (5.2.2) on the open domain. On the closed domain Losonczi and Maksa (1982) presented a description of the general solution (5.2.2) for fixed $\ell \geq 3$ and $m \geq 3$ in the case $k = 1$. But the description uses a function of two variables which is additive in one variable and satisfies a functional equation in which another biadditive function occurs, so it is not very explicit.

Let us mention here that the general solution F of (5.2.2) sometimes may be expressed in the form

$$F(p) = g(p) + c, \qquad p \in J, \tag{7.3.6}$$

where $g : J \to \mathbb{R}$ is a solution of (6.3.40). For example, if $\ell = m = 2$ and $F(p) + F(1 - p) = 1$ then (7.3.6) follows immediately from (5.2.2). Now, if $\ell = 2$, $m \geq 3$ and $\sum_{i=1}^{\ell} F(p_i) = 1$ $(P \in \Gamma_\ell^0)$, we write $P = (p, 1 - p)$, where $p \in J$. Using $\sum_{j=1}^{m} M(q_j) = M\left(\sum_{j=1}^{m} q_j\right) = 1$, where $M : J \to \mathbb{R}$ is any projection, equation (5.2.2) goes over into

$$\sum_{i=1}^{2} \sum_{j=1}^{m} F(p_i q_j) = 1 + \sum_{j=1}^{m} F(q_j) = \sum_{j=1}^{m} (M(q_j) + F(q_j))$$

or

$$\sum_{j=1}^{m} [F(pq_j) + F((1 - p)q_j) - M(q_j) - F(q_j)] = 0.$$

Thus an application of Theorem 6.4.1 yields

$$F(pq) + F((1 - p)q) - M(q) - F(q) = A\left(p, q - \frac{1}{m}\right) \tag{7.3.7}$$

where $A : J \times \mathbb{R}^k \to \mathbb{R}$ is additive in the second variable. Replacing q by $1 - q$ in (7.3.7) and adding the resulting equation to (7.3.7) we get

$$F(pq) + F(p(1 - q)) + F((1 - p)q) + F((1 - p)(1 - q))$$
$$= 1 + F(q) + F(1 - q) + \left(1 - \frac{2}{m}\right) A(p, 1), \tag{7.3.8}$$

$p, q \in J$. Since the left-hand side of (7.3.8) is symmetric in p and q we obtain $2 + \left(1 - \frac{2}{m}\right) A(p, 1) = 2 + \left(1 - \frac{2}{m}\right) A(q, 1) = $ constant, so that we again arrive at the representation (7.3.5).

This representation (7.3.5) also plays an important role in results of Chapters 8 and 9 (see Lemma 8.3.3, Remark 8.3.6, Step 1 of Theorem 8.4.3, Theorem 9.4.1) which suggests that it would be very interesting to find the general solution of (6.3.40).

7.4. Some heuristic arguments

In sections 7.2 and 7.3 the measurable solutions of the functional equation

$$\sum_{i=1}^{\ell}\sum_{j=1}^{m} F(p_i q_j) = \sum_{i=1}^{\ell} F(p_i) + \sum_{j=1}^{m} F(q_j) \qquad (5.2.2)$$

were presented. The main idea of the proof of Theorem 7.1.1 (the case $\ell \geq 2$, $m \geq 3$ or $\ell \geq 3$, $m \geq 2$) was to reduce the functional equation (5.2.2) to the equation (7.2.5), which is nearly equation (5.2.2) with $\ell = 2 = m$.

There is a heuristic argument for finding the measurable solutions of (5.2.2) (in the 1-dimensional case, with $\ell \geq 3$, $m \geq 3$). Using $\sum_{i=1}^{\ell} p_i = 1 = \sum_{j=1}^{m} q_j$ we rewrite (5.2.2) as

$$\sum_{i=1}^{\ell}\sum_{j=1}^{m} \{ F(p_i q_j) - q_j F(p_i) - p_i F(q_j) \} = 0. \qquad (7.4.1)$$

If we neglect for one moment the double sum, we arrive at the equation

$$F(pq) - qF(p) - pF(q) = 0.$$

That is, for $f(p) = F(p)/p$, we get

$$f(pq) = f(p) + f(q), \qquad (7.4.2)$$

the logarithmic functional equation. But the measurable solution of (7.4.2) on $]0, 1[$ is given by

$$f(p) = c \log p, \qquad c \in \mathbb{R}, \qquad (7.4.3)$$

so that

$$F(p) = pc \log p.$$

Since we have ignored the double sum we correct the solution F of (5.2.2) by an additive term of the form

$$A(p) + b, \qquad b \in \mathbb{R}, \qquad (7.4.4)$$

where A is a measurable additive function (see Theorem 6.4.1). Thus

$$F(p) = p\,c\,\log p + A(p) + b$$
$$= p\,c\,\log p + ap + b, \qquad a \in \mathbb{R}.$$

Since $ap + b$ is a solution of (5.2.2) if

$$a + \ell m b = a + \ell b + a + m b,$$

that is if $a = (\ell m - \ell - m)b$, we get exactly (7.2.1) and (7.2.2). Let us note here that by the given heuristic argument one could guess that

$$F(p) = pL(p) + A(p) + b, \quad A(1) = (\ell m - \ell - m)b,$$

is the general solution of (5.2.2). But because of the example presented in section 1 this is not the complete solution.

Similar heuristic arguments work for the functional equations (5.3.12) and (5.4.7) (if $\ell \geq 3$, $m \geq 3$), and these arguments actually do lead to the general solutions! To see this we rewrite (5.3.12) (in the 1-dimensional case) and (5.4.7) as

$$\sum_{i=1}^{\ell}\sum_{j=1}^{m}\{F(p_iq_j) - q_jF(p_i) - p_iF(q_j) - \lambda F(p_i)F(q_j)\} = 0 \qquad (7.4.5)$$

and

$$\sum_{i=1}^{\ell}\sum_{j=1}^{m}\{F(p_iq_j) - M_1(p_i)F(q_j) - M_2(q_j)F(p_i)\} = 0, \qquad (7.4.6)$$

respectively. We may assume that $\lambda \neq 0$ and that M_1 or M_2 is not additive (otherwise we are back to (7.4.1)). Without loss of generality we assume that M_1 is not additive. Moreover we assume for simplicity that M_1 is nonconstant. Neglecting again the double sums, we get

$$F(pq) - qF(p) - pF(q) - \lambda F(p)F(q) = 0, \qquad (7.4.7)$$

respectively

$$F(pq) - M_1(p)F(q) - M_2(q)F(p) = 0. \qquad (7.4.8)$$

With the substitution

$$f(p) = p + \lambda F(p) \qquad (7.4.9)$$

equation (7.4.7) is equivalent to

$$
\begin{aligned}
f(p)f(q) &= \{p + \lambda F(p)\}\{q + \lambda F(q)\} \\
&= pq + \lambda\{pF(q) + qF(p) + \lambda F(p)F(q)\} \\
&= pq + \lambda F(pq) \\
&= f(pq).
\end{aligned}
\tag{7.4.10}
$$

Thus $f = 0$ or f is a nonzero multiplicative function. Now taking into account the neglected double sum by adding an additive term of the form (7.4.4), we arrive at either

$$
f(p) = 0 + A(p) + b
\tag{7.4.11}
$$

or

$$
f(p) = M(p) + A'(p) + b'
\tag{7.4.12}
$$

for some nonzero multiplicative function M, for additive functions A and A', and constants $b, b' \in \mathbb{R}$. In the first case (7.4.11), $A(p) + b$ is a solution of

$$
\sum_{i=1}^{\ell}\sum_{j=1}^{m}\{f(p_iq_j) - f(p_i)f(q_j)\} = 0,
\tag{7.4.13}
$$

if and only if

$$
A(1) + \ell m b = \{A(1) + \ell b\}\{A(1) + m b\}.
\tag{7.4.14}
$$

In the second case (7.4.12) we may assume that M is neither additive nor constant and that A' is nonzero, since otherwise we are back to (7.4.11). Now $f(p) = M(p) + A'(p) + b'$ is a solution of (7.4.13), if and only if

$$
\begin{aligned}
A'(1) + \ell m b' = &\{m b' + A'(1)\}\sum_{i=1}^{\ell} M(p_i) \\
&+ \{\ell b' + A'(1)\}\sum_{j=1}^{m} M(q_j) \\
&+ \{A'(1) + \ell b'\}\{A'(1) + m b'\}
\end{aligned}
\tag{7.4.15}
$$

for all $P \in \Gamma_{\ell}^{o}$, $Q \in \Gamma_{m}^{o}$.

Suppose for the moment that $m b' + A'(1) \neq 0$. Then fixing Q in (7.4.15) we get $\sum_{i=1}^{\ell} M(p_i) = $ constant. Thus Theorem 6.4.1 yields $M(p) = B(p) + c$

for some nonzero additive function B and some constant c. But this contradicts the fact (see Lemma 6.5.2) that $\{1, M, B\}$ are linearly independent. So, $mb' + A'(1) = 0$.

Similarly $\ell b' + A'(1) = 0$ and we get $A'(1) + \ell mb' = 0$ which gives

$$A'(1) = 0 \quad \text{and} \quad b' = 0. \tag{7.4.16}$$

Combining (7.4.9), (7.4.10), (7.4.12), (7.4.14) and (7.4.16), we get that F satisfies (7.4.5) if and only if

$$\left.\begin{array}{l} F(p) = \dfrac{A(p) + b - p}{\lambda}, \qquad p \in]0, 1[\\[3ex] F(p) = \dfrac{M(p) + A'(p) - p}{\lambda}, \qquad p \in]0, 1[\end{array}\right\} \tag{7.4.17}$$

where A, A' are additive, M is multiplicative, b is a constant such that (7.4.14) holds, and where $A'(1) = 0$.

Now we come back to equation (7.4.8). We consider first the case $M_1 = M_2$, and we assume that M_1 is neither additive nor constant. In this case we rewrite (7.4.8) in the form (7.4.2) with f given by $f(p) = F(p)/M(p)$. Thus f is a logarithmic function which we denote again by L, and we get $F(p) = M(p)L(p)$. Incorporating again an additive term of the form (7.4.4) we arrive at

$$F(p) = A(p) + b + M(p)L(p), \tag{7.4.18}$$

where A is additive, L is logarithmic and b is a constant. Substituting (7.4.18) into (7.4.6) we get

$$A(1) + \ell mb = \{A(1) + mb\} \sum_{i=1}^{\ell} M(p_i) + \{A(1) + \ell b\} \sum_{j=1}^{m} M(q_j). \tag{7.4.19}$$

Similar to the considerations concerning equation (7.4.15), we get (with the aid of Lemma 6.5.2)

$$A(1) = 0, \qquad b = 0. \tag{7.4.20}$$

Finally we treat the case $M_1 \neq M_2$. We rewrite (7.4.8) as

$$F(pq) = M_1(p)F(q) + M_2(q)F(p). \tag{7.4.21}$$

Since the left side in (7.4.21) is symmetric in p and q the right side also must be symmetric in p and q, that is

$$\{M_1(q) - M_2(q)\}F(p) = \{M_1(p) - M_2(p)\}F(q). \tag{7.4.22}$$

Since $M_1 \neq M_2$, there is a q^* such that $M_1(q^*) - M_2(q^*) \neq 0$. Thus

$$F(p) = c\{M_1(p) - M_2(p)\} \tag{7.4.23}$$

where $c = \frac{F(q^*)}{\{M_1(q^*) - M_2(q^*)\}}$. As before we "correct" this solution to get

$$F(p) = A(p) + b + c\{M_1(p) - M_2(p)\}, \tag{7.4.24}$$

where A is additive and b is a constant. By Lemma 6.5.4 we get $A(1) = 0$ and $b = 0$. Combining again all the results we have that F satisfies (7.4.6), if and only if

$$F(p) = \begin{cases} A(p) + M_1(p)L(p) & \text{if } M_1 = M_2 \\ \\ A(p) + c\{M_1(p) - M_2(p)\} & \text{if } M_1 \neq M_2, \end{cases} \tag{7.4.25}$$

where $A : \mathbb{R} \to \mathbb{R}$ is an additive function with $A(1) = 0$, c is a real constant and $L : \mathbb{R}_+ \to \mathbb{R}$ is a logarithmic function.

In chapters 8 and 9 we will show that (7.4.17) and (7.4.25) indeed are the solutions of (7.4.5) and (7.4.6), respectively (for fixed $\ell \geq 3$, $m \geq 3$). But we will also see that the correct proofs are much more complicated. For this reason perhaps these heuristic arguments are useful prior to the rigorous proofs.

CHAPTER 8

ADDITIVE SUM FORM INFORMATION MEASURES OF TYPE λ

8.1. Introduction

In this chapter we derive the representation of sum form information measures $\{I_n\}$ of the form

$$I_n(P) = \sum_{i=1}^{n} F(p_i), \qquad P \in \Gamma_n^o, \tag{5.1.1}$$

where the generating function $F : J \to \mathbb{R}$ satisfies

$$\sum_{i=1}^{\ell}\sum_{j=1}^{m} F(p_i q_j) = \sum_{i=1}^{\ell} F(p_i) + \sum_{j=1}^{m} F(q_j) + \lambda \sum_{i=1}^{\ell} F(p_i) \sum_{j=1}^{m} F(q_j) \tag{5.3.12}$$

for all $P \in \Gamma_\ell^o$, $Q \in \Gamma_m^o$, for fixed $\lambda \in \mathbb{R} \setminus \{0\}$ and (ℓ, m) a fixed pair of positive integers, $\ell \geq 2, m \geq 2$. The case $\lambda = 0$ was considered in Chapter 7, where we determined the measurable solutions in all cases. In this chapter we are able to determine not only all measurable solutions of equation (5.3.12), but also the general solutions of (5.3.12) for most ℓ and m. Let us start with a result which reduces the functional equation (5.3.12) to a simpler-looking equation.

Lemma 8.1.1. *The function $f : J \to \mathbb{R}$ defined by*

$$f(x) = f(x_1, x_2, ..., x_k) = \frac{1}{k} \sum_{s=1}^{k} x_s + \lambda F(x) = \frac{1}{k}(\mathbf{1} \odot x) + \lambda F(x) \tag{8.1.1}$$

transforms the functional equation (5.3.12) into

$$\sum_{i=1}^{\ell}\sum_{j=1}^{m} f(p_i q_j) = \sum_{i=1}^{\ell} f(p_i) \sum_{j=1}^{m} f(q_j), \tag{8.1.2}$$

where $P \in \Gamma_\ell^o$, $Q \in \Gamma_m^o$.

Proof: Using (8.1.1), (5.3.12) and

$$\frac{1}{k}\sum_{i=1}^{\ell}\sum_{s=1}^{k} p_{si} = \frac{1}{k}\sum_{j=1}^{m}\sum_{r=1}^{k} q_{rj} = \frac{1}{k}\sum_{i=1}^{\ell}\sum_{j=1}^{m}\sum_{s=1}^{k} p_{si}q_{sj} = 1$$

we get (8.1.2) by the following computation:

$$\sum_{i=1}^{\ell} f(p_i) \sum_{j=1}^{m} f(q_j)$$

$$= \sum_{i=1}^{\ell}\left\{\frac{1}{k}\sum_{s=1}^{k} p_{si} + \lambda F(p_i)\right\} \sum_{j=1}^{m}\left\{\frac{1}{k}\sum_{r=1}^{k} q_{rj} + \lambda F(q_j)\right\}$$

$$= \left\{1 + \lambda\sum_{i=1}^{\ell} F(p_i)\right\}\left\{1 + \lambda\sum_{j=1}^{m} F(q_j)\right\}$$

$$= 1 + \lambda\left\{\sum_{i=1}^{\ell} F(p_i) + \sum_{j=1}^{m} F(q_j) + \lambda\sum_{i=1}^{\ell} F(p_i)\sum_{j=1}^{m} F(q_j)\right\}$$

$$= 1 + \lambda\sum_{i=1}^{\ell}\sum_{j=1}^{m} F(p_i q_j)$$

$$= \sum_{i=1}^{\ell}\sum_{j=1}^{m}\left\{\frac{1}{k}\sum_{s=1}^{k} p_{si}q_{sj} + \lambda F(p_i q_j)\right\}$$

$$= \sum_{i=1}^{\ell}\sum_{j=1}^{m} f(p_i q_j).$$

Lemma 8.1.1 shows that F is a solution of (5.3.12) if and only if f given by (8.1.1) is a solution of

$$\sum_{i=1}^{\ell}\sum_{j=1}^{m} [f(p_i q_j) - f(p_i)f(q_j)] = 0. \qquad (8.1.2)$$

Again, for technical reasons we have to distinguish three cases:

$$\begin{cases} \text{Case 1.} & \ell \geq 3, m \geq 3. \\ \text{Case 2.} & \ell \geq 3, m = 2 \text{ or } \ell = 2, m \geq 3. \\ \text{Case 3.} & \ell = m = 2. \end{cases} \qquad (8.1.3)$$

The reason is the following: In case 1 we can apply Theorem 6.4.1 twice to reduce (8.1.2) to a simpler equation where no double sum occurs. In case 2 we can apply Theorem 6.4.1 only one time, and in case 3 we can not apply Theorem 6.4.1 at all. It turns out that, just as it happened before when $\lambda = 0$, the third case is the most difficult.

In Chapter 6 (when $\lambda = 0$) we were led in case 3 to the functional equation

$$f_1(pq) + f_2(p(1-q)) + f_3((1-p)q) + f_4((1-p)(1-q)) = g(p) + h(q), \quad (6.1.1)$$

for all $p, q \in J$. In the present chapter, functional equation (8.1.3) leads in case 3 to the similar-looking functional equation

$$f(pq) + f(p(1-q)) + f((1-p)q) + f((1-p)(1-q)) = h(p)h(q), \quad (8.1.4)$$

where $h(p) = f(p) + f(1-p)$ $(p, q \in]0,1[)$. To cover also case 3 of functional equation (5.4.7), we investigate the following generalization of (8.1.4):

$$\begin{aligned} f_1(pq) + f_2(p(1-q)) + f_3((1-p)q) + f_4((1-p)(1-q)) \\ = g_1(p)h_1(q) + g_2(p)h_2(q), \end{aligned} \quad (6.2.1)$$

where $f_1, f_2, f_3, f_4, g_1, h_1, g_2, h_2 :]0,1[\to \mathbb{R}$ are measurable.

Let us now compare (6.1.1) with (6.2.1). We have seen in Theorem 6.2.8 that the solutions of (6.1.1) have the form

$$f_i(p) = \sum_{i=0}^{3} a_i p^i + \sum_{i=0}^{1} b_i p^i \log p \quad (8.1.5)$$

where $a_0, a_1, a_2, a_3, b_0, b_1 \in \mathbb{R}$. But now the structure of the functional equation (6.2.1) is much more complicated because of the more complex right-hand side of (6.2.1). (The trick we used in proving Theorem 6.2.8 doesn't work here. Differentiating (6.2.1) first with respect to p and then with respect to q, the right-hand side will not disappear as it did when $h_1 \equiv g_2 \equiv 1$.) The simple reason is the presence of products on the right-hand side of (6.2.1). Thus we will first prove in section 8.2 a general structure theorem for equations of the form (6.2.1) which states that the solutions f_i, $1 \le i \le 4$, of equation (6.2.1) are finite linear combinations of functions of the form

$$x \to x^\lambda (\log x)^k,$$

for some $\lambda \in \mathbb{R}$, $k \in \mathbb{N}$. More precisely, all f_i $(i = 1, 2, 3, 4)$ are solutions of a certain Euler differential equation.

From a mathematical point of view there is no difference in proving this structure theorem for equation (6.2.1) or for the more general functional equation

$$f_1(pq) + f_2(p(1-q)) + f_3((1-p)q) + f_4((1-p)(1-q))$$
$$= \sum_{i=1}^{N} g_i(p)h_i(q) \tag{8.1.6}$$

for all $p, q \in]0, 1[$. So we will treat equation (8.1.6) for general $N \in \mathbb{N}$, although we only need the cases $N = 1$ and $N = 2$. Section 8.2 then will be finished by results concerning the linear independence of some elementary functions. In sections 8.3 and 8.4 we then present the form of all (ℓ, m)-additive information measures of type λ, first in the cases 1 and 2, and then in the case 3 of (8.1.3).

8.2. A general structure theorem

One of the main result of this chapter is the following structure theorem for sum form equations. For technical reasons we consider not only real-valued, but also complex-valued functions.

We follow the presentation of Losonczi (1997), who kindly provided us with some of his unpublished results.

Theorem 8.2.1. *Let $f_1, f_2, f_3, f_4, g_i, h_i :]0, 1[\to \mathbf{C}$ $(i = 1, \dots, N)$ satisfy the functional equation*

$$f_1(pq) + f_2(p(1-q)) + f_3((1-p)q)$$
$$+ f_4((1-p)(1-q)) = \sum_{i=1}^{N} g_i(p)h_i(q) \tag{8.1.6}$$

for all $p, q \in]0, 1[$. If f_1, f_2, f_3, f_4 are measurable on $]0, 1[$ then there exist distinct complex numbers $\lambda_1, \dots, \lambda_M$ and natural numbers $m_1, \dots, m_M$, with

$$\sum_{i=1}^{M} m_i \le 30N + 7, \tag{8.2.1}$$

such that

$$f_i(p) = \sum_{j=1}^{M} \sum_{k=0}^{m_j-1} c_{ijk} \, p^{\lambda_j} (\log p)^k \tag{8.2.2}$$

holds for some complex constants c_{ijk} $(j = 1,\ldots,M;\ k = 0,\ldots,m_j - 1;$ $i = 1,\ldots 4)$. *If* $f_1 = f_2 = f_3 = f_4 = f$ *in (8.1.6), then* $m_1,\ldots,m_M$ *satisfy the inequality*

$$\sum_{i=1}^{M} m_i \leq 15N + 5. \tag{8.2.3}$$

Moreover, if $f_1 = f_2 = f_3 = f_4 = f$ *and* $N = 0$ *in (8.1.6) (that is,* $\sum_{i=1}^{0} g_i(p)h_i(q) := 0$), *then* f *satisfies the Euler differential equation*

$$p^5 f^{(5)}(p) + 5p^4 f^{(4)}(p) + 3p^3 f^{(3)}(p) = 0, \quad p \in]0,1[. \tag{8.2.4}$$

Proof: First, we remark that without loss of generality we may assume that the functions $g_1,\ldots g_N$ and $h_1,\ldots h_N$ are linearly independent on $]0,1[$, since otherwise the right-hand side of (8.1.6) can be written in the form $\sum_{i=1}^{N'} g_i'(p)h_i'(q)$ with $N' \leq N$ and with linearly independent functions $g_i',\ldots,g_{N'}'$ and $h_i',\ldots,h_{N'}'$. By Remark 6.2.2(e) we know that all functions occurring in (8.1.6) are differentiable infinitely often. The main idea of the proof is to find solvable differential equations for $f_1,\ldots,f_4$ by making extensive use of appropriate differential operators. Let us introduce

$$D_1 = \frac{\partial}{\partial p}, \quad D_2 = \frac{\partial}{\partial q}, \ I = \text{identity operator.} \tag{8.2.5}$$

(Here all operators are mappings from $C^\infty(]0,1[)$ into $C^\infty(]0,1[)$; in this proof D_1 and D_2 have nothing to do with the domain D_n defined in section 1.1.) Further we define

$$L_1 = pD_1 - qD_2, \quad L_2 = pD_1 + (1-q)D_2,$$
$$L_3 = (1-p)D_1 + qD_2, \quad L_4 = (1-p)D_1 - (1-q)D_2. \tag{8.2.6}$$

Moreover, we denote by

$$\mathcal{L}(p,q) \quad \text{and} \quad \mathcal{R}(p,q) \tag{8.2.7}$$

the left and right-hand sides, respectively, of equation (8.1.6). An immediate calculation shows that the operators L_i are constructed in such a manner that the application of L_i to $\mathcal{L}(p,q)$ "kills" f_i; that is, in $L_i\mathcal{L}(p,q)$ the function f_i does not appear. For example,

$$L_1\mathcal{L}(p,q) = L_1\big[f_2(p(1-q)) + f_3((1-p)q) + f_4((1-p)(1-q))\big].$$

Now we introduce the operators A_n and B_n $(n = 1,2,\ldots,6)$ given by

$$\begin{aligned}
A_1 &= L_3, \quad A_2 = L_4 + I, \\
A_3 &= L_2 + 2I, \quad A_4 = L_2 + I, \\
A_5 &= L_2, \quad A_6 = L_1 - I, \\
B_n &= A_n A_{n-1} \cdots A_1.
\end{aligned} \tag{8.2.8}$$

The proof proceeds along the following path.

Step 1. We shall show that

$$B_n\mathcal{L}(p,q) = B_n f_1(pq) \quad \text{if } n = 5,6. \tag{8.2.9}$$

Step 2. We demonstrate that

$$B_6\mathcal{L}(p,q) = -2p^4 q f_1^{(5)}(pq) - 4p^3(q+2)f_1^{(4)}(pq) - 12p^2 f_1^{(3)}(pq), \tag{8.2.10}$$

so that $p \to pB_6\mathcal{L}(p,q)$ is the result of applying an Euler differential operator to f_1.

Step 3. We prove that $B_6\mathcal{R}(p,q)$ has the decomposition

$$B_6\mathcal{R}(p,q) = \sum_{i=1}^{15N} G_i(p)H_i(q) \tag{8.2.11}$$

for appropriate functions $G_i, H_i :]0,1[\to \mathbf{C}$.

Step 4. For each fixed $q \in]0,1[$ we show that f_1 satisfies a homogeneous Euler differential equation of order between 3 and $15N + 5$ on the interval $]0,q[$ namely

$$\sum_{i=3}^{15N+5} \gamma_i(q)u^i f_1^{(i)}(u) = 0, \quad 0 < u < q < 1, \tag{8.2.12}$$

where the $\gamma_i :]0,1[\to \mathbf{C}$ satisfy

$$\sum_{i=1}^{15N+5} |\gamma_i(q)| \neq 0 \tag{8.2.13}$$

for all $q \in]0, 1[$.

Step 5. We verify representation (8.2.2) together with the condition (8.2.1).

Verification of Step 1. It is clear that $B_1 = A_1 = L_3$ "kills" f_3. Now we show that B_2 kills f_3 and also f_4. Making use of (8.2.6), (8.2.8), $L_3 = L_4 + D_2$ and $(L_4 + I)D_2 = D_2 L_4$, we can factor B_2 as

$$
\begin{aligned}
(L_4 + D_2 + I)L_4 &= (L_4 + I)L_4 + D_2 L_4 \\
&= (L_4 + I)L_4 + (L_4 + I)D_2 \\
&= (L_4 + I)(L_4 + D_2) = (L_4 + I)L_3 \\
&= A_2 A_1 = B_2.
\end{aligned}
\tag{8.2.14}
$$

The factorizations $B_2 = (L_4 + I)L_3$ and $B_2 = (L_4 + D_2 + I)L_4$ show that B_2 kills both f_3 and f_4.

We prove next that B_5 and B_6 have factorizations

$$
B_5 = UL_2 \qquad \text{and} \qquad B_6 = VL_2 \tag{8.2.15}
$$

with suitable differential operators U and V. To simplify the following calculations let us introduce the following notation. For two differential operators U, V we write

$$
U \sim V \text{ iff there is an operator } W \text{ such that } U - V = WL_2.
$$

An immediate calculation shows that $U \sim V$ and $R \sim S$ implies

$$
X(\alpha U + \beta R) \sim X(\alpha V + \beta S)
$$

for all $\alpha, \beta \in \mathbf{C}$ and for all operators X. Because of $L_4 = -L_2 + D_1 \sim D_1$ (see (8.2.6)) we obtain from (8.2.14)

$$
\begin{aligned}
B_2 = (L_4 + D_2 + I)L_4 &= (-L_2 + D_1 + D_2 + I)(-L_2 + D_1) \\
&\sim (-L_2 + D_1 + D_2 + I)D_1 = -L_2 D_1 + D_1^2 + D_2 D_1 + D_1.
\end{aligned}
\tag{8.2.16}
$$

From the immediately verified equations

$$
\begin{aligned}
L_2 D_1 &= D_1 L_2 - D_1, \\
L_2 D_2 &= D_2 L_2 + D_2,
\end{aligned}
$$

we get

$$
L_2 D_1 \sim -D_1 \text{ and } L_2 D_2 \sim D_2, \tag{8.2.17}
$$

$$L_2 D_1^2 = (D_1 L_2 - D_1) D_1$$
$$= D_1(L_2 D_1) - D_1^2 \sim D_1(-D_1) - D_1^2 \qquad (8.2.18)$$
$$= -2D_1^2,$$

and

$$L_2 D_2 D_1 = (D_2 L_2 + D_2) D_1$$
$$= D_2(L_2 D_1) + D_2 D_1 \sim D_2(-D_1) + D_2 D_1 = 0 \qquad (8.2.19)$$

where 0 is the zero operator. From (8.2.16)–(8.2.19) and (8.2.8), we get

$$B_2 \sim -L_2 D_1 + D_1^2 + D_2 D_1 + D_1 \sim -(-D_1) + D_1^2 + D_2 D_1 + D_1$$
$$\sim D_1^2 + 2D_1 + D_2 D_1,$$
$$B_3 = A_3 B_2 \sim (L_2 + 2I)(D_1^2 + 2D_1 + D_2 D_1)$$
$$\sim -2D_1^2 - 2D_1 + (2D_1^2 + 4D_1 + 2D_2 D_1) = 2D_1 + 2D_2 D_1,$$
$$B_4 = A_4 B_3 \sim (L_2 + I)(2D_1 + 2D_2 D_1)$$
$$\sim -2D_1 + (2D_1 + 2D_2 D_1) = 2D_2 D_1,$$

and finally
$$B_5 = A_5 B_4 \sim L_2(2D_2 D_1) \sim 0,$$
$$B_6 = A_6 B_5 = (L_1 + I) B_5 \sim 0.$$

Thus B_5 and B_6 satisfy (8.2.15). Since B_1 kills f_3, B_2 kills f_3 and f_4, and B_5 and B_6 kill f_2, (8.2.9) is proven.

Verification of Step 2. We first calculate $B_2 \mathcal{L}(p, q)$. Since B_2 kills f_3 and f_4, an easy calculation yields (using (8.2.14)) after simplifications

$$\left.
\begin{aligned}
&B_2 \mathcal{L}(p, q) \\
&\quad = (L_4 + I) L_3 (f_1(pq) + f_2(p(1 - q))) \\
&\quad = (L_4 + I)((1 - p)D_1 + qD_2)(f_1(pq) + f_2(p(1 - q))) \\
&\quad = (L_4 + I)(qf_1'(pq) + (1 - p - q)f_2'(p(1 - q))) \\
&\quad = (2q - 1)f_1'(pq) + q(q - p)f_1''(pq) \\
&\quad + (1 - 2q)f_2'(p(1 - q)) + (1 - q)(1 - p - q)f_2''(p(1 - q)).
\end{aligned}
\right\} \qquad (8.2.20)$$

When applying $A_6 A_5 A_4 A_3$ to $B_2 \mathcal{L}(p, q)$ we have to consider only the terms f_1' and f_1'' in (8.2.20) (since B_5 and B_6 kill f_2). This calculation results in (8.2.10). (We note that $B_5 \mathcal{L}(p, q)$ is not of Euler type, and this is the reason we use $B_6 \mathcal{L}(p, q)$.)

Verification of Step 3. We apply first B_2 to $\mathcal{R}(p, q)$ and get after some rearrangements

$$
\left.
\begin{aligned}
& B_2 \mathcal{R}(p, q) \\
& = \sum_{i=1}^{N} (L_4 + I) L_3(g_i(p) h_i(q)) \\
& = \sum_{i=1}^{N} \Bigg(g_i(p) \left[(2q - 1) h_i'(q) + (q^2 - q) h_i''(q) \right] \\
& \qquad + (1 - p) g_i'(p) (2q - 1) h_i'(q) + (1 - p)^2 g_i''(p) h_i(q) \Bigg) \\
& =: \sum_{i=1}^{3N} K_i(p) L_i(q)
\end{aligned}
\right\} \qquad (8.2.21)
$$

where

$$
\begin{aligned}
K_i(p) &= g_i(p), & L_i(q) &= (2q - 1) h_i'(q) + (q^2 - q) h_i''(q), & 1 \leq i \leq N; \\
K_i(p) &= (1 - p) g_i'(p), & L_i(q) &= (2q - 1) h_i'(q), & N + 1 \leq i \leq 2N; \\
K_i(p) &= (1 - p)^2 g_i''(p), & L_i(q) &= h_i(q), & 2N + 1 \leq i \leq 3N.
\end{aligned}
$$

To prove (8.2.11) we apply the operator $A_6 A_5 A_4 A_3$ to $B_2 \mathcal{R}(p, q)$. With the notations

$$
p D_1 = X, \quad (1 - q) D_2 = Y, \quad q D_2 = Z
$$

we get

$$
\begin{aligned}
& A_6 A_5 A_4 A_3 \\
& = (X - Z - I)(X + Y)(X + Y + I)(X + Y + 2I) \\
& = (X - Z - I)((X + Y)^3 + 3(X + Y)^2 + 2(X + Y)) \\
& = X^4 + X^3(3Y - Z + 2I) + X^2(3Y^2 + 3Y - 3YZ - 3Z - I) \\
& \quad + X(Y^3 - 4Y - 3Y^2 Z - 6YZ - 2Z - 2I) \\
& \quad + I(-Y^3 - 3Y^2 - 2Y - Y^3 Z - 3Y^2 Z - 2YZ) \\
& =: X^4 p_4(Y, Z) + X^3 p_3(Y, Z) + X^2 p_2(Y, Z) \\
& \quad + X p_1(Y, Z) + X^0 p_0(Y, Z)
\end{aligned}
$$

where p_i are differential polynomials in Y, Z (e.g. $p_4(Y, Z) = 1$, $p_3(Y, Z) =$

$3Y - Z + 2I$, etc.). Thus we obtain from (8.2.21)

$$B_6 \mathcal{R}(p, q) = A_6 A_5 A_4 A_3 A_2 B_2 \mathcal{R}(p, q)$$

$$= \sum_{i=0}^{4} X^i p_i(Y, Z) \left(\sum_{j=1}^{3N} K_j(p) L_j(q) \right)$$

$$= \sum_{j=1}^{3N} \sum_{i=0}^{4} (X^i K_j(p))(p_i(Y, Z) L_j(q)),$$

where $X^i K_j(p)$ and $p_i(Y, Z) L_j(q)$ depend only on p and q, respectively. Putting

$$G_{j+3iN}(p) = X^i K_j(p), \quad H_{j+3iN}(q) = p_i(Y, Z) L_j(q),$$

for $p, q \in]0, 1[$, $j = 1, \dots 3N$ and $i = 0, 1, 2, 3, 4$, we get exactly (8.2.11).

Verification of Step 4. We need to work both sides of the equation

$$p D_2^n B_6 \mathcal{L}(p, q) = p D_2^n B_6 \mathcal{R}(p, q), \quad n = 0, 1, \dots, 15N. \qquad (8.2.22)$$

Using (8.2.10) and (8.2.11) we get (remembering that $q^{(j)} = 0$ if $j > 1$)

$$
\left.
\begin{aligned}
&\sum_{i=1}^{15N} p G_i(p) H_i^{(n)}(q) \\
&= p(-2p^4) D_2^n \big(q f_1^{(5)}(pq) \big) + p(-4p^3) D_2^n \big((q+2) f_1^{(4)}(pq) \big) \\
&\quad + p(-12p^2) D_2^n f_1^{(3)}(pq) \\
&= -2p^5 \sum_{i=0}^{n} \binom{n}{i} q^{(n-i)} \big(f_1^{(5+i)}(pq) \big)^{(i)} p^i \\
&\quad - 4p^4 \sum_{i=0}^{n} \binom{n}{i} (q+2)^{(n-i)} \big(f_1^{(4+i)}(pq) \big)^{(i)} p^i \\
&\quad - 12p^{n+3} f_1^{(n+3)}(pq) \\
&= -2p^{n+5} q f_1^{(n+5)}(pq) - 2n p^{n+4} f_1^{(n+4)}(pq) \\
&\quad - 4p^{n+4}(q+2) f_1^{(n+4)}(pq) \\
&\quad - 4n p^{n+3} f_1^{(n+3)}(pq) - 12 p^{n+3} f_1^{(n+3)}(pq) \\
&= -2p^{n+5} q f_1^{(n+5)}(pq) - p^{n+4}(2n + 8 + 4q) f_1^{(n+4)}(pq) \\
&\quad - 4(n+3) p^{n+3} f_1^{(n+3)}(pq).
\end{aligned}
\right\} \qquad (8.2.23)
$$

Denoting the last line by $-E_n(p, q)$ we obtain

$$1 \cdot E_n(p, q) + \sum_{i=1}^{15N} pG_i(p)H_i^{(n)}(q) = 0, \qquad (8.2.24)$$

for all $p, q \in]0, 1[$ and $n = 0, \ldots, 15N$. This equation can be written as a linear homogeneous system

$$A \cdot (1, pG_1(p), \ldots, pG_{15N}(p))^T = 0 \qquad (8.2.25)$$

for the unknowns $1, pG_1(p), \ldots, pG_{15N}(p)$, where A is a matrix given by

$$A = \begin{pmatrix} E_0(p, q) & H_1(q) & \cdots & H_{15N}(q) \\ E_1(p, q) & H_1'(q) & \cdots & H_{15N}'(q) \\ \vdots & \vdots & \vdots & \vdots \\ E_{15N}(p, q) & H_1^{(15N)}(q) & \cdots & H_{15N}^{(15N)}(q) \end{pmatrix}$$

(v^T is the transpose of the row-vector v). Since the system has nontrivial solutions we have $\det A = 0$. This implies that the rows of A form a linearly dependent vector-system. Thus the linear dependence for the first coordinates gives

$$\sum_{i=0}^{15N} \alpha_i(q)E_i(p, q) = 0, \qquad (8.2.26)$$

where for any fixed $q \in]0, 1[$ not all coefficients $\alpha_i(q)$ are zero; that is

$$\sum_{i=0}^{15N} |\alpha_i(q)| \neq 0 \qquad (8.2.27)$$

for all $q \in]0, 1[$. Substituting $E_i(p, q)$ (that is (8.2.23)) into (8.2.26), we obtain by simple index transformations

$$\sum_{i=5}^{15N+5} \alpha_{i-5}(q)2qp^i f_1^{(i)}(pq) + \sum_{i=4}^{15N+4} \alpha_{i-4}(q)(2i + 4q)p^i f_1^{(i)}(pq)$$

$$+ \sum_{i=3}^{15N+3} \alpha_{i-3}(q)4ip^i f_1^{(i)}(pq) = 0.$$

That is,

$$\sum_{i=3}^{15N+5} \beta_i(q)p^i f_1^{(i)}(pq) = 0, \qquad (8.2.28)$$

where

$$\beta_i(q) := 2q\alpha_{i-5}(q) + (2i + 4q)\alpha_{i-4}(q) + 4i\alpha_{i-3}(q) \tag{8.2.29}$$

$(i = 3, \ldots, 15N + 5)$, with $\alpha_{-2} = \alpha_{-1} = \alpha_{15N+1} = \alpha_{15N+2} = 0$.

We claim that $\sum_{i=3}^{15N+5} |\beta_i(q)| \neq 0$ for all $q \in]0, 1[$. Otherwise there would be a $q_0 \in]0, 1[$ such that $\beta_i(q_0) = 0$ for all $i = 3, \ldots, 15N + 5$. But then (8.2.29) implies $\alpha_i(q_0) = 0$ for all $i = 0, \ldots, 15N$ contradicting (8.2.27). Substituting

$$u = pq, \qquad \gamma_i(q) := \beta_i(q)q^{-i}$$

into (8.2.28) we obtain the desired Euler differential equation (8.2.12) (with (8.2.13)). Note that $pB_6\mathcal{L}(p, q) = 0$ (see (8.2.10)) represents the case $N = 0$ in (8.2.28).

Verification of Step 5. For a fixed $q_0 \in]0, 1[$ substitute $f_1(u) = u^\lambda$ ($\lambda \in \mathbb{R}$) into (8.2.12) to get the characteristic equation

$$c_{q_0}(\lambda) = \sum_{i=3}^{15N+5} \gamma_i(q_0)i! \binom{\lambda}{i} = 0 \tag{8.2.30}$$

of the Euler equation (8.2.12). Let $A = \{\mu_1, \ldots, \mu_S\}$ be the set of (distinct) zeros of (8.2.30), and let $n_1, \ldots, n_S$ be the respective multiplicities of these zeros. Then

$$\sum_{i=1}^{S} n_i \leq 15N + 5, \tag{8.2.31}$$

where we may assume $\mu_1 = 0$, $\mu_2 = 1$, $\mu_3 = 2$ (since $\lambda(\lambda - 1)(\lambda - 2)$ is a common factor of all summands in (8.2.30)). The functions

$$p \to p^{\mu_j}(\log p)^k \quad (j = 1, \ldots, S; \ k = 0, \ldots, n_j - 1) \tag{8.2.32}$$

are linearly independent on $]0, q_0[$, hence the solution of (8.2.12) on $]0, q_0[$ has the form

$$f_1(u) = \sum_{j=1}^{S} \sum_{k=0}^{n_j-1} c_{jk} u^{\mu_j}(\log u)^k \tag{8.2.33}$$

for all $u \in]0, q_0[$, where the c_{jk} are constants.

Taking another fixed value $q_1 > q_0$, the solution of (8.2.12) on $]0, q_1[$ has the representation

$$f_1(u) = \sum_{j=1}^{S'} \sum_{k=0}^{n'_j - 1} c'_{jk}\, u^{\mu'_j} (\log u)^k, \qquad u \in]0, q_1[\tag{8.2.34}$$

where $\mu'_1 = 0$, $\mu'_2 = 1$, $\mu'_3 = 2$, $\mu'_4, \ldots, \mu'_{S'}$ are distinct complex numbers, $n'_1, \ldots, n'_{S'}$ are natural numbers and c'_{jk} are constants. Since the functions (8.2.32) are linearly independent on any interval $]0, t[$, $t > 0$, the functions in (8.2.33) and (8.2.34) are identical not only on the interval $]0, min(q_0, q_1)[=]0, q_0[$, but also on each interval $]0, t[$, and thus also on $]0, 1[$. This means that f_1 has the representation (8.2.33) for all $u \in]0, 1[$. Replacing p by $1 - p$ in (8.1.6), f_3 will play the role of f_1 and $g_i(1 - p)$ will play the role of $g_i(p)$. Thus f_3 satisfies the same differential equation as f_1 and has the form

$$f_3(u) = \sum_{j=1}^{S} \sum_{k=0}^{n_j - 1} d_{jk}\, u^{\mu_j} (\log u)^k, \qquad u \in]0, 1[, \tag{8.2.35}$$

with some constants d_{jk}. Replacing in (8.1.6) first q by $1 - q$ and then both p by $1 - p$ and q by $1 - q$, we get that f_2 and f_4 have the representations

$$f_2(u) = \sum_{j=1}^{T} \sum_{k=0}^{p_j - 1} e_{jk}\, u^{\nu_j} (\log u)^k,$$

$$f_4(u) = \sum_{j=1}^{T} \sum_{k=0}^{p_j - 1} h_{jk}\, u^{\nu_j} (\log u)^k, \tag{8.2.36}$$

for all $u \in]0, 1[$, where $B = \{\nu_1 = 0,\, \nu_2 = 1,\, \nu_3 = 2,\, \nu_4, \ldots, \nu_T\}$ contains distinct complex numbers, $p_1, \ldots, p_T$ are natural numbers satisfying $\sum_{j=1}^{T} p_j \le 15N + 5$, and e_{jk}, h_{jk} are constants.

Now let $\lambda_1, \ldots, \lambda_M$ be the distinct elements of $A \cup B$ and define m_j $(j = 1, \ldots M)$ by

$$m_j = \begin{cases} \max\{n_k, p_\ell\} & \text{if } \lambda_j \in A \cap B \text{ and } \lambda_j = \mu_k = \nu_\ell \\ n_k & \text{if } \lambda_j \in A \setminus B \text{ and } \lambda_j = \mu_k \\ p_\ell & \text{if } \lambda_j \in B \setminus A \text{ and } \lambda_j = \nu_\ell \end{cases}.$$

Then we get $\lambda_1 = 0$, $\lambda_2 = 1$, $\lambda_3 = 2$ and

$$\sum_{j=1}^{M} m_j \le 2(15N + 5) - 3 = 30N + 7.$$

Moreover, every f_i has the representation (8.2.2). If we assume that $f_1 = f_2 = f_3 = f_4 = f$ in (8.2.2), then we get from (8.2.31) with $M = S$ and $n_j = m_j$ exactly (8.2.3). Finally, to prove (8.2.4) we put $q = \frac{1}{2}$ into (8.2.10) so that $pB_6\mathcal{L}\left(p, \frac{1}{2}\right) = 0$ (which represents the case $N = 0$ in (8.2.28)) together with the substitution $u = \frac{p}{2}$ goes over into

$$u^5 f^{(5)}(u) + 5u^4 f^{(4)}(u) + 3u^3 f^{(3)}(u) = 0,$$

for all $u \in \,]0, \frac{1}{2}[$. But from the argument above we know that this equation is in fact valid on $]0, 1[$. This completes the proof.

Remark 8.2.2 (a) To illustrate Theorem 8.2.1 let us determine once more the measurable solutions of equation

$$f(pq) + f(p(1-q)) + f((1-p)q) + f((1-p)(1-q)) = 0 \qquad (6.2.18)$$

for all $p, q \in]0, 1[$. From Theorem 8.2.1 we know that f satisfies the Euler differential equation (8.2.4). Its characteristic polynomial $c(\lambda)$ is given by (see (8.2.30))

$$\begin{aligned} c(\lambda) &= \lambda(\lambda - 1)(\lambda - 2)\big[(\lambda - 3)(\lambda - 4) + 5(\lambda - 3) + 3\big] \\ &= \lambda^2(\lambda - 1)(\lambda - 2)^2. \end{aligned}$$

Thus f has the form (see (8.2.33))

$$f(p) = a + b\log p + cp + p^2(d + e\log p), \quad p \in]0, 1[. \qquad (8.2.37)$$

Substitution of (8.2.37) into (6.2.18) yields $b = d = e = 0$ and $4a + c = 0$, so that we get the same solutions found in Lemma 6.2.3.

(b) If $\ell = 2$, $m \ge 3$ or $\ell \ge 3$, $m = 2$, then it can be shown easily that a function f satisfying (8.1.2) has the form (8.2.2), also. The proof is the same as the proof of Theorem 7.1.1. By a substitution similar to (7.1.7) and (7.1.8) we can reduce (8.1.2) to a functional equation of type (8.1.6). Thus Theorem 8.2.1 can be applied and we get the result. We don't pursue this line here because we are able to give the general solution (without assuming measurability) of (8.1.2) for $\ell = 2$, $m \geq 3$ or $\ell \geq 3$, $m = 2$, and we can prove our results more directly. For the interested reader we refer to Losonczi (1997).

In Theorem 8.2.1 we used the following well-known result from the theory of differential equations:

Let $M, m_1, m_2, \ldots, m_M \in \mathbb{N}$ and let $\lambda_1, \ldots, \lambda_M$ be distinct complex numbers. Then the functions

$$p^{\lambda_j} (\log p)^k \quad (j = 1, \ldots, M; k = 0, \ldots, m_j - 1)$$

are linearly independent on $]0, 1[$. That is, if

$$\sum_{j=1}^{M} \sum_{k=0}^{m_j - 1} c_{jk}\, p^{\lambda_j} (\log p)^k = 0 \tag{8.2.38}$$

holds for some complex constants c_{jk}, then $c_{jk} = 0$ ($1 \leq j \leq M$, $0 \leq k \leq m_j - 1$). The next theorem is similar to this result and also known in the theory of (complex-valued) differential equations, even in a more general setting. Since we will use it rather often, we present and comment on this result for the convenience of the reader. (To shorten the notation we write $\log^k p$ for $(\log p)^k$). For a proof we refer to Reich (1992) or Ebanks and Losonczi (1992).

Theorem 8.2.3. *Let $M, m_1, m_2, \ldots, m_M \in \mathbb{N}$ and let $\lambda_1, \ldots, \lambda_M$ be distinct complex numbers. Suppose that*

$$\sum_{j=1}^{M} \sum_{k=0}^{m_j - 1} a_{jk} \big[p^{\lambda_j} \log^k p \pm (1 - p)^{\lambda_j} \log^k (1 - p) \big] = 0, \quad p \in]0, 1[, \tag{8.2.39}$$

holds for some complex constants a_{jk} ($1 \leq j \leq M$, $0 \leq k \leq m_j - 1$) either with the + or with the − sign. Then for any $1 \leq j \leq M$ we have:
(i) if $\lambda_j \notin \mathbb{N} \cup \{0\}$, then $a_{jk} = 0$ ($0 \leq k \leq m_j - 1$), and
(ii) if $\lambda_j \in \mathbb{N} \cup \{0\}$ and $m_j > 1$, then $a_{jk} = 0$ ($1 \leq k \leq m_j - 1$).

Remark 8.2.4. (a) Theorem 8.2.3 can also be expressed in the following way. If $M^* = \{ j \mid 1 \leq j \leq M, \lambda_j \in \mathbb{N} \cup \{0\} \}$, then the relation (8.2.39) reduces to

$$\sum_{j \in M^*} a_{j0} \big(p^{\lambda_j} \pm (1 - p)^{\lambda_j} \big) = 0. \tag{8.2.40}$$

(b) A special case of Theorem 8.2.3 is that the existence of complex nonzero constants $a_1, \ldots, a_M$ with the property

$$\sum_{j=1}^{M} a_j \left(p^{\lambda_j} + (1-p)^{\lambda_j} \right) = 0, \quad p \in]0,1[, \qquad (8.2.41)$$

implies that $\lambda_1, \ldots, \lambda_M \in \mathbb{N} \cup \{0\}$.

(c) The importance of Theorem 8.2.3 is that relations of the form

$$\sum_{j=1}^{M} \sum_{k=0}^{m_j - 1} c_{jk} A_{jk}(p) = 0,$$

where

$$A_{jk}(p) = p^{\lambda_j} \log^k p \pm (1-p)^{\lambda_j} \log^k (1-p),$$

can be reduced to relations of the form (8.2.41) where λ_j are not arbitrary elements of $\mathbf{C}$ but must belong to $\mathbb{N} \cup \{0\}$.

Let us now give some results concerning the linear dependence and linear independence of certain functions A_k, $k \in \mathbb{N}$, where $A_\alpha :]0,1[\to \mathbb{R}$, $\alpha \in \mathbf{C}$, is defined by

$$A_\alpha(p) = p^\alpha + (1-p)^\alpha. \qquad (8.2.42)$$

Theorem 8.2.5. (a) *If $n \geq 7$ is an odd integer and for all $p \in]0,1[$*

$$\alpha_n A_n(p) + \alpha_{n-1} A_{n-1}(p) + \alpha_{n-2} A_{n-2}(p) + \alpha_{n-3} A_{n-3}(p) = 0 \qquad (8.2.43)$$

holds with some constants $\alpha_n, \alpha_{n-1}, \alpha_{n-2}, \alpha_{n-3} \in \mathbf{C}$ then

$$\alpha_n = \alpha_{n-1} = \alpha_{n-2} = \alpha_{n-3} = 0.$$

(b) *The equation*

$$\sum_{k=0}^{5} \beta_k A_k(p) = 0, \quad p \in]0,1[, \qquad (8.2.44)$$

holds with some constants $\beta_k \in \mathbf{C}$ ($k = 0,1,\ldots,5$) if and only if

$$\left.
\begin{aligned}
\beta_3 &= -4\beta_2 - 10(\beta_1 + 2\beta_0), \\
\beta_4 &= 5\beta_2 + 15(\beta_1 + 2\beta_0), \\
\beta_5 &= -2\beta_2 - 6(\beta_1 + 2\beta_0)
\end{aligned}
\right\} \qquad (8.2.45)$$

is satisfied. (Here $\beta_1 + 2\beta_0$ and β_2 are arbitrary.)

(c) *The equation*

$$\sum_{k=0}^{3} \gamma_k A_k(p) = 0, \quad p \in]0,1[, \tag{8.2.46}$$

holds with some constants $\gamma_k \in \mathbf{C}$ ($k = 0,1,2,3$) if and only if

$$\gamma_2 = -3(\gamma_1 + 2\gamma_0), \quad \gamma_3 = 2(\gamma_1 + 2\gamma_0). \tag{8.2.47}$$

(d) *The functions 1 and $A_\alpha, \alpha \in \mathbf{C}$, are linearly independent on $]0,1[$ if and only if $\alpha \notin \{0,1\}$.*

Proof. (a) With $p = \frac{1}{2} - t$ and $1 - p = \frac{1}{2} + t$ $(t \in]0, \frac{1}{2}[)$ and by the binomial theorem we get

$$\begin{aligned}
A_k(p) &= \left(\frac{1}{2} - t\right)^k + \left(\frac{1}{2} + t\right)^k \\
&= \sum_{l=0}^{k} \binom{k}{l} \frac{1 + (-1)^l}{2^{k-l}} t^l = \sum_{m=0}^{[k/2]} \binom{k}{2m} 2^{2m+1-k} t^{2m}.
\end{aligned} \tag{8.2.48}$$

Putting (8.2.48) into (8.2.43) we get

$$\sum_{m=0}^{\frac{n-1}{2}} t^{2m} 2^{2m+1-n} \left(\alpha_n \binom{n}{2m} + 2\alpha_{n-1} \binom{n-1}{2m} \right)$$

$$+ \sum_{m=0}^{\frac{n-3}{2}} t^{2m} 2^{2m+3-n} \left(\alpha_{n-2} \binom{n-2}{2m} + 2\alpha_{n-3} \binom{n-3}{2m} \right) = 0.$$

The vanishing of the coefficients of $t^{n-1}, 2^{-2}t^{n-3}, 2^{-4}t^{n-5}, 2^{-6}t^{n-7}$ implies the linear homogeneous system

$$\begin{pmatrix} \binom{n}{1} & 2 & 0 & 0 \\ \binom{n}{3} & \binom{n-1}{2} & \binom{n-2}{1} & \binom{n-3}{0} \\ \binom{n}{5} & \binom{n-1}{4} & \binom{n-2}{3} & \binom{n-3}{2} \\ \binom{n}{7} & \binom{n-1}{6} & \binom{n-2}{5} & \binom{n-3}{4} \end{pmatrix} \begin{pmatrix} \alpha_n \\ 2\alpha_{n-1} \\ 2^2\alpha_{n-2} \\ 2^3\alpha_{n-3} \end{pmatrix} = \begin{pmatrix} 0 \\ 0 \\ 0 \\ 0 \end{pmatrix}. \tag{8.2.49}$$

Denoting the coefficient matrix by C_n, calculation shows that

$$\det C_n = 4725^{-1} n(n-1)(n-2)^2 (n-3)^2 (n-4)^2 (n-5)(n-6) \neq 0,$$

since $n \geq 7$. Thus the system (8.2.49) has only trivial solutions; that is, $\alpha_n = \alpha_{n-1} = \alpha_{n-2} = \alpha_{n-3} = 0$.

(b) Substitution of (8.2.48) into (8.2.44) results in

$$
\begin{aligned}
2\beta_0 + \beta_1 + \sum_{m=0}^{1} t^{2m} 2^{2m-2} \left(2\beta_2 \binom{2}{2m} + \beta_3 \binom{3}{2m} \right) \\
+ \sum_{m=0}^{2} t^{2m} 2^{2m-4} \left(2\beta_4 \binom{4}{2m} + \beta_5 \binom{5}{2m} \right) = 0.
\end{aligned}
\tag{8.2.50}
$$

The left-hand side in (8.2.50) is an even polynomial of degree ≤ 4, and (8.2.50) is valid if and only if the coefficients of $t^4, 2^{-2}t^2$ and $2^{-4}t^0$ are zero, that is

$$5\beta_5 + 2\beta_4 = 0,$$

$$10\beta_5 + 12\beta_4 + 12\beta_3 + 8\beta_2 = 0,$$

$$\beta_5 + 2\beta_4 + 4\beta_3 + 8\beta_2 + 16(\beta_1 + 2\beta_0) = 0.$$

But it is easy to see that this system is equivalent to (8.2.45).

(c) The statement concerning equation (8.2.46) is a special case of statement (b). Putting $\gamma_k = \beta_k$, $k = 0, 1, 2, 3$ and $\beta_4 = \beta_5 = 0$ into (8.2.45) we get immediately (8.2.47).

(d) If $aA_\alpha(p) + b = 0$ for some $a, b \in \mathbf{C}$, we differentiate this equation to get

$$a\alpha \left[p^{\alpha-1} - (1-p)^{\alpha-1} \right] = 0,$$

for $p \in]0, 1[$. Thus $a = 0$ unless either $\alpha = 0$ or $\alpha - 1 = 0$. Clearly $a = 0$ implies $b = 0$. Hence 1 and A_α are linearly independent unless $\alpha \in \{0, 1\}$.

8.3. Solution of (5.3.12) in the case m ≥ 3 or ℓ ≥ 3

In this section we derive all λ-parametric (ℓ, m)-additive sum form information measures, for $\lambda \neq 0$ and for pairs of integers (ℓ, m) satisfying $\ell \geq 2$, $m \geq 3$ or $\ell \geq 3$, $m \geq 2$. The case $\ell = m = 2$ will be treated in section 8.4. To give the reader a better insight into the structure of the proofs we start with two lemmas.

Lemma 8.3.1. *Let* $f : J \rightarrow \mathbb{R}$ *satisfy (8.1.2) for fixed integers* ℓ, m *and for all* $P \in \Gamma_\ell^o$ *and* $Q \in \Gamma_m^o$.

(a) *If* $\ell \geq 3, m \geq 2$ *then there are two possibilities. Either*

$$
\sum_{j=1}^{m} f(q_j) = \text{constant} \qquad \text{for all} \quad Q \in \Gamma_m^o,
\tag{8.3.1}
$$

or there is an additive function $A : \mathbb{R}^k \to \mathbb{R}$ with $A(1) = 0$ such that

$$h(p) := f(p) - A(p) \tag{8.3.2}$$

satisfies

$$\sum_{j=1}^{m} [h(pq_j) - h(p)h(q_j)] = 0. \tag{8.3.3}$$

(b) *If $\ell \geq 2$, $m \geq 3$ then there are two possibilities. Either*

$$\sum_{i=1}^{\ell} f(p_i) = \text{constant} \qquad \text{for all} \quad P \in \Gamma_\ell^o, \tag{8.3.4}$$

or there is an additive function $A' : \mathbb{R}^k \to \mathbb{R}$ with $A'(1) = 0$ such that

$$h'(p) := f(p) - A'(p) \tag{8.3.5}$$

satisfies

$$\sum_{i=1}^{\ell} [h'(p_iq) - h'(p_i)h'(q)] = 0 \tag{8.3.6}$$

Proof: Without loss of generality we treat part (a). Let us keep $Q \in \Gamma_m^o$ momentarily fixed in (8.1.2). Defining

$$H(p) := \sum_{j=1}^{m} \{f(pq_j) - f(p)f(q_j)\}, \qquad p \in J,$$

we get

$$\sum_{i=1}^{\ell} H(p_i) = 0.$$

Now we use the fact that $\ell \geq 3$. Theorem 6.4.1 yields the existence of a function $B : \mathbb{R}^k \times \Gamma_m^o \to \mathbb{R}$ which is additive in the first variable and

$$\sum_{j=1}^{m} \{f(pq_j) - f(p)f(q_j)\} = B\left(p - \frac{1}{\ell}, Q\right), \qquad p \in J, \ Q \in \Gamma_m^o, \tag{8.3.7}$$

where Q is now free to vary again.

Let $P = (p_1, p_2, ..., p_m) \in \Gamma_m^o$, substitute xp_i for p $(i = 1, 2, ..., m; x \in J)$ in (8.3.7), and add these m equations. Using the fact that $\sum_{i=1}^{m} xp_i = x$ and the additivity of B, we get

$$\sum_{i=1}^{m} \sum_{j=1}^{m} \{f(xp_iq_j) - f(xp_i)f(q_j)\} = B\left(x - \frac{m}{\ell}, Q\right), \qquad x \in J. \qquad (8.3.8)$$

Next we put $p = x$ and $Q = P \in \Gamma_m^o$ in (8.3.7) to obtain

$$\sum_{i=1}^{m} \{f(xp_i) - f(x)f(p_i)\} = B\left(x - \frac{1}{\ell}, P\right), \qquad p \in J. \qquad (8.3.9)$$

By (8.3.8) and (8.3.9) we get

$$\sum_{i=1}^{m} \sum_{j=1}^{m} \{f(xp_iq_j) - f(x)f(p_i)f(q_j)\}$$

$$= \sum_{i=1}^{m} \sum_{j=1}^{m} \{f(xp_iq_j) - f(xp_i)f(q_j) + f(xp_i)f(q_j) - f(x)f(p_i)f(q_j)\}$$

$$= B\left(x - \frac{1}{\ell}, P\right) \sum_{j=1}^{m} f(q_j) + B\left(x - \frac{m}{\ell}, Q\right). \qquad (8.3.10)$$

The left hand side of (8.3.10) is symmetric in P and Q, therefore the right hand side must be also symmetric. That is,

$$B\left(x - \frac{1}{\ell}, P\right) \sum_{j=1}^{m} f(q_j) + B\left(x - \frac{m}{\ell}, Q\right)$$

$$= B\left(x - \frac{1}{\ell}, Q\right) \sum_{i=1}^{m} f(p_i) + B\left(x - \frac{m}{\ell}, P\right). \qquad (8.3.11)$$

Letting $x = \frac{1}{\ell}$ in (8.3.11) we see that

$$\frac{1-m}{\ell} B(1, Q) = \frac{1-m}{\ell} B(1, P), \qquad P, Q \in \Gamma_m^o,$$

which implies

$$B(1, P) = a \quad \text{(constant)}, \qquad P \in \Gamma_m^o. \qquad (8.3.12)$$

Rewriting (8.3.11) with (8.3.12) we have (using the additivity of B)

$$
\begin{aligned}
B(x, Q) &\left\{ 1 - \sum_{i=1}^{m} f(p_i) \right\} + \frac{a}{\ell} \sum_{i=1}^{m} f(p_i) \\
&= B(x, P) \left\{ 1 - \sum_{i=1}^{m} f(q_j) \right\} + \frac{a}{\ell} \sum_{j=1}^{m} f(q_j).
\end{aligned}
\tag{8.3.13}
$$

If $\sum_{i=1}^{m} f(p_i)$ is constant for all $P \in \Gamma_m^o$ then (8.3.1) is satisfied. If not, then there exists $P' \in \Gamma_m^o$ such that $\sum_{i=1}^{m} f(p'_i) \neq 1$. Substituting P' into (8.3.13) we obtain

$$
B(x, Q) = A(x) \left\{ 1 - \sum_{j=1}^{m} f(q_j) \right\} + b \sum_{j=1}^{m} f(q_j) + c,
\tag{8.3.14}
$$

where b, c are constants and $A : \mathbb{R}^k \to \mathbb{R}$ is an additive function. Using (8.3.14) and the additivity of B and A, we get

$$
0 = -B(x+y, Q) + B(x, Q) + B(y, Q) = b \sum_{j=1}^{m} f(q_j) + c.
$$

Hence (8.3.14) reduces to

$$
B(x, Q) = A(x) \left\{ 1 - \sum_{j=1}^{m} f(q_j) \right\}.
\tag{8.3.15}
$$

Putting $x = 1$ into (8.3.14) and using (8.3.12) and the nonconstancy of $\sum_{j=1}^{m} f(q_j)$, we get

$$
A(1) = a = B(1, Q) = 0.
\tag{8.3.16}
$$

By (8.3.15) and (8.3.16) we rewrite (8.3.7) as

$$
\sum_{j=1}^{m} \{ f(pq_j) - f(p)f(q_j) \} = A(p) \left\{ 1 - \sum_{j=1}^{m} f(q_j) \right\}.
\tag{8.3.17}
$$

Defining

$$
h(p) := f(p) - A(p), \qquad p \in J,
$$

equation (8.3.17) converts into (8.3.3) (since $A(1) = 0$).

Remark 8.3.2. (a) If f in Lemma 8.3.1 is measurable then A and A' are identically zero and h and h' are measurable, too. Indeed, let us assume that f is measurable in case (a). Then (8.3.9) shows that B is measurable in its first variable. Thus from (8.3.15) we conclude that A is measurable, that is A has the form

$$A(p) = A(1) \odot p, \qquad p \in J.$$

Because of $A(1) = 0$ we get $A \equiv 0$, so that $h = f$.

(b) Let us add a further remark which will be needed later. An arbitrary additive function B satisfies (8.1.2), if and only if $B(1) = B(1)^2$, that is $B(1) = 0$ or $B(1) = 1$. Thus, if B is in addition measurable, then $B(p) = 0$ or $B(p) = a \odot p$ with $1 \odot a = 1$ for some constant $a \in \mathbb{R}^k$. (In the last case we first conclude $B(p) = \sum_{i=1}^{n} a_i p_i$ where $a_1, a_2, ..., a_n \in \mathbb{R}$. The condition $B(1) = 1$ yields $\sum_{i=1}^{n} a_i = 1 = 1 \odot a$.)

Lemma 8.3.3. *Let $\ell \geq 3$. If the function $f : J \to \mathbb{R}$ satisfies*

$$\sum_{i=1}^{\ell} \sum_{j=1}^{2} \{f(p_i q_j) - f(p_i)f(q_j)\} = 0, \qquad P \in \Gamma_\ell^o, \ Q \in \Gamma_2^o, \qquad (8.3.18)$$

and for some constant d

$$f(p) + f(1 - p) = d, \qquad p \in J, \qquad (8.3.19)$$

then f has the form

$$f(p) = g(p) + c \qquad (8.3.20)$$

where g is a solution of

$$g(pq) + g(p(1 - q)) + g((1 - p)q) + g((1 - p)(1 - q)) = 0, \quad p, q \in J, \quad (6.3.40)$$

and c is a constant.

Proof: We substitute (8.3.19) into (8.3.18) to arrive at

$$\sum_{i=1}^{\ell} \left\{ \sum_{j=1}^{2} f(p_i q_j) - d\, f(p_i) \right\} = 0 \qquad (8.3.21)$$

for all $P \in \Gamma_\ell^o$, $Q \in \Gamma_2^o$. Now we put $q = q_1$ and $1 - q = q_2$, $q \in J$ into (8.3.21) and get

$$\sum_{i=1}^{\ell} \{f(p_i q) + f(p_i(1 - q)) - d\, f(p_i)\} = 0. \qquad (8.3.22)$$

By Theorem 6.4.1 there is a function $A : \mathbb{R}^k \times J \to \mathbb{R}$ which is additive in its first variable, satisfying

$$f(pq) + f(p(1-q)) = d\,f(p) + A\left(p - \frac{1}{\ell}, q\right) \tag{8.3.23}$$

for all $p, q \in J$. Replacing p by $1 - p$ in (8.3.23) and adding the resulting equation to (8.3.23) we obtain

$$
\begin{aligned}
f(pq) &+ f(p(1-q)) + f((1-p)q) + f((1-p)(1-q)) \\
&= d\,f(p) + d\,f(1-p) + A(p) \\
&\quad - \frac{1}{\ell}A(1,q) + A(1-p,q) - \frac{1}{\ell}A(1,q) \\
&= d^2 + \left(1 - \frac{2}{\ell}\right)A(1,q).
\end{aligned} \tag{8.3.24}
$$

Since the left side in (8.3.24) is symmetric in p and q we get from (8.3.24)

$$d^2 + \left(1 - \frac{2}{\ell}\right)A(1,q) = d^2 + \left(1 - \frac{2}{\ell}\right)A(1,p)$$

or, since $1 - \frac{2}{\ell} \neq 0$,

$$A(1,p) = A(1,q) = a \quad \text{(constant)}. \tag{8.3.25}$$

Substituting (8.3.25) into (8.3.24) we have

$$
\begin{aligned}
f(pq) &+ f(p(1-q)) + f((1-p)q) + f((1-p)(1-q)) \\
&= d^2 + \left(1 - \frac{2}{\ell}\right)a.
\end{aligned} \tag{8.3.26}
$$

Putting

$$4c = d^2 + \left(1 - \frac{2}{\ell}\right)a, \tag{8.3.27}$$

equation (8.3.26) goes over into (6.3.40), where

$$g(p) = f(p) - c, \qquad p \in J.$$

Remark 8.3.4. (a) Every solution of the system (8.3.18) and (8.3.19) is (because of Theorem 6.4.1) equivalent to the system (8.3.23) and (8.3.19) and can thus be characterized by solutions g of (6.3.40) (using (8.3.25), (8.3.27) and (8.3.20)). Since g satisfies not only (6.3.40) but also the equations (8.3.19) and (8.3.23) (rewritten with (8.3.20)) we can say that g is a "special" solution of (6.3.40) which differs from the solution of the system (8.3.18) and (8.3.19) only by a constant. Unfortunately, even the general solution of (6.3.40) with the additional information (8.3.19) and (8.3.23) is not known.

(b) Nevertheless, Lemma 8.3.3 is important because we get immediately the measurable solutions of the system (8.3.18) and (8.3.19). If f is measurable, g is a measurable solution of (6.3.40) so that g has the form

$$g(p) = A \odot p + c$$

where $A \in \mathbb{R}^k$ and $c \in \mathbb{R}$.

Now we are ready to prove the main results in this section. To give the solution $F : J \to \mathbb{R}$ of (5.3.12) with $\lambda \neq 0$ it is sufficient to give the general solution $f : J \to \mathbb{R}$ of (8.1.2) where f is given by (8.1.1) (see Lemma 8.1.1).

Theorem 8.3.5. *Let* $f : J \to \mathbb{R}$ *satisfy the functional equation (8.1.2) for a fixed pair of integers* $\ell \geq 2$, $m \geq 2$.
(a) *The general solution of (8.1.2) for* $\ell \geq 3$, $m \geq 3$ *is given by*

$$f(p) = A(p) + c, \qquad p \in J, \tag{8.3.28}$$

or

$$f(p) = M(p) + B(p), \qquad p \in J, \tag{8.3.29}$$

where c *is a constant,* $M : \mathbb{R}_+^k \to \mathbb{R}$ *is multiplicative and* $A, B : \mathbb{R}^k \to \mathbb{R}$ *are additive functions satisfying*

$$B(1) = 0, \quad A(1) + \ell m c = (A(1) + \ell c)(A(1) + m c) \tag{8.3.30}$$

(b) *The general solution of (8.1.2) for* $\ell \geq 3$, $m = 2$ *or* $\ell = 2$, $m \geq 3$ *with the additional condition* $f(p) + f(1 - p)$ *nonconstant is also given by (8.3.28)-(8.3.30).*

Proof: (a) We first treat the case $\ell \geq 3$, $m \geq 3$. Using Lemma 8.3.1 we first consider the case (8.3.1) or (8.3.4), that is

$$\sum_{j=1}^{m} f(q_j) = \text{constant} \quad \text{or} \quad \sum_{i=1}^{\ell} f(p_i) = \text{constant}$$

for all $Q \in \Gamma_m^o$ or $P \in \Gamma_l^o$. Since $\ell \geq 3$ and $m \geq 3$, we get from Theorem 6.4.1

$$f(p) = A(p) + c$$

where $A : \mathbb{R}^k \to \mathbb{R}$ is additive and c is a constant. This is exactly solution (8.3.28). Substituting (8.3.28) into (8.1.2) we get the second condition in (8.3.30).

Now we consider the case that $\sum_{j=1}^m f(q_j)$ and $\sum_{i=1}^\ell f(p_i)$ are nonconstant. By Lemma 8.3.1 there is an additive function $A : \mathbb{R}^k \to \mathbb{R}$ with $A(1) = 0$ such that

$$h(p) := f(p) - A(p) \qquad (8.3.2)$$

satisfies

$$\sum_{j=1}^m \{h(pq_j) - h(p)h(q_j)\} = 0 \qquad (8.3.3)$$

for all $p \in J$ and $Q \in \Gamma_m^o$. Since $m \geq 3$, keeping p temporarily fixed and applying again Theorem 6.4.1, we get

$$h(pq) - h(p)h(q) = B\left(p, q - \frac{1}{m}\right), \qquad p, q \in J, \qquad (8.3.31)$$

where $B : J \times \mathbb{R}^k \to \mathbb{R}$ is additive in the second variable. If $B(p, q - \frac{1}{m}) = 0$ for all $p, q \in J$, then h is multiplicative and (8.3.2) yields

$$f(p) = h(p) + A(p), \qquad A(1) = 0, \qquad p \in J.$$

Thus we arrive this time at the representation (8.3.29) and (8.3.30) for f.

Next, we assume that there are $p^*, q^* \in J$ such that

$$r := B\left(p^*, q^* - \frac{1}{m}\right) \neq 0. \qquad (8.3.32)$$

Now let $x \in J$. In the following computation we use (8.3.31) and the associativity of the multiplication (in the form $h(p^*q^* \cdot x) = h(p^* \cdot q^*x)$) to arrive at

$$\left.\begin{aligned}
B&\left(p^*q^*, x - \frac{1}{m}\right) + h(x)B\left(p^*, q^* - \frac{1}{m}\right)\\
&= h(p^*q^* \cdot x) - h(p^*q^*)h(x) + h(x)h(p^*q^*)\\
&\quad - h(x)h(p^*)h(q^*)\\
&= h(p^* \cdot q^*x) - h(p^*)h(q^*)h(x)\\
&= h(p^* \cdot q^*x) - h(p^*)h(q^*x) + h(p^*)h(q^*x)\\
&\quad - h(p^*)h(q^*)h(x)\\
&= B\left(p^*, q^*x - \frac{1}{m}\right) + h(p^*)B\left(q^*, x - \frac{1}{m}\right).
\end{aligned}\right\} \tag{8.3.33}$$

From the first and last line in (8.3.33) we get

$$\left.\begin{aligned}
h(x) = \frac{1}{r}\Bigg\{ &B\left(p^*, q^*x - \frac{1}{m}\right)\\
&+ h(p^*)B\left(q^*, x - \frac{1}{m}\right) - B\left(p^*q^*, x - \frac{1}{m}\right)\Bigg\}
\end{aligned}\right\} \tag{8.3.34}$$

for all $x \in J$. Since B is additive in the second variable, we get from (8.3.34) the representation

$$h(x) = C(x) + b, \qquad x \in J \tag{8.3.35}$$

where

$$C(x) = \frac{1}{r}\{B(p^*, q^*x) + h(p^*)B(q^*, x) - B(p^*q^*, x)\}, \qquad x \in \mathbb{R}^k,$$

is additive and b is given by

$$b := \frac{1}{rm}\{-B(p^*, 1) - h(p^*)B(q^*, 1) + B(p^*q^*, 1)\}.$$

Thus (8.3.2) and (8.3.35) imply

$$\begin{aligned}
f(p) &= h(p) + A(p)\\
&= C(p) + b + A(p)\\
&= D(p) + b, \qquad p \in J
\end{aligned} \tag{8.3.36}$$

where $D = C + A : \mathbb{R}^k \to \mathbb{R}$ is additive. Again we get solution (8.3.28).

(b) Suppose first that $\ell \geq 3$ and $m = 2$. As in part (a), if $\sum_{i=1}^{\ell} f(p_i)$ is constant then we get (8.3.28). Now assume that $\sum_{i=1}^{\ell} f(p_i)$ is nonconstant.

By Lemma 8.3.1 again, we arrive at (8.3.2) and (8.3.3). That is, the function h, defined by (8.3.2), satisfies

$$h(pq) + h(p(1-q)) = h(p)\{h(q) + h(1-q)\}, \qquad p, q \in J. \qquad (6.4.9)$$

By Theorem 6.4.6 we have

$$h(p) = B(p) \qquad \text{or} \qquad h(p) = M(p), \qquad p \in J, \qquad (8.3.37)$$

where $M : \mathbb{R}_+^k \to \mathbb{R}$ is multiplicative and $B : \mathbb{R}^k \to \mathbb{R}$ is additive. Hence we get from (8.3.2) and (8.3.37)

$$f(p) = A'(p) := A(p) + B(p), \qquad p \in J, \qquad (8.3.38)$$

or

$$f(p) = M(p) + A(p), \qquad A(1) = 0, \qquad p \in J, \qquad (8.3.39)$$

where A', B are additive. Again (8.3.38) and (8.3.39) have the form (8.3.28) (with $c = 0$) and (8.3.29), respectively.

The case $\ell = 2$, $m \geq 3$ can be treated in exactly the same way as the case $\ell \geq 3$, $m = 2$ (using part (b) of Lemma 8.3.1 instead of part (a)). Thus the proof is complete.

Remark 8.3.6. (a) Theorem 8.3.5 shows that the general solution of (8.1.2) is known with the exceptions of Case 3 of (8.1.3) and Case 2 in the special case that

$$f(p) + f(1-p) = \text{constant}, \qquad p \in J.$$

By Lemma 8.3.3 we know that in the latter case the general solution of (8.1.2) can be described as

$$f(p) = g(p) + c$$

where g is a solution of (6.3.40) and c is a constant.

(b) In section four of Chapter 5 we used the general solution of (8.1.2) in the case $\ell \geq 3$, $m \geq 3$. Thus this gap is now closed.

(c) Using Lemma 8.1.1 and Theorem 8.3.5 we get immediately that the general solution $F : J \to \mathbb{R}$ of functional equation (5.3.12) with $\lambda \neq 0$ in Case 1 and Case 2 of (8.1.3) (with the additional condition

$$\frac{1}{k}(1 \odot p) + \lambda F(p) + \frac{1}{k}(1 \odot (1-p)) + \lambda F(1-p) \neq \text{constant},$$

or equivalently $F(p) + F(1-p) \neq$ constant) is given by

$$F(p) = \frac{1}{\lambda}\left(A(p) + c - \frac{1}{k}(1 \odot p)\right), \quad p \in J, \qquad (8.3.40)$$

or

$$F(p) = \frac{1}{\lambda}\left(M(p) + B(p) - \frac{1}{k}(1 \odot p)\right), \quad p \in J, \qquad (8.3.41)$$

where c is a constant, $M : \mathbb{R}_+^k \to \mathbb{R}$ is multiplicative and $A, B : \mathbb{R}^k \to \mathbb{R}$ are additive functions satisfying (8.3.30).

The following result is an immediate consequence of Theorem 8.3.5, Remark 8.3.4 (b) and the fact that a measurable additive function $B : \mathbb{R}^k \to \mathbb{R}$ satisfying $B(1) = 0$ is identically zero (since $B(p) = B(1) \odot p = 0$).

Theorem 8.3.7. *Let $f : J \to \mathbb{R}$ be a measurable solution of (8.1.2) for fixed $\ell \geq 2$, $m \geq 2$. If $\ell \geq 3$ or $m \geq 3$, then*

$$f(p) = a \odot p + c, \quad p \in J,$$

or

$$f(p) = p^\alpha, \quad p \in J,$$

where $a, \alpha \in \mathbb{R}^k, c \in \mathbb{R}$, and

$$a \odot 1 + \ell m c = (a \odot 1 + \ell c)(a \odot 1 + m c).$$

From the above results we derive some characterizations for information measures where we make use of our notations (1.1.10)–(1.1.12).

Theorem 8.3.8. *Let $\{I_n\}$ have the sum form for some generating function $F : J \to \mathbb{R}$. Then $I_n : \Gamma_n^\circ \to \mathbb{R}$ is λ-parametric (ℓ, m)-additive for fixed $\ell \geq 3$, $m \geq 3$ $(\lambda \neq 0)$, if and only if I_n is given by*

$$I_n(p_1, p_2, ..., p_n) = \frac{1}{\lambda}\left(\sum_{i=1}^{n} M(p_i) - 1\right)$$

$$= \frac{1}{\lambda}\left(\sum_{i=1}^{n}\prod_{j=1}^{k} M_j(\pi_{ji}) - 1\right), \qquad (8.3.42)$$

where $M : \mathbb{R}_+^k \to \mathbb{R}$ (and each $M_j : \mathbb{R}_+ \to \mathbb{R}$, $1 \leq j \leq k$) is multiplicative, or by

$$I_n(p_1, p_2, ..., p_n) = \frac{1}{\lambda}[cn + A(1) - 1], \qquad (8.3.43)$$

for some additive $A : \mathbb{R}^k \to \mathbb{R}$ and constant c satisfying (8.3.30).

Proof: By Remark 8.3.6 (c) we know that F has one of the forms given in (8.3.40) or (8.3.41). Moreover we write the additive function $A : \mathbb{R}^k \to \mathbb{R}$ and the multiplicative function $M : \mathbb{R}_+^k \to \mathbb{R}$ in the forms

$$A(x_1, x_2, ..., x_k) = \sum_{j=1}^{k} A_j(x_j)$$

$$M(x_1, x_2, ..., x_k) = \prod_{j=1}^{k} M_j(x_j),$$

where $A_j : \mathbb{R} \to \mathbb{R}$ and $M : \mathbb{R}_+ \to \mathbb{R}$ are additive and multiplicative, respectively.

In the case (8.3.40), we get

$$\lambda I_n(P) = \lambda \sum_{i=1}^{n} F(p_i)$$

$$= \sum_{i=1}^{n} \left[A(\pi_{1i}, ..., \pi_{ji}, ..., \pi_{ki}) + c - \frac{1}{k}(1 \odot p_i) \right]$$

$$= \sum_{i=1}^{n} \sum_{j=1}^{k} A_j(\pi_{ji}) + nc - \frac{1}{k} \sum_{i=1}^{n} \sum_{j=1}^{k} \pi_{ji}$$

$$= \sum_{j=1}^{k} A_j \left(\sum_{i=1}^{n} \pi_{ji} \right) + nc - \frac{k}{k}$$

$$= \sum_{j=1}^{k} A_j(1) + nc - 1$$

$$= A(1) + nc - 1,$$

which gives (8.3.43).

If F is given by (8.3.41), then we arrive at

$$
\begin{aligned}
\lambda I_n(P) &= \lambda \sum_{i=1}^{n} F(p_i) \\
&= \sum_{i=1}^{n} \left(M(p_i) + B(p_i) - \frac{1}{k} \left(1 \odot p_i \right) \right) \\
&= \sum_{i=1}^{n} M(p_i) + B(1) - 1 \\
&= \sum_{i=1}^{n} M(p_i) - 1 \\
&= \sum_{i=1}^{n} \prod_{j=1}^{k} M_j(\pi_{ji}) - 1.
\end{aligned}
$$

Conversely, I_n given by (8.3.42) or (8.3.43) is λ-parametric (ℓ, m)-additive. The proof of the theorem is now complete.

If in Theorem 8.3.8 the generating function is measurable then the entropies given by (8.3.42) take the form

$$
\begin{aligned}
I_n(P) &= \frac{1}{\lambda} \left(\sum_{i=1}^{n} p_i^\alpha - 1 \right) \\
&= \frac{2^{1-\alpha} - 1}{\lambda} H_n^\alpha(P), \qquad \alpha \in \mathbb{R}^k,
\end{aligned}
\tag{8.3.44}
$$

where we have defined $H_n^\alpha : \Gamma_n^o \to \mathbb{R}$ completely analogous to the expression of H_n^α depending upon one probability distribution.

Note that (8.3.44) can be rewritten as

$$
I_n(P) = \frac{2^{1-\alpha_1} 2^{1-\alpha_2} \cdots 2^{1-\alpha_k} - 1}{\lambda} \left(\sum_{i=1}^{n} \pi_{1i}^{\alpha_1} \pi_{2i}^{\alpha_2} \cdots \pi_{ki}^{\alpha_k} - 1 \right)
$$

where $\alpha = (\alpha_1, ..., \alpha_k) \in \mathbb{R}^k$.

Note also that, because of (8.3.30), solution (8.3.43) collapses further in presence of other (ℓ, m)-additivities. If ℓ and m are varied, then (8.3.30) forces c, $A(1) \in \{0, 1\}$ with $cA(1) = 0$, leading to $I_n = 0$, $I_n = \frac{1}{\lambda}(n - 1)$, or $I_n = -\frac{1}{\lambda}$. The second of these, $I_n = \frac{1}{\lambda}(n - 1)$, is included in the following theorem as $\frac{1}{\lambda} H_n^o$, while the other two (degenerate solutions) are excluded by the requirement that $\{I_n\}$ be nonconstant.

In view of Theorem 7.2.1 and the above results we have the following characterization theorem.

Theorem 8.3.9. *Let* $I_n : \Gamma_n^o \to \mathbb{R}$ *have the sum form for some measurable generating function* $F : J \to \mathbb{R}$. *Then* $\{I_n\}$ *is nonconstant,* λ-*parametric* (ℓ, m)-*additive for all* $\ell \geq 2$, $m \geq 3$ *or for all* $\ell \geq 3$, $m \geq 2$, *if and only if* I_n *has the form*

$$I_n(p_1, p_2, ..., p_n) = I_n(P_1, P_2, ..., P_k) =$$

$$I_n(P) = \begin{cases} \displaystyle\sum_{i=1}^{k} c_i H_n(P_i) + \sum_{i=1}^{k} \sum_{\substack{j=1 \\ i \neq j}}^{k} c_{ij} K_n(P_i, P_j) & \text{if } \lambda = 0 \\[4mm] \dfrac{2^{1-\alpha}-1}{\lambda} H_n^\alpha(P) & \text{if } \lambda \neq 0, \end{cases}$$

where $\alpha \in \mathbb{R}^k$ *and* c_{ij}, c_i *are constants* $(i, j = 1, 2, ..., k; i \neq j)$.

Using our definition for the generalized Shannon entropy (see (7.1.10)), Theorem 8.3.9 with the additional assumption $I_2(\frac{1}{2}, \frac{1}{2}) = 1$ (so that $\lambda = 2^{1-\alpha} - 1$) leads to

$$I_n(P) = \begin{cases} H_n(P) & \text{if } \lambda = 0 \\[2mm] H_n^\alpha(P) & \text{if } \lambda \neq 0 . \end{cases}$$

This is analogous to a result for information measures depending upon one probability distribution.

8.4. Solution of (5.3.12) in the case $\ell = m = 2$

In this section we determine the measurable solutions $f :]0, 1[\to \mathbb{R}$ of (8.1.2) if $\ell = m = 2$, that is

$$f(pq) + f(p(1-q)) + f((1-p)q) + f((1-p)(1-q))$$
$$= \{f(p) + f(1-p)\}\{f(q) + f(1-q)\} \qquad (8.4.1)$$

for all $p, q \in]0, 1[$. We already remarked that the general solution of (8.4.1) is not known. Moreover we will see that even when $k = 1$ and when f is measurable, it is very hard to obtain the solutions of (8.4.1). Let us start with some simple but tedious calculations related to equation (8.4.1). In the following we denote by $L_f(p,q)$ the left-hand side and by $R_f(p,q)$ the right-hand side of (8.4.1):

$$L_f(p,q) := f(pq) + f(p(1-q)) + f((1-p)q) + f((1-p)(1-q)), \quad (8.4.2)$$

$$R_f(p,q) := (f(p) + f(1-p))(f(q) + f(1-q)), \quad (8.4.3)$$

$p,q \in]0,1[$. Moreover we introduce the notations

$$\left.\begin{aligned} T_f(p) &:= f(p) + f(1-p), \\ A_k(p) &:= p^k + (1-p)^k, \end{aligned}\right\} \quad (8.4.4)$$

and

$$A_{jk}(p) := p^{\lambda_j} \log^k p + (1-p)^{\lambda_j} \log^k (1-p), \quad (8.4.5)$$

where $\lambda_j \in \mathbf{C}$ and $k \in \mathbf{N} \cup \{0\}$.

Lemma 8.4.1. *Let $f :]0,1[\to \mathbf{C}$ satisfy (8.4.1) for all $p,q \in]0,1[$, and suppose f has the form*

$$f(p) = \sum_{j=1}^{M} \sum_{k=0}^{m_j-1} c_{jk}\, p^{\lambda_j} \log^k p, \quad p \in]0,1[, \quad (8.4.6)$$

where $M, m_1, \ldots, m_M \in \mathbf{N}$, where $\lambda_1, \ldots, \lambda_M$ are distinct complex numbers, and where c_{jk} are complex constants, $1 \le j \le M$, $0 \le k \le m_j - 1$. Then we have

$$L_f(p,q) = \sum_{j=1}^{M} \sum_{l=0}^{m_j-1} A_{jl}(p) d_{jl}(q) \quad (8.4.7)$$

with

$$d_{jl}(q) := \sum_{k=l}^{m_j-1} c_{jk} \binom{k}{l} A_{jk-l}(q), \quad (8.4.8)$$

and

$$L_f(p,q) - R_f(p,q) = \sum_{j=1}^{M} \sum_{l=0}^{m_j-1} A_{jl}(p) D_{jl}(q) = 0 \quad (8.4.9)$$

with

$$D_{jl}(q) = d_{jl}(q) - c_{jl} \sum_{r=1}^{M} \sum_{s=0}^{m_r-1} c_{rs} A_{rs}(q), \quad (8.4.10)$$

for $1 \leq j \leq M,\ 0 \leq l \leq m_j - 1.$

Proof: By the binomial theorem we have

$$(\log p + \log q)^k = \sum_{l=0}^{k} \binom{k}{l} \log^l p \, \log^{k-l} q, \qquad (8.4.11)$$

$p, q \in]0,1[,\ k \in \mathbb{N}$. Moreover an index transformation yields (for $u_{kl} \in \mathbb{C}$)

$$\sum_{k=0}^{m_j-1} \sum_{l=0}^{k} u_{kl} = \sum_{l=0}^{m_j-1} \sum_{k=l}^{m_j-1} u_{kl}. \qquad (8.4.12)$$

Using (8.4.11) and (8.4.12) we obtain

$$
\begin{aligned}
L_f(p,q) &= \sum_{j=1}^{M} \sum_{k=0}^{m_j-1} \Bigg(c_{jk} p^{\lambda j} \Big[q^{\lambda j} (\log p + \log q)^k \\
&\quad + (1-q)^{\lambda j} (\log p + \log(1-q))^k \Big] \\
&\quad + c_{jk}(1-p)^{\lambda j} \Big[q^{\lambda j} (\log(1-p) + \log q)^k \\
&\quad + (1-q)^{\lambda j} (\log(1-p) + \log(1-q))^k \Big] \Bigg) \\
&= \sum_{j=1}^{M} \sum_{k=0}^{m_j-1} \sum_{l=0}^{k} c_{jk} \binom{k}{l} \Big[p^{\lambda j} q^{\lambda j} \log^l p \, \log^{k-l} q \\
&\quad + p^{\lambda j}(1-q)^{\lambda j} \log^l p \, \log^{k-l}(1-q) \\
&\quad + (1-p)^{\lambda j} q^{\lambda j} \log^l(1-p) \, \log^{k-l} q \\
&\quad + (1-p)^{\lambda j}(1-q)^{\lambda j} \log^l(1-p) \, \log^{k-l}(1-q) \Big] \\
&= \sum_{j=1}^{M} \sum_{l=0}^{m_j-1} \sum_{k=l}^{m_j-1} c_{jk} \binom{k}{l} \Big[p^{\lambda j} \log^l p + (1-p)^{\lambda j} \log^l(1-p) \Big] \cdot \\
&\quad \cdot \Big[q^{\lambda j} \log^{k-l} q + (1-q)^{\lambda j} \log^{k-l}(1-q) \Big] \\
&= \sum_{j=1}^{M} \sum_{l=0}^{m_j-1} A_{jl}(p) \sum_{k=l}^{m_j-1} c_{jk} \binom{k}{l} A_{j,k-l}(q) \\
&= \sum_{j=1}^{M} \sum_{l=0}^{m_j-1} A_{jl}(p) d_{jl}(q),
\end{aligned}
$$

which is (8.4.7) with (8.4.8).

Because of (8.4.7) and

$$R_f(p,q) = \left(\sum_{j=1}^{M} \sum_{l=0}^{m_j-1} c_{jl} A_{jl}(p) \right) \left(\sum_{r=1}^{M} \sum_{s=0}^{m_r-1} c_{rs} A_{rs}(q) \right)$$

we arrive at (8.4.9) and (8.4.10).

Lemma 8.4.2. *Let* $f :]0,1[\to \mathbf{C}$ *satisfy (8.4.1) for all* $p,q \in]0,1[$, *and suppose* f *is a polynomial of the form*

$$f(p) = \sum_{k=0}^{n} a_k p^k, \tag{8.4.13}$$

$p \in]0,1[$, $a_n \neq 0$, $a_k \in \mathbf{C}$, $0 \leq k \leq n$. *Then we have*

$$L_f(p,q) = \sum_{k=0}^{n} p^k \left[a_k A_k(q) + (-1)^k \sum_{i=k}^{n} a_i \binom{i}{k} A_i(q) \right], \tag{8.4.14}$$

$$R_f(p,q) = T_f(p) T_f(q) = T_f(q) \cdot \sum_{k=0}^{n} b_k p^k, \tag{8.4.15}$$

where

$$T_f(p) = \sum_{k=0}^{n} b_k p^k = \sum_{k=0}^{n} a_k \left[p^k (1 + (-1)^k) + \sum_{l=0}^{k-1} \binom{k}{l} (-1)^l p^l \right], \tag{8.4.16}$$

and

$$a_n (1 + (-1)^n) A_n(q) = a_n (1 + (-1)^n) T_f(q). \tag{8.4.17}$$

If in addition n *is an odd integer, then the following four equations hold.*

$$\left. \begin{aligned} 2a_{n-1} & A_{n-1}(q) + na_n A_n(q) \\ & = T_f(q) b_{n-1} \\ & = T_f(q)(2a_{n-1} + na_n) \quad (n \geq 1), \end{aligned} \right\} \tag{8.4.18}$$

$$\left. \begin{aligned} 2a_{n-3} & A_{n-3}(q) + a_{n-2}(n-2) A_{n-2}(q) \\ & \quad + a_{n-1} \binom{n-1}{2} A_{n-1}(q) + a_n \binom{n}{3} A_n(q) \\ & = T_f(q) b_{n-3} \\ & = T_f(q) \Big[2a_{n-3} + a_{n-2}(n-2) \\ & \qquad + a_{n-1} \binom{n-1}{2} + a_n \binom{n}{3} \Big], \end{aligned} \right\} \tag{8.4.19}$$

$$4a_{n-1} + n^2 a_n = b_{n-1}^2 = (2a_{n-1} + na_n)^2 \quad (n \geq 1), \qquad (8.4.20)$$

and

$$
\left.
\begin{aligned}
2a_{n-1} &\binom{n-1}{2} + n\binom{n}{3} a_n \\[1em]
&= b_{n-1}b_{n-3} \\[0.5em]
&= (2a_{n-1} + na_n)\left[2a_{n-3} + a_{n-2}(n-2)\right. \\[0.5em]
&\quad \left. + a_{n-1}\binom{n-1}{2} + a_n\binom{n}{3}\right] \quad (n \geq 3).
\end{aligned}
\right\} \qquad (8.4.21)
$$

Proof: Since f is a polynomial of degree n we get by the Taylor formula

$$f(x+h) = \sum_{i=0}^{n} \frac{f^{(i)}(x)}{i!} h^i \qquad (8.4.22)$$

for all $x, h, x+h \in]0,1[$. Moreover, remembering that

$$\left(p^i\right)^{(k)} = \binom{i}{k} k!\, p^{i-k}$$

$(i, k \in \mathbb{N}, i \geq k)$ we have

$$f^{(k)}(p) = \sum_{i=k}^{n} a_i \binom{i}{k} k! p^{i-k}, \qquad (8.4.23a)$$

and

$$f^{(k)}(1-p) = \sum_{i=k}^{n} a_i \binom{i}{k} k!(-1)^k (1-p)^{i-k}. \qquad (8.4.23b)$$

Using (8.4.22) and (8.4.23) we obtain

$$
\begin{aligned}
L_f(p,q) &= f(pq) + f(p(1-q)) + f(q-pq) + f((1-q)-p(1-q)) \\
&= \sum_{k=0}^{n} p^k \Bigg[a_k \big(q^k + (1-q)^k \big) \\
&\quad + \frac{f^{(k)}(q)}{k!} (-1)^k q^k + \frac{f^{(k)}(1-q)}{k!} (-1)^k (1-q)^k \Bigg] \\
&= \sum_{k=0}^{n} p^k \Bigg[a_k A_k(q) + (-1)^k \left(\frac{f^{(k)}(q)}{k!} q^k + \frac{f^{(k)}(1-q)}{k!} (1-q)^k \right) \Bigg] \\
&= \sum_{k=0}^{n} p^k \Bigg[a_k A_k(q) + \frac{(-1)^k q^k}{k!} \sum_{i=k}^{n} a_i \binom{i}{k} k! q^{i-k} \\
&\quad + \frac{(-1)^k (1-q)^k}{k!} \sum_{i=k}^{n} a_i \binom{i}{k} k! (1-q)^{i-k} \Bigg] \\
&= \sum_{k=0}^{n} p^k \Bigg[a_k A_k(q) + (-1)^k \sum_{i=k}^{n} a_i \binom{i}{k} A_i(q) \Bigg],
\end{aligned}
$$

which is (8.4.14).

Substituting (8.4.13) into the right-hand side of (8.4.1) we get (using (8.4.4))

$$
\begin{aligned}
R_f(p,q) \\
&= T_f(p) T_f(q) \\
&= T_f(q) \sum_{k=0}^{n} a_k (p^k + (1-p)^k) \\
&= T_f(q) \sum_{k=2}^{n} a_k \left(p^k + \sum_{l=0}^{k} \binom{k}{l} (-1)^l p^l \right) \\
&= T_f(q) \sum_{k=2}^{n} a_k \Bigg[p^k (1 + (-1)^k) + \sum_{l=0}^{k-1} \binom{k}{l} (-1)^l p^l \Bigg].
\end{aligned}
$$

Thus $T_f(p)$ is a polynomial of the form $\sum_{k=0}^{n} b_k p^k$ where the coefficients b_k can be determined from (8.4.16).

By comparison of the coefficients of p^n, p^{n-1} and p^{n-3} in

$$L_f(p,q) = R_f(p,q)$$

we obtain the equations (see (8.4.14)-(8.4.16))

$$a_n A_n(q) + (-1)^n a_n \binom{n}{n} A_n(q) = a_n(1 + (-1)^n) T_f(q), \qquad (8.4.24)$$

$$\left.\begin{array}{l} a_{n-1} A_{n-1}(q) \\[2mm] + (-1)^{n-1} \left[a_{n-1} \binom{n-1}{n-1} A_{n-1}(q) + a_n \binom{n}{n-1} A_n(q) \right] \\[2mm] = T_f(q) b_{n-1} \\[2mm] = T_f(q) \left[a_{n-1}(1 + (-1)^{n-1}) + \binom{n}{n-1}(-1)^{n-1} a_n \right], \\[2mm] (n \geq 1) \end{array}\right\} \qquad (8.4.25)$$

and

$$\left.\begin{array}{l} a_{n-3} A_{n-3}(q) + (-1)^{n-3} \left[a_{n-3} \binom{n-3}{n-3} A_{n-3}(q) \right. \\[2mm] \quad + a_{n-2} \binom{n-2}{n-3} A_{n-2}(q) \\[2mm] \quad + a_{n-1} \binom{n-1}{n-3} A_{n-1}(q) + a_n \binom{n}{n-3} A_n(q) \Big] \\[2mm] = T_f(q) b_{n-3} \\[2mm] = T_f(q) \left[(1 + (-1)^{n-3}) a_{n-3} + (-1)^{n-3} \left(a_{n-2} \binom{n-2}{n-3} \right. \right. \\[2mm] \quad \left. \left. + a_{n-1} \binom{n-1}{n-3} + a_n \binom{n}{n-3} \right) \right], \quad (n \geq 3) \end{array}\right\} \qquad (8.4.26)$$

respectively. Equation (8.4.24) gives (8.4.17), and if n is odd then (8.4.25)–(8.4.26) are exactly (8.4.18)–(8.4.19).

Substituting (8.4.16) and

$$A_i(q) = q^i + \sum_{l=0}^{i} \binom{i}{l}(-1)^l q^l = (1 + (-1)^i) q^i + \sum_{l=0}^{i-1} \binom{i}{l}(-1)^l q^l$$

into (8.4.18) and comparing the coefficients of q^{n-1} and q^{n-3}, we arrive at

$$2a_{n-1}(1+(-1)^{n-1}) + na_n \binom{n}{n-1}(-1)^{n-1} = b_{n-1}^2$$

$$= (2a_{n-1} + na_n)\left[a_{n-1}(1+(-1)^{n-1}) + \binom{n}{n-1}(-1)^{n-1}a_n\right]$$

and

$$2a_{n-1}\binom{n-1}{n-3}(-1)^{n-3} + na_n\binom{n}{3}(-1)^{n-3} = b_{n-1}b_{n-3}$$

$$= (2a_{n-1} + na_n)\left((1+(-1)^n)a_{n-3} + (-1)^{n-3}\left[a_{n-2}\binom{n-2}{n-3}\right.\right.$$

$$\left.\left. + a_{n-1}\binom{n-1}{n-3} + a_n\binom{n}{n-3}\right]\right),$$

respectively. Since n is odd, these last two equations are exactly (8.4.20) and (8.4.21).

Now we are ready to prove the main results of Section 8.4.

Theorem 8.4.3. *If $f :]0,1[\to \mathbf{C}$ is a measurable solution of (8.4.1), then f has one of the forms*

$$f(p) = p^\alpha, \tag{8.4.26}$$

$$f(p) = (1-s^2)p + \frac{s^2 - s}{2}, \tag{8.4.27}$$

$$f(p) = \frac{t^2 - 1}{3}p^3 - \frac{t^2 - t - 2}{2}p^2 + \frac{t^2 - 3t + 2}{6}p, \tag{8.4.28}$$

$$f(p) = \frac{u^2 - 1}{15}p^5 - \frac{u^2 - 3u - 4}{6}p^4 + \frac{u^2 - 5u + 4}{270}(40p^3 - 15p^2 + 2p) \tag{8.4.29}$$

for all $p \in]0,1[$, where $\alpha, s, t, u \in \mathbf{C}$.

If f is real-valued then the solutions are again given by (8.4.26)–(8.4.29) where $\alpha, s, t, u \in \mathbb{R}$.

Proof: A straightforward (but tedious) calculation shows that f given by (8.4.26) to (8.4.29) satisfies (8.4.1). The proof of the converse is rather long, so we give first an outline of the steps involved.

Step 1. Let $f :]0,1[\to \mathbf{C}$ be a measurable solution of (8.4.1). If

$$T_f(p) = f(p) + f(1-p) = \text{constant}, \quad p \in]0,1[, \tag{8.4.30}$$

then f has the form (8.4.27).

Step 2. If

$$T_f(p) \neq \text{constant}, \quad p \in \,]0, 1[, \tag{8.4.31}$$

then either there exists $\alpha \notin \mathbb{N} \cup \{0\}$ such that

$$f(p) = p^\alpha + P(p), \quad p \in \,]0, 1[, \tag{8.4.32}$$

or

$$f(p) = P(p), \quad p \in \,]0, 1[, \tag{8.4.33}$$

where in either case P is a polynomial.

Step 3. In case (8.4.32) $P = 0$ so that f has the form (8.4.26).

Step 4. In case (8.4.33) let

$$f(p) = \sum_{k=0}^{n} a_k p^k, \tag{8.4.13}$$

$p \in \,]0, 1[$, $a_n \neq 0$, $a_k \in \mathbf{C}$, $0 \le k \le n$. Then the continuous extension $\overline{f} : [0, 1] \to \mathbf{C}$ of f satisfies

$$\overline{f}(0) = 0, \qquad \overline{f}(1) = 1. \tag{8.4.34}$$

Step 5. In (8.4.13) if n is even, or if n is odd and $a_{n-1} = 0$, then $f(p) = p^n$. Hence f has the form (8.4.26) again.

Step 6. In the case $a_{n-1} \neq 0$ (and n odd with $a_n \neq 0$) f is a polynomial of degree 5, 3, or 1.

Step 7. The three cases of Step 6 lead to the explicit forms of f given by (8.4.28)-(8.4.29).

Verification of Step 1. Let $f(p) + f(1-p) = \text{constant} = 1 - s$ for some $s \in \mathbf{C}$. Because of $L_f(p, q) = (1-s)^2$ the function $g :]0, 1[\to \mathbf{C}$ defined by

$$g(p) = f(p) - \frac{(1-s)^2}{4}$$

satisfies (6.3.40). Thus g has the form $g(p) = 4ap - a$ for some $a \in \mathbf{C}$ (see Corollary 6.3.4) and we have

$$f(p) = 4ap - a + \frac{(1-s)^2}{4}, \quad p \in \,]0, 1[. \tag{8.4.35}$$

The condition $f(p) + f(1-p) = 1 - s$ yields $4a = 1 - s^2$ so that (8.4.35) goes over into (8.4.27).

Verification of Step 2. If T_f is not constant then $T_f \neq 0$. Since equation (8.4.1) is of the form $L_f(p,q) = T_f(p)T_f(q)$ and since f and T_f are measurable, Theorem 8.2.1 is applicable (with $N = 1$) and we get that f has the form (8.4.6) where $M, m_1, \ldots, m_M \in \mathbb{N}$ and $\lambda_1, \ldots, \lambda_M, c_{jk} \in \mathbb{C}$, $1 \leq j \leq M, 0 \leq k \leq m_j - 1$. By Lemma 8.4.1 we have

$$\sum_{j=1}^{M} \sum_{l=0}^{m_j-1} A_{jl}(p) D_{jl}(q) = 0 \qquad (8.4.9)$$

for all $q \in]0,1[$, where D_{jl} is given by (8.4.10). If $m_j = 1$ for all $1 \leq j \leq M$ then (8.4.9) has the form

$$\sum_{j=1}^{M} A_{j0}(p) D_{j0}(q) = 0 \qquad (8.4.36)$$

But (8.4.36) is valid also for $m_j \geq 2$. Indeed, applying Theorem 8.2.3 to (8.4.9), we find that

$$D_{jl}(q) = 0 \qquad (1 \leq l \leq m_j - 1, \ q \in]0,1[) \qquad (8.4.37)$$

for all $1 \leq j \leq M$ with $m_j \geq 2$. Note that $D_{jl}(q) = 0$ means that

$$D_{jl}(q) = \sum_{k=l}^{m_j-1} c_{jk} \binom{k}{l} A_{j,k-l}(q) - c_{jl} \sum_{r=1}^{M} \sum_{s=0}^{m_r-1} c_{rs} A_{rs}(q) = 0$$

(see (8.4.8) and (8.4.10) for the definition of D_{jl}). Now we prove that

$$c_{jl} = 0 \qquad (1 \leq l \leq m_j - 1) \qquad (8.4.38)$$

for $1 \leq j \leq M$. If $m_j = 1$ there is nothing to prove, so assume $m_j \geq 2$. Let $1 \leq j \leq M$ and put $l = m_j - 1 > 0$ into (8.4.37) to arrive at (using (8.4.8) and (8.4.10))

$$c_{j,m_j-1} A_{j0}(q)$$

$$- c_{j,m_j-1} \left(c_{j0} A_{j0}(q) + \sum_{\substack{r=1 \\ (r,s) \neq (j,0)}}^{M} \sum_{s=0}^{m_r-1} c_{rs} A_{rs}(q) \right)$$

$$= c_{j,m_j-1} \left[(1 - c_{j0}) A_{j0}(q) - \sum_{\substack{r=1 \\ (r,s) \neq (j,0)}}^{M} \sum_{s=0}^{m_r-1} c_{rs} A_{rs}(q) \right]$$

$$= 0$$

From (8.4.39) we claim that $c_{j,m_j-1} = 0$. For if this were not the case then the bracket in (8.4.39) would be zero for all $q \in]0,1[$. But then Theorem 8.2.3 yields

$$c_{js} = 0 \text{ for } s = 1, \ldots, m_j - 1.$$

In particular we have $c_{j,m_j-1} = 0$ contradicting our assumption. Suppose now that we have proved

$$c_{j,m_j-1} = c_{j,m_j-2} = \ldots = c_{jt} = 0 \qquad (t \geq 2). \qquad (8.4.40)$$

Then we show that

$$c_{j,t-1} = 0$$

is valid, too. Now (8.4.37) and (8.4.40) lead to

$$0 = D_{j,t-1}(q) = c_{j,t-1} \binom{t-1}{t-1} A_{j0}(q)$$

$$- c_{j,t-1} \left[c_{j0} A_{j0}(q) + \sum_{\substack{r=1 \\ (r,s) \neq (j,0)}}^{M} \sum_{s=0}^{m_r-1} c_{rs} A_{rs}(q) \right] \qquad (8.4.41)$$

$$= c_{j,t-1} \left[(1 - c_{j0}) A_{j0}(q) - \sum_{\substack{r=1 \\ (r,s) \neq (j,0)}}^{M} \sum_{s=0}^{m_r-1} c_{rs} A_{rs}(q) \right]$$

$$= 0.$$

If $c_{j,t-1} \neq 0$ then the bracket in (8.4.41) would be zero, so that Theorem 8.2.3 implies

$$c_{js} = 0, \quad s = 1, \ldots, t-1$$

(note that already $c_{js} = 0$ for $s = t, t+1, \ldots, m_j - 1$). But this would give the contradiction $c_{j,t-1} = 0$. Thus (8.4.40) implies $c_{j,t-1} = 0$ and (8.4.38) is proven by letting t run over $m_j - 1$, $m_j - 2, \ldots, 2$.

From (8.4.38) we know that nearly all coefficients c_{jk} in (8.4.6) are zero, that is f has the form

$$f(p) = \sum_{j=1}^{M} c_{j0} \, p^{\lambda j}. \qquad (8.4.42)$$

We now consider two cases.

Case 1. If there exists $j \in \mathbb{N}$ with $\lambda_j \notin \mathbb{N} \cup \{0\}$ such that $c_{j0} \neq 0$ then we conclude from (8.4.36)

$$\sum_{j=1}^{M} A_{j0}(p) D_{j0}(q)$$

$$= \sum_{j=1}^{M} A_{j0}(p) \left(c_{j0} A_{j0}(q) - c_{j0} \sum_{r=1}^{M} c_{r0} A_{r0}(q) \right) \qquad (8.4.43)$$

$$= \sum_{j=1}^{M} c_{j0} \left(A_{j0}(q) - \sum_{r=1}^{M} c_{r0} A_{r0}(q) \right) A_{j0}(p) = 0.$$

Since $\lambda_j \notin \mathbb{N} \cup \{0\}$ the coefficient of $A_{j0}(p)$ in (8.4.43) is zero by Theorem 8.2.3 again. That is (because of $c_{j0} \neq 0$)

$$A_{j0}(q) - \sum_{r=1}^{M} c_{r0} A_{r0}(q) = A_{j0}(q)(1 - c_{j0}) - \sum_{\substack{r=1 \\ r \neq j}}^{M} c_{r0} A_{r0}(q) = 0. \qquad (8.4.44)$$

Now Theorem 8.2.3 implies $c_{j0} = 1$ (since $\lambda_j \notin \mathbb{N} \cup \{0\}$) and $c_{r0} = 0$ for those r, $1 \leq r \leq M, r \neq j$ with $\lambda_j \notin \mathbb{N} \cup \{0\}$. Thus (8.4.42) goes over into (8.4.32) with $\alpha = \lambda_j$.

Case 2. If for all $j \in \mathbb{N}$ with $\lambda_j \notin \mathbb{N} \cup \{0\}$ we have $c_{j0} = 0$, then $f(p) = \sum_{j \in F} c_{j0} p^{\lambda_j}$ where $F = \{j : 1 \leq j \leq M, \lambda_j \in \mathbb{N} \cup \{0\}\}$; that is, f is a polynomial. Thus Step 2 is verified.

Verification of Step 3. If f has the representation

$$f(p) = p^{\alpha} + P(p) = p^{\alpha} + \sum_{k=0}^{n} a_k p^k$$

$(\alpha_k \in \mathbf{C}, 0 \leq k \leq n)$ substitution of this representation into $L_f(p, q) = R_f(p, q)$ yields

$$A_{\alpha}(p) A_{\alpha}(q) + \sum_{k=0}^{n} a_k A_k(p) A_k(q)$$

$$= A_{\alpha}(p) A_{\alpha}(q) + \left(\sum_{k=0}^{n} a_k A_k(p) \right) \left(\sum_{l=0}^{n} a_l A_l(q) \right)$$

$$+ A_{\alpha}(p) \sum_{k=0}^{n} a_k A_k(q) + A_{\alpha}(q) \sum_{k=0}^{n} a_k A_k(p),$$

or equivalently

$$A_\alpha(p)\left(\sum_{k=0}^{n} a_k A_k(q)\right)$$

$$+ \sum_{k=0}^{n} A_k(p)\left[a_k A_\alpha(q) - a_k A_k(q) + a_k \sum_{l=0}^{n} a_l A_l(q)\right] = 0. \qquad (8.4.45)$$

Because of $\alpha \notin \mathbb{N} \cup \{0\}$ Theorem 8.2.3 implies

$$\sum_{k=0}^{n} a_k A_k(q) = 0, \qquad q \in]0,1[, \qquad (8.4.46)$$

so that (8.4.45) is simplified to

$$\sum_{k=0}^{n} a_k A_k(p) A_k(q) = 0. \qquad (8.4.47)$$

This means that $P(p) = \sum_{k=0}^{n} a_k p^k$ is a measurable solution of (6.3.40), and Corollary 6.3.4 implies the representation

$$P(p) = 4ap - a, \qquad p \in]0,1[, \qquad (8.4.48)$$

for some constant a. Putting P given by (8.4.48) into (8.4.46) we get

$$4a - 2a = 0.$$

Hence $a = 0$ so that $P = 0$ (see (8.4.48)) and $f(p) = p^\alpha$, $p \in]0,1[$.

Verification of Step 4. In case (8.4.33) f is a polynomial satisfying (8.4.1) for all $p, q \in]0,1[$ and therefore its continuous extension $\overline{f}$ will satisfy (8.4.1) for all $p, q \in [0,1]$, too. Putting $q = 0$ into (8.4.1) we get

$$[\overline{f}(p) + \overline{f}(1-p)][\overline{f}(1) + \overline{f}(0) - 1] - 2\overline{f}(0) = 0.$$

Since $\{\overline{f}(p)+\overline{f}(1-p), 1\}$ is linearly independent (note that $p \to \overline{f}(p)+\overline{f}(1-p)$ is not constant on $[0,1]$ since T_f is not), we get (8.4.34).

Verification of Step 5. If f has the representation (8.4.13) with $a_n \neq 0$, we can apply Lemma 8.4.2. If n is even then (8.4.17) implies

$$2a_n A_n(q) = 2a_n T_f(q).$$

Moreover if n is odd and $a_{n-1} = 0$ then (8.4.18) gives the same result. Since $a_n \neq 0$ we get in both cases $T_f(q) = A_n(q)$ so that (8.4.1) goes over into

$$L_f(p, q) = (p^n + (1-p)^n)(q^n + (1-q)^n).$$

Therefore $g(p) := f(p) - p^n$ satisfies (6.4.30), hence

$$f(p) = p^n + 4ap - a, \quad p \in]0, 1[, \tag{8.4.49}$$

for some constant a. But the condition $T_f(q) = f(q) + f(1-q) = A_n(q)$ results in $a = 0$ so that f in (8.4.49) has the form (8.4.26) (with $\alpha = n$).

Verification of Step 6. Now we know that the solution f of (8.4.1) has the form (8.4.13) where n is odd and $a_n \cdot a_{n-1} \neq 0$. We prove that the assumption $n \geq 7$ leads to $a_n a_{n-1} = 0$ which is a contradiction. (Hence $n < 7$ and Step 6 is proven.) Let us now suppose $n \geq 7$. By Lemma 8.4.2 we know that the equations (8.4.18)-(8.4.21) hold. Moreover, from (8.4.20) we conclude that

$$b_{n-1} = 2a_{n-1} + na_n \neq 0. \tag{8.4.50}$$

For, if $2a_{n-1} + na_n = 0$ then (8.4.20) implies $4a_{n-1} + n^2 a_n = 0$. Thus $-2a_{n-1} = na_n = \frac{n^2}{2} a_n$, or $na_n \left(1 - \frac{n}{2}\right) = 0$. But this implies the contradiction $n = 2$. Now we eliminate $T_f(q)$ from (8.4.18) and (8.4.19) and get

$$2a_{n-3} A_{n-3}(q) + a_{n-2}(n-2) A_{n-2}(q)$$

$$+ a_{n-1} \binom{n-1}{2} A_{n-1}(q) + a_n \binom{n}{3} A_n(q) \tag{8.4.51}$$

$$= \frac{b_{n-3}}{b_{n-1}} (2a_{n-1} A_{n-1}(q) + na_n A_n(q)).$$

But the ratio $\frac{b_{n-3}}{b_{n-1}}$ can be determined from (8.4.20) and (8.4.21) which give

$$\frac{b_{n-3}}{b_{n-1}} = \frac{b_{n-3} b_{n-1}}{b_{n-1}^2} = \frac{2a_{n-1} \binom{n-1}{2} + n \binom{n}{3} a_n}{4a_{n-1} + n^2 a_n}. \tag{8.4.52}$$

Substitution of (8.4.52) into (8.4.51) and rearrangements of the terms leads to

$$\alpha_n A_n(q) + \alpha_{n-1} A_{n-1}(q) + \alpha_{n-2} A_{n-2}(q) + \alpha_{n-3} A_{n-3}(q) = 0$$

for all $q \in]0, 1[$, where

$$\alpha_{n-3} = 2a_{n-3} b_{n-1}^2, \alpha_{n-2} = (n-2)a_{n-2} b_{n-1}^2,$$

$$\alpha_{n-1} = n \binom{n}{3} a_n a_{n-1}, \alpha_n = -2 \binom{n}{3} a_n a_{n-1}. \tag{8.4.53}$$

Now Theorem 8.2.5 (a) implies $\alpha_n = \alpha_{n-1} = \alpha_{n-2} = \alpha_{n-3} = 0$. But then $\alpha_n = -2\binom{n}{3} a_n a_{n-1} = 0$ is a contradiction since $a_n a_{n-1} \neq 0$.

Verification of Step 7. We have shown that f has the form

$$f(p) = \sum_{k=0}^{n} a_k p^k, \tag{8.4.13}$$

where $a_k \in \mathbf{C}$, $a_n a_{n-1} \neq 0$ and $n \in \{1,3,5\}$. Now (8.4.18), (8.4.20) and (8.4.50) imply

$$
\begin{aligned}
& f(p) + f(1-p) \\
& = T_f(p) \\
& = (4a_{n-1} + n^2 a_n)^{-\frac{1}{2}} (2a_{n-1} A_{n-1}(p) + n a_n A_n(p))
\end{aligned}
\tag{8.4.54}
$$

If we substitute (8.4.54) and (8.4.13) into (8.4.1) we get an equation of the form

$$\sum_{k=0}^{n} \gamma_k A_k(p) = 0,$$

from which we conclude (using Theorem 8.2.5) the explicit form of the coefficients a_k, $0 \leq k \leq n$, $n \in \{1,3,5\}$.

Case n = 1. In this case we have

$$f(p) = a_1 p + a_0, \quad a_1 a_0 \neq 0.$$

But (8.4.34) gives immediately $a_0 = 0$, a contradiction. Hence this case cannot occur.

Case n = 3. Now f has the representation

$$f(p) = a_3 p^3 + a_2 p^2 + a_1 p + a_0 \tag{8.4.55}$$

together with $a_2 a_3 \neq 0$. Using first (8.4.20) and then (8.4.21) with $n = 3$ we get

$$4a_2 + 9a_3 = (2a_2 + 3a_3)^2 \tag{8.4.56}$$

and

$$(2a_2 + 3a_3) = (2a_2 + 3a_3)(2a_0 + a_1 + a_2 + a_3) \tag{8.4.57}$$

where $2a_2 + 3a_3 \neq 0$ by (8.4.50). Thus (8.4.57) gives

$$2a_0 + a_1 + a_2 + a_3 = 1$$

which simplifies because of $\overline{f}(0) = 0$ (that is $a_0 = 0$) to

$$a_1 + a_2 + a_3 = 1. \tag{8.4.58}$$

(Equation (8.4.58) follows also from $\overline{f}(1) = 1$ and $\overline{f}(0) = 0$). Putting $2a_2 + 3a_3 = t + 1$, $t \in \mathbf{C}$, condition (8.4.56) leads to the linear system

$$2a_2 + 3a_3 = t + 1$$
$$4a_2 + 9a_3 = t^2 + 2t + 1$$

in the unknowns a_2 and a_3, with solutions

$$a_2 = \frac{1}{2}(-t^2 + t + 2) = -\frac{1}{2}(t+1)(t-2), \qquad a_3 = \frac{t^2 - 1}{3}.$$

The condition (8.4.58) leads to

$$a_1 = \frac{1}{6}(t^2 - 3t + 2) = \frac{1}{6}(t-1)(t-2)$$

so that f given by (8.4.55) has the representation (8.4.28). The excluded values for t, which result from the conditions $a_2 a_3 \neq 0$ and $4a_2 + 9a_3 \neq 0$, are $t = 1$, $t = -1$ and $t = 2$. In these cases we get solutions of the form (8.4.26) with $\alpha = 2$, $\alpha = 1$ and $\alpha = 3$, respectively.

Case n = 5. Since f has the representation

$$f(p) = a_5 p^5 + a_4 p^4 + a_2 p^3 + a_2 p^2 + a_1 p + a_0 \tag{8.4.59}$$

(together with $a_4 a_5 \neq 0$) we get from (8.4.20) and (8.4.50) (with $n = 5$)

$$4a_4 + 25a_5 = (2a_4 + 5a_5)^2 \neq 0. \tag{8.4.60}$$

As in the case $n = 3$ we set $2a_4 + 5a_5 = u + 1$, $u \in \mathbf{C}$, and get from (8.4.60) the linear system

$$\begin{aligned} 2a_4 + 5a_5 &= u + 1 \\ 4a_4 + 25a_5 &= u^2 + 2u + 1, \end{aligned} \tag{8.4.61}$$

with the solutions

$$a_5 = \frac{u^2 - 1}{15} = \frac{(u-1)(u+1)}{15}, \tag{8.4.62}$$

$$a_4 = \frac{1}{6}(-u^2 + 3u + 4) = -\frac{1}{6}(u-4)(u+1). \tag{8.4.63}$$

To determine the missing coefficients we put (8.4.20) into (8.4.18) (with $n = 5$) to arrive at

$$f(p) + f(1 - p) = (4a_4 + 25a_5)^{-\frac{1}{2}}(2a_4 A_4(p) + 5a_5 A_5(p)), \qquad (8.4.64)$$

$p \in]0, 1[$. Substituting (8.4.59) and (8.4.64) into (8.4.1) we obtain

$$\sum_{k=0}^{5} a_k A_k(q) A_k(p)$$

$$= \left(\frac{5a_5 A_5(q) + 2a_4 A_4(q)}{25a_5 + 4a_4} \right) (5a_5 A_5(p) + 2a_4 A_4(p)),$$

$(p, q \in]0, 1[)$, or rearranging the terms,

$$\sum_{k=0}^{5} \beta_k A_k(p) = 0, \quad p \in]0, 1[, \qquad (8.4.65)$$

where

$$\beta_k := a_k A_k(q) \qquad (k = 0, 1, 2, 3), \qquad (8.4.66)$$

$$\beta_4 := \frac{-5a_4 a_5}{25a_5 + 4a_4}[2A_5(q) - 5A_4(q)], \qquad (8.4.67)$$

$$\beta_5 := \frac{2a_4 a_5}{25a_5 + 4a_4}[2A_5(q) - 5A_4(q)]. \qquad (8.4.68)$$

Using the identity

$$2A_5(q) - 5A_4(q) = -4A_3(q) + A_2(q), q \in]0, 1[,$$

(verified by direct calculation, or by an application of Theorem 8.2.5 (b) with $\beta_1 = \beta_0 = 0$) we apply Theorem 8.2.5 (b) to (8.4.65)-(8.4.68) and get from (8.2.45)

$$a_3 A_3(q) = -4a_2 A_2(q) - 10[a_1 A_1(q) + 2a_0 A_0(q)] \qquad (8.4.69)$$

and

$$\frac{-5a_4 a_5}{25a_5 + 4a_4}[-4A_3(q) + A_2(q)]$$
$$= 5a_2 A_2(q) + 15[a_1 A_1(q) + 2a_0 A_0(q)]. \qquad (8.4.70)$$

The last two equations are of the form (8.2.46) and we can apply Theorem 8.2.5 (c) both to equation (8.4.69) and (8.4.70). Using (8.2.47) two times we arrive at

$$4a_2 = -3(10a_1 + 40a_0), \tag{8.4.71}$$

$$a_3 = 2(10a_1 + 40a_0), \tag{8.4.72}$$

and

$$\frac{-5a_4a_5}{25a_5 + 4a_4} - 5a_2 = -3(-15a_1 - 60a_0), \tag{8.4.73}$$

$$\frac{20a_4a_5}{25a_5 + 4a_4} = 2(-15a_1 - 60a_0). \tag{8.4.74}$$

The comparison of (8.4.71) and (8.4.74) gives (together with the condition $a_0 = 0$ which results from $\overline{f}(0) = 0$ (see Step 4))

$$a_2 = \frac{5a_4a_5}{25a_5 + 4a_4} \quad \text{and} \quad a_1 = -\frac{2}{3}\frac{a_4a_5}{25a_5 + 4a_4},$$

while the comparison of (8.4.72) and (8.4.74) yields

$$a_3 = -\frac{40}{3}\frac{a_4a_5}{25a_5 + 4a_4}.$$

Now, substituting

$$\frac{a_4a_5}{25a_5 + 4a_4} = -\frac{1}{6 \cdot 15}\frac{(u+1)^2(u-1)(u-4)}{(u+1)^2} \tag{8.4.75}$$
$$= -\frac{1}{90}(u-4)(u-1)$$

(see (8.4.61)–(8.4.63)) into the last three equations we obtain, together with (8.4.62) and (8.4.63), the solution (8.4.29).

The restrictions $a_4a_5 \neq 0$ and $2a_4 + 5a_5 = 0$ imply $u \neq 1$, $u \neq -1$, $u \neq 4$ (see (8.4.61)–(8.4.63)). But these restrictions can be omitted, since $u = 1$ and $u = 4$ in (8.4.29) yield solution (8.4.26) with $\lambda = 4$ and $\lambda = 5$. If $u = -1$, (8.4.29) goes over into

$$f(p) = \frac{1}{27}(40p^3 - 15p^2 + 2p)$$

which is (8.4.28) with $t = \frac{7}{3}$. Thus (8.4.26)–(8.4.29) are all solutions of (8.4.1) for all $p \in \mathbf{C}$.

(b) The solution (8.4.26) is real valued if and only if $\alpha \in \mathbb{R}$. Moreover, a polynomial of a real variable with complex coefficients is real valued if and only if its coefficients are real. Thus the solutions (8.4.27)-(8.4.29) are real valued if and only if $s, t, u \in \mathbb{R}$.

Now we present (in the 1-dimensional case) all normalized, λ-parametric (2,2)-additive sum form information measures with measurable generating function. Because of Theorem 8.4.3, the class of these information measures is wider than the entropies of degree α ($\alpha \in \mathbb{R}\backslash\{1\}$) if $\lambda = 2^{1-\alpha} - 1$. The following result is due to Losonczi (1993a).

Theorem 8.4.4. *Let $\lambda = 2^{1-\alpha} - 1$, $\alpha \in \mathbb{R}\backslash\{1\}$. Suppose that $I_n : \Gamma_n^o \to \mathbb{R}$ has the sum form for some measurable generating function $F :]0, 1[\to \mathbb{R}$ and that $\{I_n\}$ is nonconstant λ-parametric (2,2)-additive satisfying the normalization condition $I_2 \left(\frac{1}{2}, \frac{1}{2}\right) = 1$. Then $\{I_n\}$ has one of the following forms:*

$$I_n(P) = H_n^\alpha(P), \tag{8.4.76}$$

$$I_n(P) = -\frac{1}{2}(\lambda + 1)H_n^0(P) - \frac{1}{2}(3\lambda + 1), \tag{8.4.77}$$

$$I_n(P) = -(4\lambda + 2)H_n^3(P) + (4\lambda + 3)H_n^2(P), \tag{8.4.78}$$

or

$$
\begin{aligned}
I_n(P) = & -3(48\lambda + 82 + 35\lambda^{-1})H_n^5(P) + \frac{7}{2}(96\lambda + 170 + 75\lambda^{-1})H_n^4(P) \\
& - 2(128\lambda + 232 + 105\lambda^{-1})H_n^3(P) \\
& + \frac{1}{2}(128\lambda + 232 + 105\lambda^{-1})H_n^2(P) + 1.
\end{aligned}
\tag{8.4.79}
$$

Proof. By (5.1.1), (8.1.1) and Theorem 8.4.3 we get

$$I_n(P) = \frac{1}{\lambda}\sum_{i=1}^{k}[f(p_i) - p_i] \tag{8.4.80}$$

where f is given by (8.4.26)-(8.4.29). Using the fact that $H_n^5, H_n^4, \ldots, H_n^0$ have the generating functions $F_5(p) = -\frac{16}{15}(p^5 - p)$, $F_4(p) = -\frac{8}{7}(p^4 - p)$, $F_3(p) = -\frac{4}{3}(p^3 - p)$, $F_2(p) = -2(p^2 - p)$, $F_0(p) = 1 - p$, respectively, we get (8.4.76)-(8.4.79) from (8.4.80) and (8.4.26)-(8.4.29) by putting $s = -\lambda$, $t = -4\lambda - 1$ and $u = -48\lambda - 41$.

Remark 8.4.5. We give some historical remarks concerning Theorems 8.3.5 and 8.4.3. If $k = 1$ and $\ell \geq 3, m \geq 3$ then Theorem 8.3.5 was established by Losonczi (1985b) (see Kannappan and Sahoo (1985a, 1986a) for the measurable solutions). If $k = 2$ and $\ell \geq 3, m \geq 3$ then the same result is due to Kannappan and Sahoo (1986d). For arbitrary k, Theorem 8.3.5 was established by Sahoo in his thesis (1986) and also was found independently by Kannappan and Sander (unpublished). In the case $\ell = 2, m \geq 3$ or $\ell \geq 3, m = 2$ the general solution in the 1-dimensional case is essentially due to Kannappan and Sahoo (1991) and also Maksa (1987)). The present version of Theorem 8.3.5 is due to the authors.

In Losonczi (1993b) Theorem 8.4.3 was proved in a more general version. In this chapter we presented a more direct proof. We also want to point out that nontrivial results like Theorem 6.2.1 and Theorem 8.2.1 and many auxiliary result like Theorem 8.2.3, Theorem 8.2.5, Lemma 8.4.1 and Lemma 8.4.2 were used to prove Theorem 8.4.3. In particular, Theorem 8.2.3 was a key tool in these investigations, and it will prove to be essential again in the next chapter.

CHAPTER 9

ADDITIVE SUM FORM INFORMATION MEASURES
OF MULTIPLICATIVE TYPE

9.1. Introduction

In this chapter we consider sum form information measures $\{I_n\}$ of the form

$$I_n(P) = \sum_{i=1}^{n} F(p_i), \qquad P \in \Gamma_n^o \quad (n = 2, 3, \ldots), \tag{5.1.1}$$

where the generating function $F : J \to \mathbb{R}$ satisfies

$$\sum_{i=1}^{\ell} \sum_{j=1}^{m} F(p_i q_j) = \sum_{i=1}^{\ell} M_1(p_i) \sum_{j=1}^{m} F(q_j) + \sum_{j=1}^{m} M_2(q_j) \sum_{i=1}^{\ell} F(p_i) \tag{5.4.7}$$

for all $P \in \Gamma_\ell^o$, $Q \in \Gamma_m^o$, where (ℓ, m) is a fixed pair of positive integers, and where M_1, $M_2 : J \to \mathbb{R}$ are multiplicative. Recall that information measures of the form (5.1.1) satisfying (5.4.7) are (ℓ, m)-additive sum form information measures of multiplicative type (M_1, M_2). In particular, if $M_1(p) = p^\alpha$, $M_2(p) = p^\beta$ we speak of (ℓ, m)-additive sum form information measures of degree (α, β) (since the information measures depend upon the *a priori* chosen parameters α, β).

In Chapter 7, we have already treated the case where M_1 and M_2 are projections. Now we handle the case where M_1 and M_2 are multiplicative maps but not both projections. Note that a multiplicative function M is not a projection if either $M \equiv 0$ or $M \equiv 1$ (that is, M is constant) or M is nonconstant and not additive (see Lemma 1.2.10).

The results in this chapter are in some respect analogous to the results given in Chapter 8. If $\ell \geq 3$ and $m \geq 3$, then the general solution of (5.4.7) is known. If $\ell = m = 2$ then all measurable triplets (F, M_1, M_2) in the 1-dimensional case are known. As in sections 7.2 and 8.4, the behavior of the solutions of equation (5.4.7) is again somewhat surprising, since as in section 8.4 new solutions arise, namely polynomials of odd degree up to 5 (which are different from the solutions in Theorem 8.4.3).

To prove all these results we will make use of a sequence of lemmas which will be given in section 9.2. Finally, in sections 9.3 and 9.4, the solutions of (5.4.7) in the cases $\ell \geq 3$, $m \geq 3$, respectively $\ell = m = 2$, are presented.

Remark 9.1.1. We note here that results in the cases $\ell \geq 3$, $m = 2$, and $\ell = 2$, $m \geq 3$ are known. However, the authors refrain from presenting these proofs as they are tedious and in some cases we do not have a characterization.

9.2. Preliminaries

A necessary part of our analysis of sum form functional equations is the following.

Lemma 9.2.1. *Let* $M : J \to \mathbb{R}$ *be constant multiplicative function (that is, either $M = 0$ or $M = 1$), and suppose that $E : J \times \mathbb{R}^k \to \mathbb{R}$ is additive in the second variable. If $h, k : J \to \mathbb{R}$ satisfy*

$$h(pq) - M(p)h(q) - M(q)h(p) = E(p, q) - k(p) \qquad (9.2.1)$$

for all $p, q \in J$, then

$$h(p) = M(p)L(p) + A(p) + b, \qquad (9.2.2)$$

where $L : \mathcal{P} \to \mathbb{R}$ is logarithmic, $A : \mathbb{R}^k \to \mathbb{R}$ is additive, and b is a constant. Moreover, if h and E are measurable, then A and L are measurable, too.

Proof: First, suppose $M = 0$. Then (9.2.1) becomes

$$h(pq) = E(p, q) - k(p). \qquad (9.2.3)$$

Replacing q by $1 - q$ here, adding the resulting equation to (9.2.3), and using the additivity of E, we arrive at the Pexider equation

$$h(pq) + h(p - pq) = E(p, 1) - 2k(p), \qquad p, q \in J.$$

Thus h has the form given by (9.2.2). Clearly A is measurable if h is.
    Now suppose $M = 1$, in which case (9.2.1) becomes

$$h(pq) - h(q) - h(p) = E(p, q) - k(p), \qquad p, q \in J. \qquad (9.2.4)$$

In this case we replace q by $r + s$, with $(r, s) \in D_2$, and use (9.2.4) three times together with additivity of E to get

$$
\begin{aligned}
h(p(r + s)) - h(r + s) - h(p) &= E(p, r + s) - k(p) \\
&= E(p, r) + E(p, s) - k(p) \\
&= h(pr) - h(r) - h(p) + h(ps) - h(s) - (h - k)(p).
\end{aligned}
$$

That is, defining $L : J \to \mathbb{R}$ and $C : D_2 \to \mathbb{R}$ by

$$
L(p) := (h - k)(p), \qquad p \in J, \tag{9.2.5}
$$

$$
C(r, s) := h(r + s) - h(r) - h(s), \qquad (r, s) \in D_2,
$$

we have

$$
C(pr, ps) = C(r, s) - L(p), \qquad (r, s) \in D_2, \, p \in J.
$$

Applying this last equation three times, we find that

$$
\begin{aligned}
C(r, s) - L(pq) &= C(pqr, pqs) \\
&= C(qr, qs) - L(p) \\
&= C(r, s) - L(q) - L(p)
\end{aligned}
$$

for all $(r, s) \in D_2$ and $p, q \in J$. Thus L is logarithmic: $L(pq) = L(p) + L(q)$ for all $p, q \in J$.

With this information, substituting $h(p) = k(p) + L(p)$ from (9.2.5) into (9.2.4) we arrive after simplification at

$$
k(pq) - k(q) = E(p, q), \qquad p, q \in J. \tag{9.2.6}
$$

Defining $K : D_2 \to \mathbb{R}$ by

$$
K(r, s) := k(r + s) - k(r) - k(s), \qquad (r, s) \in D_2, \tag{9.2.7}
$$

and using (9.2.6) three times, we deduce that

$$
\begin{aligned}
K(pr, ps) &= k(p(r + s)) - k(pr) - k(ps) \\
&= E(p, r + s) + k(r + s) - [E(p, r) + k(r)] \\
&\qquad - [E(p, s) + k(s)] \\
&= k(r + s) - k(r) - k(s) \\
&= K(r, s),
\end{aligned}
$$

where we have used the additivity of E once again. Hence K in (9.2.7) is a Cauchy difference which is homogeneous of degree zero (that is, satisfies (3.2.5) with $M = 1$). Therefore, by Theorem 3.3.1, there exists a constant b such that

$$K(r, s) = -b[1 + 1 - 1] = -b, \qquad (r, s) \in D_2.$$

Comparing this with (9.2.7), we see that

$$k(r + s) = k(r) + k(s) - b, \qquad (r, s) \in D_2,$$

which implies that

$$k(p) = A(p) + b, \qquad p \in J, \tag{9.2.8}$$

for some additive map $A : \mathbb{R}^k \to \mathbb{R}$. Combining this with (9.2.5), we have (9.2.2).

Finally, if h and E are measurable, then (9.2.1) shows that k is measurable also. Then (9.2.8) and (9.2.5), respectively, give the measurability of A and L. This concludes the proof of Lemma 9.2.1.

Remark 9.2.2. Because the left hand side of (9.2.1) is symmetric in p and q, the right hand side of (9.2.1) may be replaced by $E(p, q) - k(q)$ if E is additive in the first variable. We arrive at the same representation (9.2.2) for h.

Lemma 9.2.3. *Let $M_1, M_2 : J \to \mathbb{R}$ be different multiplicative functions, at least one of which is not a projection, and let $\ell \geq 3$, $m \geq 3$ be fixed integers. If $F : J \to \mathbb{R}$ satisfies (5.4.7) for all $P \in \Gamma_\ell^o$ and $Q \in \Gamma_m^o$, then*

$$\sum_{j=1}^{m}\{F(q_j) - a[M_1(q_j) - M_2(q_j)]\} = d \tag{9.2.9}$$

for some constants a, d. Moreover $d = 0$ in any of the following cases: (i) if $\ell = m$, (ii) if $\ell \neq m$, M_1 and M_2 are constant, (iii) if $\ell \neq m$, $\ell \geq 3$, $m \geq 2$, M_1 is constant and M_2 additive, (iv) if $\ell \neq m$, $\ell \geq 2$, $m \geq 3$, M_1 additive and M_2 constant.

Proof: Without loss of generality we assume that M_1 is not a projection. Now let $\gamma \in \mathbb{R}$ be arbitrary. Defining

$$G(p) = F(p) - \gamma[M_1(p) - M_2(p)], \qquad p \in J, \tag{9.2.10}$$

we see that G also satisfies equation (5.4.7), that is,

$$\sum_{i=1}^{\ell}\sum_{j=1}^{m}[G(p_iq_j) - M_1(p_i)G(q_j) - M_2(q_j)G(p_i)] = 0 \qquad (9.2.11)$$

for all $P \in \Gamma_{\ell}^o, Q \in \Gamma_m^o$. Keeping $Q \in \Gamma_m^o$ temporarily fixed in (9.2.11) and defining

$$H(p) = \sum_{j=1}^{m}[G(pq_j) - M_1(p)G(q_j) - M_2(q_j)G(p)], \quad p \in J,$$

we get $\sum_{i=1}^{\ell} H(p_i) = 0$. Thus Theorem 6.4.1 yields

$$\sum_{j=1}^{m}[G(pq_j) - M_1(p)G(q_j) - M_2(q_j)G(p)] = A\left(p - \frac{1}{\ell}, Q\right) \qquad (9.2.12)$$

where $A : \mathbb{R}^k \times \Gamma_m^o \to \mathbb{R}$ is additive in the first variable.

Let $P = (p_1, \ldots, p_m) \in \Gamma_m^o$ and substitute in (9.2.12) p_i for p ($i = 1, 2, \ldots, m$). Adding the equations so obtained we get

$$\sum_{i=1}^{m}\sum_{j=1}^{m} G(p_iq_j)$$
$$= \sum_{i=1}^{m}\sum_{j=1}^{m}[M_1(p_i)G(q_j) + M_2(q_j)G(p_i)] + A\left(1 - \frac{m}{\ell}, Q\right). \qquad (9.2.13)$$

Since the left side of (9.2.13) is symmetric in p_i, q_j we obtain from the right side

$$\sum_{j=1}^{m} G(q_j) \sum_{i=1}^{m}[M_1(p_i) - M_2(p_i)] + \left(1 - \frac{m}{\ell}\right) A(1, Q)$$
$$= \sum_{i=1}^{m} G(p_i) \sum_{j=1}^{m}[M_1(q_j) - M_2(q_j)] + \left(1 - \frac{m}{\ell}\right) A(1, P). \qquad (9.2.14)$$

By Lemma 6.5.3 (i) there exists $P' = (p_1', \ldots, p_m') \in \Gamma_m^o$ such that

$$b := \sum_{j=1}^{m}[M_1(p_j') - M_2(p_j')] \neq 0. \qquad (9.2.15)$$

Now choosing the arbitrary constant γ in (9.2.10) to be

$$\gamma := b^{-1} \sum_{i=1}^{m} F(p_i')$$

we see that

$$\sum_{i=1}^{m} G(p_i') = 0. \tag{9.2.16}$$

Substituting P' for $(p_1, p_2, \ldots, p_m)$ into (9.2.14), we get (see (9.2.16) and (9.2.15))

$$b \sum_{j=1}^{m} G(q_j) + \left(1 - \frac{m}{\ell}\right) A(1, Q) = \left(1 - \frac{m}{\ell}\right) A(1, P'). \tag{9.2.17}$$

If $m = \ell$ in (9.2.17) then we get $\sum_{j=1}^{m} G(q_j) = 0$, for $Q \in \Gamma_m^o$. Thus (9.2.9) is satisfied with $d = 0$ and $a = \gamma$.

If $m \neq \ell$ in (9.2.17), then we get

$$A(1, Q) = \delta \sum_{j=1}^{m} G(q_j) + \epsilon \tag{9.2.18}$$

where δ and ϵ are real constants. (Note that $\delta = -b\left(1 - \frac{m}{\ell}\right)^{-1} \neq 0$). Substituting (9.2.18) into (9.2.14) and using $\left(1 - \frac{m}{\ell}\right)\delta = -b$, (9.2.14) can be rewritten as

$$\left(\sum_{i=1}^{m}[M_1(p_i) - M_2(p_i)] - b\right) \sum_{j=1}^{m} G(q_j)$$
$$= \left(\sum_{j=1}^{m}[M_1(q_j) - M_2(q_j)] - b\right) \sum_{i=1}^{m} G(p_i). \tag{9.2.19}$$

At this point let us consider the case that M_1 is nonconstant and not additive. By Lemma 6.5.3 (ii) we know that

$$\sum_{i=1}^{m}[M_1(p_i'') - M_2(p_i'')] \neq b$$

for some $P'' \in \Gamma_m^o$. Substituting $P = P''$ in (9.2.19) we get

$$\sum_{j=1}^{m} G(q_j) = \alpha \sum_{j=1}^{m}[M_1(q_j) - M_2(q_j)] - \alpha b$$

for some constant α. Combining this with (9.2.10), we have (9.2.9) with $a = \gamma + \alpha$ and $d = -\alpha b$.

On the other hand, consider the case that M_1 is constant or additive. Since M_1 is not a projection by hypothesis, we must have either $M_1 = 0$ or $M_1 = 1$. Now (9.2.19) becomes

$$\left(b' - \sum_{i=1}^{m} M_2(p_i) \right) \sum_{j=1}^{m} G(q_j) = \left(b' - \sum_{j=1}^{m} M_2(q_j) \right) \sum_{i=1}^{m} G(p_i), \qquad (9.2.20)$$

where $b' = -b$ if $M_1 = 0$, and $b' = m - b$ if $M_1 = 1$. The rest of the proof divides into cases depending on the form of M_2.

If M_2 is nonconstant and not additive, then we claim that there is a $\overline{P} \in \Gamma_m^o$ for which

$$\sum_{i=1}^{m} M_2(\overline{p}_i) \neq b'. \qquad (9.2.21)$$

Indeed, suppose to the contrary that $\sum_{i=1}^{m} M_2(p_i) = b'$ for all $P \in \Gamma_m^o$. Since $m \geq 3$, Theorem 6.4.1 yields $M_2(p) = A_1(p) + a_1$ for some additive $A_1 : \mathbb{R}^k \to \mathbb{R}$ and constant a_1. Moreover, $A_1 \neq 0$ since M_2 is nonconstant, and $a_1 \neq 0$ since M_2 is not additive. But then $M_2 = A_1 + a_1$ contradicts Lemma 6.5.2. Thus (9.2.21) is established.

Continuing, put $P = \overline{P}$ into (9.2.20) to obtain

$$\sum_{j=1}^{m} G(q_j) = \beta \left(b' - \sum_{j=1}^{m} M_2(q_j) \right) = \beta \sum_{j=1}^{m} [M_1(q_j) - M_2(q_j)] - \beta b$$

for some constant β. Again, combining this with (9.2.10) we arrive at (9.2.9) with $a = \gamma + \beta$ and $d = -\beta b$.

The only cases remaining are that M_2 is constant or additive (that is, $M_2 = 1$ or $M_2 = 0$ or M_2 is a projection), while either $M_1 = 0$ or $M_1 = 1$. (Also, recall that $M_1 \neq M_2$.) In all these cases, (9.2.19) provides no further information, so we return to (9.2.12). To handle these cases, we substitute now in (9.2.12) $\frac{p_i}{\ell}$ for p ($i = 1, 2, ..., m$), where $P = (p_1, p_2, ..., p_m) \in \Gamma_m^o$. Adding the m resulting equations, we obtain

$$\sum_{i=1}^{m} \sum_{j=1}^{m} G\left(\frac{p_i q_j}{\ell} \right) = \sum_{i=1}^{m} \sum_{j=1}^{m} \left[M_1\left(\frac{p_i}{\ell} \right) G(q_j) + M_2(q_j) G\left(\frac{p_i}{\ell} \right) \right]$$

$$+ \left(\frac{1 - m}{\ell} \right) A(1, Q).$$

By the symmetry of the left hand side in p_i and q_j, this leads to

$$
\sum_{i=1}^{m} \sum_{j=1}^{m} \left[M_1 \left(\frac{p_i}{\ell} \right) G(q_j) + M_2(q_j) G \left(\frac{p_i}{\ell} \right) \right]
$$
$$
+ \left(\frac{1-m}{\ell} \right) A(1, Q)
$$
$$
= \sum_{i=1}^{m} \sum_{j=1}^{m} \left[M_1 \left(\frac{q_j}{\ell} \right) G(p_i) + M_2(p_i) G \left(\frac{q_j}{\ell} \right) \right] \tag{9.2.22}
$$
$$
+ \left(\frac{1-m}{\ell} \right) A(1, P).
$$

This is the key to the remaining four cases.

First, suppose $M_1 = 1$ and $M_2 = 0$. Now (9.2.22) becomes

$$
m \sum_{j=1}^{m} G(q_j) + \left(\frac{1-m}{\ell} \right) A(1, Q) = m \sum_{i=1}^{m} G(p_i) + \left(\frac{1-m}{\ell} \right) A(1, P).
$$

Substituting $\sum_{j=1}^{m} G(q_j) = \delta^{-1} [A(1, Q) - \epsilon]$ from (9.2.18) into the previous equation, we get

$$
\left(\frac{m}{\delta} + \frac{1-m}{\ell} \right) A(1, Q) = \left(\frac{m}{\delta} + \frac{1-m}{\ell} \right) A(1, P). \tag{9.2.23}
$$

Furthermore, $\frac{m}{\delta} = \frac{b}{\delta} = \frac{m}{\ell} - 1$ (cf. (9.2.15)) in this case, so (9.2.23) yields the constancy of $A(1, Q)$. Thus (9.2.18) gives the constancy of $\sum_{j=1}^{m} G(q_j)$, which by (9.2.16) means $\sum_{j=1}^{m} G(q_j) = 0$. Therefore (9.2.9) with $a = \gamma$ and $d = 0$ follows from (9.2.10).

Next, suppose $M_1 = 0$ and $M_2 = 1$. Now (9.2.22) reads as

$$
m \sum_{i=1}^{m} G \left(\frac{p_i}{\ell} \right) + \left(\frac{1-m}{\ell} \right) A(1, Q)
$$
$$
= m \sum_{j=1}^{m} G \left(\frac{q_j}{\ell} \right) + \left(\frac{1-m}{\ell} \right) A(1, P). \tag{9.2.24}
$$

Moreover, (9.2.12) in this case with $p = \frac{1}{\ell}$ yields $\sum_{j=1}^{m} G \left(\frac{q_j}{\ell} \right) = mG \left(\frac{1}{\ell} \right)$. Using this in (9.2.24) we obtain again the constancy of $A(1, Q)$, and as above this leads to (9.2.9) with $a = \gamma$ and $d = 0$.

In the case that $M_1 = 0$ and M_2 is a projection, (9.2.22) takes the form

$$\sum_{i=1}^{m} G\left(\frac{p_i}{\ell}\right) + \left(\frac{1-m}{\ell}\right) A(1,Q) = \sum_{j=1}^{m} G\left(\frac{q_j}{\ell}\right) + \left(\frac{1-m}{\ell}\right) A(1,P).$$

In this case (9.2.12) with $p = \frac{1}{\ell}$ gives $\sum_{j=1}^{m} G\left(\frac{q_j}{\ell}\right) = G\left(\frac{1}{\ell}\right)$, and the analysis is concluded as in the previous paragraph.

Finally, in the case $M_1 = 1$ and M_2 is a projection, (9.2.22) yields

$$m \sum_{j=1}^{m} G(q_j) + \sum_{i=1}^{m} G\left(\frac{p_i}{\ell}\right) + \left(\frac{1-m}{\ell}\right) A(1,Q)$$

$$= m \sum_{i=1}^{m} G(p_i) + \sum_{j=1}^{m} G\left(\frac{q_j}{\ell}\right) + \left(\frac{1-m}{\ell}\right) A(1,P)$$

while (9.2.12) with $p = \frac{1}{\ell}$ becomes

$$\sum_{j=1}^{m} \left[G\left(\frac{q_j}{\ell}\right) - G(q_j) \right] - G\left(\frac{1}{\ell}\right) = 0.$$

Combining these last two equations, we arrive at

$$(m-1) \sum_{j=1}^{m} G(q_j) + \left(\frac{1-m}{\ell}\right) A(1,Q)$$

$$= (m-1) \sum_{i=1}^{m} G(p_i) + \left(\frac{1-m}{\ell}\right) A(1,P).$$

By (9.2.18) we can rewrite this as

$$\left(\frac{m-1}{\delta} + \frac{1-m}{\ell}\right) A(1,Q) = \left(\frac{m-1}{\delta} + \frac{1-m}{\ell}\right) A(1,P). \tag{9.2.25}$$

Since $\frac{m-1}{\delta} = \frac{b}{\delta} = \frac{m}{\ell} - 1$ (cf. (9.2.15) again) in this case, we conclude from (9.2.25) that $A(1,Q)$ is constant, and hence (9.2.9) holds again with $a = \gamma$ and $d = 0$.

Since there are no more cases to consider, the lemma is proven if $\ell \geq 3$ and $m \geq 3$. A careful examination of the proof yields that the case $\ell \geq 3$, $m = 2$ is included in our proof if M_1 is nonconstant and M_2 is a projection. (Note that the assumption $\ell \geq 3$ already implies (9.2.12) and (9.2.22). Only these two equations were used to investigate the case that M_1 is constant and M_2 is a projection.) Thus (iii) is proved. Since the statement (iv) can be shown similarly this concludes the proof of the lemma.

Lemma 9.2.4. *Let* $M : J \to \mathbb{R}$ *be multiplicative but not a projection. Further let* $\ell \geq 3$ *and* $m \geq 2$ *be fixed integers. If* $F : J \to \mathbb{R}$ *and* M *satisfy*

$$\sum_{i=1}^{\ell}\sum_{j=1}^{m} F(p_i q_j) = \sum_{i=1}^{\ell} M(p_i) \sum_{j=1}^{m} F(q_j) + \sum_{j=1}^{m} M(q_j) \sum_{i=1}^{\ell} F(p_i) \qquad (9.2.26)$$

for all $P \in \Gamma_\ell^o$ *and* $Q \in \Gamma_m^o$, *then there exist a constant* c *and an additive function* $A : \mathbb{R}^k \to \mathbb{R}$ *such that the map* $h : J \to \mathbb{R}$ *defined by*

$$h(p) := F(p) - A(p) + c \qquad (9.2.27)$$

satisfies

$$\sum_{j=1}^{m}\{h(pq_j) - M(p)h(q_j) - M(q_j)h(p)\} = 0, \qquad (9.2.28)$$

where

$$c = \begin{cases} \dfrac{A(1)}{m\ell} & \text{if } M \equiv 0, \\[2mm] \dfrac{A(1)}{m\ell}(m + \ell - 1) & \text{if } M \equiv 1, \\[2mm] A(1) = 0 & \text{otherwise.} \end{cases} \qquad (9.2.29)$$

Moreover, if F *and* M *are measurable, then* A *and* h *are also measurable.*

Proof: The proof is very similar to the proof of Lemma 8.3.1. Thus we refer for some details to corresponding considerations in the proof of Lemma 8.3.1.

Keeping $Q \in \Gamma_m^o$ temporarily fixed in (9.2.26), and defining

$$H(p) := \sum_{j=1}^{m}\{F(pq_j) - M(p)F(q_j) - F(p)M(q_j)\}, \quad p \in J, \qquad (9.2.30)$$

we get

$$\sum_{i=1}^{\ell} H(p_i) = 0.$$

Because of Theorem 6.4.1 there exists a function $B : \mathbb{R}^k \times \Gamma_m^o \to \mathbb{R}$ which is additive in the first variable and which satisfies

$$\sum_{j=1}^{m} \{F(pq_j) - M(p)F(q_j) - F(p)M(q_j)\}$$
$$= B\left(p - \frac{1}{\ell}, Q\right), \quad p \in J.$$

(9.2.31)

By the same procedure of obtaining equation (8.3.10) from (8.3.7) we now arrive at

$$\sum_{i=1}^{m}\sum_{j=1}^{m} \{F(xp_iq_j) - F(x)M(p_i)M(q_j)\}$$
$$- \sum_{i=1}^{m}\sum_{j=1}^{m} M(x)\{M(p_i)F(q_j) + M(q_j)F(p_i)\}$$
$$= B\left(x - \frac{1}{\ell}, P\right)\sum_{j=1}^{m} M(q_j) + B\left(x - \frac{m}{\ell}, Q\right).$$

(9.2.32)

Using the symmetry of the left-hand side in p_i and q_j in equation (9.2.32) we get

$$B\left(x - \frac{1}{\ell}, P\right)\sum_{j=1}^{m} M(q_j) + B\left(x - \frac{m}{\ell}, Q\right)$$
$$= B\left(x - \frac{1}{\ell}, Q\right)\sum_{i=1}^{m} M(p_i) + B\left(x - \frac{m}{\ell}, P\right).$$

(9.2.33)

Letting $x = \frac{1}{\ell}$ in (9.2.33) we get

$$B(1, P) = a \text{ (constant).}$$

(9.2.34)

Now (9.2.33) with (9.2.34) yields

$$B(x, Q)\left\{1 - \sum_{i=1}^{m} M(p_i)\right\} + \frac{a}{\ell}\sum_{i=1}^{m} M(p_i)$$
$$= B(x, P)\left\{1 - \sum_{i=1}^{m} M(q_j)\right\} + \frac{a}{\ell}\sum_{j=1}^{m} M(q_j).$$

(9.2.35)

Case 1. Suppose that M is constant (either $M = 0$ or $M = 1$). Since in these cases $\sum_{i=1}^{m} M(p_i) = $ constant (namely 0 or m but different from 1) (9.2.35) results in

$$B(x, Q) = B(x, P) = c(x) \quad \text{(say)}. \tag{9.2.36}$$

Here the additivity of B implies that $c : \mathbb{R}^k \to \mathbb{R}$ is additive. Thus substitution of (9.2.34) and (9.2.36) in (9.2.31) gives (note that $a = c(1)$)

$$\sum_{j=1}^{m} \{F(pq_j) - M(p)F(q_j) - F(p)M(q_j)\} = c(p) - \frac{c(1)}{\ell}, \quad p \in J. \tag{9.2.37}$$

Now, if $M = 0$ then (9.2.37) reduces to

$$\sum_{j=1}^{m} F(pq_j) = c(p) - \frac{c(1)}{\ell};$$

that is (using $c(p) = \sum_{j=1}^{m} c(pq_j)$),

$$\sum_{j=1}^{m} \left\{ F(pq_j) - c(pq_j) + \frac{c(1)}{\ell m} \right\} = 0.$$

Thus the function h in (9.2.27) (with $A(p) = c(p)$ and $c = \frac{c(1)}{\ell m}$) satisfies (9.2.28) and (9.2.29).

If $M = 1$ in (9.2.37) then the desired result follows from the following equivalences (with $A(p) := \frac{c(p)}{1-m}$ and $c := \frac{A(1)}{\ell m}(\ell + m - 1)$):

$$\sum_{j=1}^{m} \{F(pq_j) - F(q_j) - F(p)\}$$

$$= \left(\frac{c(p)}{1-m} - \frac{c(1)}{\ell(1-m)} \right)(1-m)$$

$$= \left(A(p) - \frac{A(1)}{\ell} \right)(1-m)$$

$$\Leftrightarrow \sum_{j=1}^{m} (\{F(pq_j) - A(pq_j)\} - \{F(q_j) - A(q_j)\} - \{F(p) - A(p)\})$$

$$= \frac{A(1)}{\ell}(\ell + m - 1) = mc$$

$$\Leftrightarrow \sum_{j=1}^{m} \{h(pq_j) - h(q_j) - h(p)\} = 0,$$

where h is again defined by (9.2.27).

Case 2. Now suppose M is nonconstant and not additive in (9.2.35). This implies that, given any constant γ there is a $P' \in \Gamma_m^o$ such that

$$\sum_{i=1}^{m} M(p_i') \neq \gamma. \qquad (9.2.38)$$

For, if $m \geq 3$ we may argue as in the justification of (9.2.21) in the preceding lemma, while for $m = 2$ we argue as follows. Since M is nonconstant and not additive, the equation $M(p) + M(1 - p) = \gamma$ would contradict Theorem 6.4.8. Substituting P' into (9.2.35) and using (9.2.38) for $\gamma = 1$, we obtain

$$B(x, Q) = A(x)\left\{1 - \sum_{j=1}^{m} M(q_j)\right\} + b\sum_{j=1}^{m} M(q_j) + c \qquad (9.2.39)$$

where b, c are constants and $A : \mathbb{R}^k \to \mathbb{R}$ is an additive function. Using this and the additivity of B and A, we get

$$0 = -B(x + y, Q) + B(x, Q) + B(y, Q) = b\sum_{j=1}^{m} M(q_j) + c.$$

Thus (9.2.39) reduces to

$$B(x, Q) = A(x)\left\{1 - \sum_{j=1}^{m} M(q_j)\right\}. \qquad (9.2.40)$$

Putting $x = 1$ into (9.2.40) and using $B(1, Q) = a$ (see (9.2.34)), we get $a = A(1)\left\{1 - \sum_{j=1}^{m} M(q_j)\right\}$. Thus $A(1) = 0$, since otherwise (9.2.38) is contradicted for $\gamma = 1$. Now

$$a = B(1, Q) = 0 = A(1). \qquad (9.2.41)$$

Using (9.2.40) and (9.2.41) we rewrite (9.2.31) as

$$\sum_{j=1}^{m} [F(pq_j) - M(p)F(q_j) - F(p)M(q_j)] = A(p)\left\{1 - \sum_{j=1}^{m} M(q_j)\right\}. \qquad (9.2.42)$$

With

$$h(p) := F(p) - A(p) - A(1) = F(p) - A(p),$$

equation (9.2.42) goes over into (9.2.28) (since $A(1) = 0$).

Finally, if F and M are measurable then h and A are measurable, too. The fact that A is measurable follows from the equations (9.2.31) and either (9.2.36) or (9.2.40). Then the measurability of h follows from (9.2.27). Thus the lemma is established.

For completeness sake we present the following lemma which can be proved in the same manner as Lemma 9.2.4.

Lemma 9.2.5. *Let $M : J \to \mathbb{R}$ be multiplicative but not a projection. Further let $\ell \geq 2$ and $m \geq 3$ be fixed integers. If $F : J \to \mathbb{R}$ and M satisfy (9.2.26) for all $P \in \Gamma_\ell^o$ and $Q \in \Gamma_m^o$, then there exist a constant c and an additive function $A : \mathbb{R}^k \to \mathbb{R}$ such that h defined by (9.2.27) with (9.2.29) satisfies*

$$\sum_{i=1}^{\ell} \{h(p_i q) - M(p_i)h(q) - M(q)h(p_i)\} = 0. \tag{9.2.43}$$

Moreover, if F and M are measurable, then A and h are also measurable.

9.3. Solution of (5.4.7) in the case $m \geq 3$ or $\ell \geq 3$

Now we are ready to present the general solution of (5.4.7) for $F : J \to \mathbb{R}$, where $M_1, M_2 : J \to \mathbb{R}$ are multiplicative and where M_1 or M_2 is not a projection (provided $\ell \geq 3$ and $m \geq 3$). In Kannappan and Sander (1989) the solution was obtained if M_1 and M_2 are multiplicative, nonconstant and not additive. By refinements of these results we get the following result.

Theorem 9.3.1. *Let $M_1, M_2 : J \to \mathbb{R}$ be multiplicative, and suppose M_1 or M_2 is not a projection. Moreover let $F : J \to \mathbb{R}$ satisfy the functional equation (5.4.7) for fixed $\ell \geq 3$, $m \geq 3$. Then, and only then,*

$$F(p) = \begin{cases} A(p) + M(p)L(p) - c & \text{if } M_1 = M_2 = M \quad (9.3.1) \\[2mm] B(p) + b[M_1(p) - M_2(p)] & \text{if } M_1 \neq M_2 \quad (9.3.2) \end{cases}$$

for all $p \in J$, where $L : \mathcal{P} \to \mathbb{R}$ is a logarithmic function, A and B are additive functions, b is a constant, c is given by (9.2.29) and $B(1) = 0$. Moreover, if F, M_1, M_2 are measurable then A, B and L are measurable and B is the zero function.

Proof: It is easy to check that F, given by (9.3.1), (9.3.2) and (9.2.29) satisfies equation (5.4.7). Let us now treat the converse. We first consider the case that $M := M_1 = M_2$, where M is multiplicative but not a projection.

If $M = M_1 = M_2$ then equation (5.4.7) goes over into (9.2.26). Thus Lemma 9.2.4 or 9.2.5 shows that there is a constant c_2 and an additive function $A : \mathbb{R}^k \to \mathbb{R}$ such that the map h defined by

$$h(p) := F(p) - A(p) + c_2 \qquad (9.3.3)$$

satisfies (9.2.28) and c_2 is given by (9.2.29). An application of Theorem 6.4.1 leads to

$$h(pq) - M(p)h(q) - M(q)h(p) = E\left(p, q - \frac{1}{m}\right), \qquad m \geq 3, \qquad (9.3.4)$$

where $E : J \times \mathbb{R}^k \to \mathbb{R}$ is additive in the second variable. First assume that M is nonconstant (and not additive, by hypothesis) in (9.3.4). Then we get

$$
\begin{aligned}
& E\left(pq, r - \frac{1}{m}\right) + M(r)E\left(p, q - \frac{1}{m}\right) \\
& = h(pqr) - M(pq)h(r) - M(r)h(pq) \\
& \quad + M(r)h(pq) - M(rp)h(q) - M(rq)h(p) \\
& = h(pqr) - M(p)h(qr) - M(qr)h(p) \\
& \quad + M(p)h(qr) - M(pq)h(r) - M(pr)h(q) \\
& = E\left(p, qr - \frac{1}{m}\right) + M(p)E\left(q, r - \frac{1}{m}\right).
\end{aligned}
\qquad (9.3.5)
$$

We want to show that $E(p, q - \frac{1}{m}) = 0$ for all $p, q \in J$. Suppose that there exists $p_o, q_o \in J$ such that $s := E(p_o, q_o - \frac{1}{m}) \neq 0$. Substituting $p = p_o$, $q = q_o$ into (9.3.5) we get

$$M(r) = \beta(r) + a, \qquad r \in J, \qquad (9.3.6)$$

where $\beta : J \to \mathbb{R}$ is the additive function

$$\beta(r) = s^{-1}\left\{E(p_o, q_o r) + M(p_o)E(q_o, r) - E(p_o q_o, r)\right\}$$

and a is the constant

$$a = s^{-1}\left\{E\left(p_o, -\frac{1}{m}\right) + M(p_o)E\left(q_o, -\frac{1}{m}\right) - E\left(p_o q_o, -\frac{1}{m}\right)\right\}.$$

Since M is nonconstant and not additive, the representation (9.3.6) contradicts Lemma 6.5.2, part (i). Thus (9.3.4) reduces to

$$h(pq) = M(p)h(q) + M(q)h(p), \qquad p, q \in J,$$

that is, the function $L : J \to \mathbb{R}$ defined by

$$L(p) := \frac{h(p)}{M(p)}, \qquad p \in J, \tag{9.3.7}$$

is logarithmic. Because of (9.3.3) (with $c_2 = A(1) = 0$ by (9.2.29)) and (9.3.7) we arrive at

$$F(p) = A(p) + M(p)L(p),$$

which is (9.3.1) (with (9.2.29)).

Now suppose M is constant in (9.3.4). Then Lemma 9.2.1 yields

$$h(p) = M(p)L_1(p) + A_1(p) + c_1 \tag{9.3.8}$$

where $A_1 : \mathbb{R}^k \to \mathbb{R}$ is additive, $L_1 : \mathcal{P} \to \mathbb{R}$ is logarithmic and c_1 is a constant. Thus we get from (9.3.3) that

$$F(p) = A_2(p) + M(p)L_1(p) - (c_2 - c_1) \tag{9.3.9}$$

for all $p \in J$, where $A_2 := A + A_1$ is additive. Substitution of (9.3.9) into (9.2.26) yields

$$c_2 - c_1 = \frac{A_2(1)}{m\ell} \quad \text{if } M \equiv 0,$$

and

$$c_2 - c_1 = \frac{A_2(1)(m + \ell - 1)}{m\ell} \qquad \text{if } M \equiv 1,$$

so that (9.3.1) is also satisfied in this case.

Now we turn to the case $M_1 \neq M_2$. From Lemma 9.2.3 we get

$$\sum_{j=1}^{m} \{F(q_j) - a[M_1(q_j) - M_2(q_j)]\} = d, \tag{9.2.9}$$

for some constants a, d. Now Theorem 6.4.1 yields

$$F(p) = B(p) + a[M_1(p) - M_2(p)] + \frac{d - B(1)}{m}, \tag{9.3.10}$$

where $B : \mathbb{R}^k \to \mathbb{R}$ is additive. By Lemma 9.2.3 we know that $d = 0$ in the cases listed there. But then Lemma 6.5.4 applied to (9.3.10) (with $d = 0$) implies $B(1) = 0$ or $\frac{1}{m}B(1) = 0 = \frac{1}{\ell}B(1)$ or $B(1)(1 - \frac{1}{m}) = 0 = B(1)(1 - \frac{1}{\ell})$, so that we get $B(1) = 0$ in all these cases, and F has the representation (9.3.2). We still have to consider the cases in which Lemma 9.2.3 does not give $d = 0$, that is when at least one of the functions M_1, M_2 is nonconstant and not additive. In these cases, by the last statement of Lemma 6.5.4 we deduce from (9.3.10) that $d - B(1) = 0$ and $B(1) = 0$, so that again (9.3.10) reduces to (9.3.2) with $B(1) = 0$.

Finally, if F, M_1 and M_2 are measurable, then the additive functions A and B in (9.3.1) and (9.3.2) are measurable (and thus the logarithmic function L is measurable, too). Of course, B is measurable since $B(p) = F(p) - b\{M_1(p) - M_2(p)\}$. In this case $B(p) = B(1) \odot p = 0$ since $B(1) = 0$. Now we show that A is measurable. From Lemma 9.2.5 and Lemma 9.2.4 we get that the function h defined in (9.3.3) is measurable. The rest follows from the proof of Theorem 9.3.1 (see the equations (9.3.7), (9.3.8), and (9.3.9)).

In Theorem 9.3.1 we have characterized the solutions of the functional equation (5.4.7) for fixed $\ell \geq 3$, $m \geq 3$ and then we have determined all measurable solutions.

Let us now present some characterization theorems for sum form information measures of multiplicative type (M_1, M_2), which are easy consequences of equation (5.1.1), Theorem 9.3.1, and Theorem 7.2.1. Again we use our notation (1.1.10)-(1.1.12).

Theorem 9.3.2. Let $I_n : \Gamma_n^o \to \mathbb{R}$ have the sum form for a generating function $F : J \to \mathbb{R}$, let $\{I_n\}$ be (ℓ, m)-additive for fixed $\ell \geq 3$, $m \geq 3$ and let $\{I_n\}$ be of type (M_1, M_2) with multiplicative functions M_1, M_2 where M_1 or M_2 is not a projection. Then $\{I_n\}$ is given by

$$I_n(P) = \begin{cases} \frac{\ell m - n}{\ell m} b & \text{if } M_1 = M_2 = 0 \\[2ex] \sum_{i=1}^{n}\sum_{j=1}^{k} L_j(\pi_{ji}) + \frac{\ell m - (\ell + m)n + n}{\ell m} b & \text{if } M_1 = M_2 = 1 \\[2ex] \sum_{i=1}^{n}\prod_{s=1}^{k} M_{1s}(\pi_{si})\sum_{j=1}^{k} L_j(\pi_{ji}) & \text{if } M_1 = M_2 \notin \{0, 1\} \\[2ex] b\left\{\sum_{i=1}^{n}\left[\prod_{s=1}^{k} M_{1s}(\pi_{si}) - \prod_{s=1}^{k} M_{2s}(\pi_{si})\right]\right\} & \text{if } M_1 \neq M_2 \end{cases}$$

where b is a constant, $L_j : \mathbb{R}_+ \to \mathbb{R}$, $1 \leq j \leq k$, are logarithmic and M_{1s}, $M_{2s} :]0,1[\to \mathbb{R}$ are multiplicative with $M_1(x) = \prod_{s=1}^{k} M_{1s}(x_s)$ and $M_2(x) = \prod_{s=1}^{k} M_{2s}(x_s)$, $x = (x_1, x_2, ..., x_k) \in J$ $(1 \leq s \leq k)$.

Theorem 9.3.3. *Let $I_n : \Gamma_n^o \to \mathbb{R}$ have the sum form for some measurable generating function $F : J \to \mathbb{R}$ and let $\alpha = (\alpha_1, ..., \alpha_k) \in \mathbb{R}^k$, $\beta = (\beta_1, ..., \beta_k) \in \mathbb{R}^k$. If $\{I_n\}$ satisfies*

$$I_{\ell m}(P \star Q) = \sum_{i=1}^{\ell} p_i^{\alpha} I_m(Q) + \sum_{j=1}^{m} q_j^{\beta} I_{\ell}(P), \qquad P \in \Gamma_{\ell}^o, Q \in \Gamma_m^o$$

for all $\ell \geq 3$, $m \geq 3$ then $\{I_n\}$ has the form

$$I_n(P) = \begin{cases} \displaystyle\sum_{i=1}^{k} c_i H_n(P_i) + \sum_{\substack{i=1 \\ i \neq j}}^{k}\sum_{j=1}^{k} c_{ij} K_n(P_i, P_j) & \text{if } \alpha, \beta \in U \\[2em] \displaystyle\sum_{i=1}^{n} \pi_{1i}^{\alpha_1} \cdots \pi_{ki}^{\alpha_k} \sum_{j=1}^{k} c_j \log \pi_{ji} & \text{if } \alpha = \beta \in \mathbb{R}^k \setminus U \\[2em] \displaystyle b\left\{ \sum_{i=1}^{n} \pi_{1i}^{\alpha_1} \cdots \pi_{ki}^{\alpha_k} - \sum_{i=1}^{n} \pi_{1i}^{\beta_1} \cdots \pi_{ki}^{\beta_k} \right\} & \text{if } \begin{matrix}\alpha \neq \beta \\ \alpha \notin U \text{ or } \beta \notin U\end{matrix} \end{cases}$$

where b, c_i, c_{ij} $(i, j = 1, 2, ..., k)$ are constants.

Remark 9.3.4. It should be noted that if one assumes $\ell \geq 3$, $m \geq 2$ or $m \geq 3$, $\ell \geq 2$ in Theorem 9.3.3 instead of $\ell \geq 3$, $m \geq 3$, then the assertation of Theorem 9.3.3 is still valid. (See Lemma 9.2.3 and Lemma 9.2.5.)

Let us now define the entropies of degree (α, β), depending upon k probability distributions by

$$H_n^{(\alpha,\beta)}(P) = \begin{cases} \displaystyle\sum_{i=1}^{n} p_i \odot c \odot \log p_i, & 1 \odot c \odot 1 = -1 & \text{if } \alpha, \beta \in U \\[2em] \displaystyle\sum_{i=1}^{n} p_i^{\alpha}(b \odot \log p_i), & b \odot 1 = -2^{\alpha-1} & \text{if } \alpha = \beta \in \mathbb{R}^k \setminus U \\[2em] \displaystyle(2^{1-\alpha} - 2^{1-\beta})^{-1}\left\{ \sum_{i=1}^{n} p_i^{\alpha} - \sum_{i=1}^{n} p_i^{\beta} \right\} & \text{if } \begin{matrix}\alpha \neq \beta \\ \alpha \notin U \text{ or } \beta \notin U\end{matrix} \end{cases}$$

where b is a $1 \times k$ vector and $c = (c_{rs})$ is a $k \times k$-matrix. This is motivated by requiring also $I_2(\frac{1}{2}, \frac{1}{2}) = 1$ in Theorem 9.3.2, so that the entropies given in Theorem 9.3.2 go over into the above expression. Thus $\{H_n^{(\alpha,\beta)}\}$ is a natural generalization of the entropies of degree (α, β), $\alpha, \beta \in \mathbb{R}$.

9.4. Solution of (5.4.7) in the case $\ell = m = 2$

Now we consider the functional equation (5.4.7) in the case that $\ell = m = 2$, that is where $F : J \to \mathbb{R}$ satisfies

$$\sum_{i=1}^{2}\sum_{j=1}^{2} F(p_i q_j) = \sum_{i=1}^{2} M_1(p_i) \sum_{j=1}^{2} F(q_j) + \sum_{j=1}^{2} M_2(q_j) \sum_{i=1}^{2} F(p_i) \qquad (9.4.1)$$

for all $(p_1, p_2), (q_1, q_2) \in \Gamma_2^o$, where $M_1, M_2 : J \to \mathbb{R}$ are multiplicative functions.

In section 9.3 we saw that it was possible to determine the general solution of (5.4.7) if M_1 and M_2 are multiplicative functions, at least one of which is not a projection, if $\ell \geq 3$ and $m \geq 3$. But, as in the case of additive or λ-parametric information measures, it seems very difficult to obtain the form of such F in the case of (2,2)-additivity. Remember that in the case that both M_1 and M_2 are projections in (9.4.1) we obtained the measurable solutions F of (9.4.1) in Chapter 7 (see Theorem 7.2.1).

In this section we assume measurability of F, M_1 and M_2 so that

$$M_1 \equiv 0 \text{ or } M_1(p) = p^\alpha, \quad \text{and} \quad M_2 \equiv 0 \text{ or } M_2(p) = p^\beta, \qquad (9.4.2)$$

for all $p \in J$ and for some $\alpha, \beta \in \mathbb{R}^k$. If

$$\left.\begin{array}{ll} & M_1 = M_2 = 0, \\ \text{or} & M_1 = 0, \quad M_2(p) = p^\beta, \\ \text{or} & M_2 = 0, \quad M_1(p) = p^\alpha, \\ \text{or} & M_1(p) = p^\alpha, \quad M_2(p) = p^\beta, \quad \alpha \neq \beta, \quad p \in J \end{array}\right\} \qquad (9.4.3)$$

then we present the measurable solution of the functional equation (9.4.1). But if

$$M_1(p) = p^\alpha = M_2(p) \qquad (9.4.4)$$

then we assume that $k = 1$ (that is, $M_1 :]0, 1[\to \mathbb{R}$ and $\alpha \in \mathbb{R}$). The cases in (9.4.3) can be treated very easily whereas the cases in (9.4.4) are very difficult and yield new, surprising solutions.

Let us start with our first result, which is similar to the results in Lemma 8.3.3 and Theorem 9.3.2.

Theorem 9.4.1. *Let $M_1, M_2 : J \to \mathbb{R}$ be measurable functions of one of the forms given by (9.4.3).*
(a) If F satisfies (9.4.1) then F has the form

$$F(p) = g(p) + b[M_1(p) - M_2(p)] \qquad (9.4.5)$$

where $b \in \mathbb{R}$ and g is a solution of (6.3.40) with $g(p) + g(1-p) = 0$, $p \in J$.
(b) If F is measurable then F satisfies (9.4.1) if and only if

$$F(p) = \begin{cases} a \odot (4p - 1) & \text{if } M_1 = M_2 = 0 \\ b\{M_1(p) - M_2(p)\} + c \odot p & \text{otherwise} \end{cases} \qquad (9.4.6)$$

where $b \in \mathbb{R}$ and $a, c \in \mathbb{R}^k$ such that $c \odot 1 = 0$.

Proof: If $M_1 = M_2 = 0$ in (9.3.1), then equation (9.3.1) goes over into equation (6.3.40) so that F has the form (9.4.5). If F is measurable then Corollary 6.3.4 implies that $F(p) = a \odot (4p - 1)$ for some constant vector $a \in \mathbb{R}^k$. Now we consider the remaining cases of M_1, M_2 given in (9.4.3) (in which we always have $M_1 \neq M_2$). By letting $p = p_1$, $q = q_1$ in (9.4.1) we obtain

$$\begin{aligned} F(pq) + F(p(1-q)) &+ F((1-p)q) + F((1-p)(1-q)) \\ &= \{M_1(p) + M_1(1-p)\}\{F(q) + F(1-q)\} \qquad (9.4.7) \\ &+ \{M_2(q) + M_2(1-q)\}\{F(p) + F(1-p)\} \end{aligned}$$

for all $p, q \in J$. Since the left-hand side of (9.4.7) is symmetric in p and q, the right hand side must be symmetric too, and thus F satisfies

$$\begin{aligned} \{M_1(p) &+ M_1(1-p)\}\{F(q) + F(1-q)\} \\ &+ \{M_2(q) + M_2(1-q)\}\{F(p) + F(1-p)\} \\ &= \{M_1(q) + M_1(1-q)\}\{F(p) + F(1-p)\} \\ &+ \{M_2(p) + M_2(1-p)\}\{F(q) + F(1-q)\}. \end{aligned}$$

By fixing q in this equation, say $q = \frac{1}{2}$ we get

$$F(p) + F(1-p) = b\{M_1(p) + M_1(1-p) - M_2(p) - M_2(1-p)\},$$

where b is given by

$$b = F\left(\frac{1}{2}\right)\left[M_1\left(\frac{1}{2}\right) - M_2\left(\frac{1}{2}\right)\right]^{-1}.$$

(Note that the denominator of b is different from zero in each remaining case of (9.4.3)). This means that the function

$$g(p) := F(p) - b\{M_1(p) - M_2(p)\}, \qquad p \in J, \tag{9.4.8}$$

satisfies the relation

$$g(p) + g(1 - p) = 0, \qquad p \in J. \tag{9.4.9}$$

Furthermore, since equation (9.4.7) is linear in F, and since an easy calculation shows that $b\{M_1(p) - M_2(p)\}$ is a solution of (9.4.7), we deduce that the map g defined in (9.4.8) satisfies (6.3.40) (because of (9.4.9)). From (9.4.8) we get the representation (9.4.5). If F is in addition measurable then g is given by

$$g(p) = c \odot (4p - 1) = c \odot 4p - c \odot 1 \tag{9.4.10}$$

for some vector $c \in \mathbb{R}^k$. But the condition (9.4.9) yields $c \odot \{4p - 1 + 4(1 - p) - 1\} = 0$ so that we have

$$c \odot 1 = 0. \tag{9.4.11}$$

Thus (9.4.10) and (9.4.11) can be rewritten as

$$g(p) = c' \odot p, \qquad c' \odot 1 = 0 \tag{9.4.12}$$

for some $c' \in \mathbb{R}^k$. Finally, (9.4.8) together with (9.4.12) yields (9.4.6). Since an immediate calculation verifies the converse we are finished.

Note that $c \in \mathbb{R}^k$ in (9.4.6) together with $c \odot 1 = 0$ yields $c = 0$ if $k = 1$. On the other hand, for sum form information measures with generating function $F(p) = b\{M_1(p) - M_2(p)\} + c \odot p$ the term $c \odot p$ is of no effect since

$$\sum_{i=1}^{n} c \odot p_i = c \odot 1 = 0.$$

Theorem 9.4.2. *Let $\{I_n : \Gamma_n^\circ \to \mathbb{R}\}$ be $(2,2)$-additive and have the sum form for some measurable generating function $F : J \to \mathbb{R}$. Then $\{I_n\}$ is of type*

(M_1, M_2) *with measurable, multiplicative functions* M_1, M_2, *given by (9.4.3) if, and only if*

$$I_n(P) = \begin{cases} (n-4) \odot a & \text{if } M_1 = M_2 = 0 \\ b \sum_{i=1}^{n} \{M_1(p_i) - M_2(p_i)\} & \text{otherwise} \end{cases} \qquad (9.4.13)$$

where $a \in \mathbb{R}^k$ *and* $b \in \mathbb{R}$.

Proof: Since $I_n(P) = \sum_{i=1}^{n} F(p_i)$, where F is given by (9.4.6), we obtain (with $a = (a_1, ..., a_k) \in \mathbb{R}^k$)

$$I_n(P) = \sum_{i=1}^{n} a \odot \{4p_i - 1\}$$

$$= \sum_{j=1}^{k} a_j \sum_{i=1}^{n} \{4p_{ji} - 1\}$$

$$= \sum_{j=1}^{k} (-a_j) \{n - 4\}$$

and

$$I_n(P) = b \sum_{i=1}^{n} \{M_1(p_i) - M_2(p_i)\} + \sum_{i=1}^{n} c \odot p_i$$

$$= b \sum_{i=1}^{n} \{M_1(p_i) - M_2(p_i)\},$$

respectively. This completes the proof.

We remark that the first case in (9.4.13) can also be written in the form

$$I_n(P) = I_n(P_1, ..., P_k)$$
$$= \{(n-1) - 3\} \odot a = \{H_n^o(P) - 3\} \odot a \qquad (9.4.14)$$
$$= \sum_{j=1}^{k} \{H_n^o(P_j) - 3\} a_j.$$

For the reminder of this section we assume that $k = 1$ and consider the functional equation (9.4.7) with $M_1(p) = M_2(p) = p^\alpha$, $p \in]0, 1[$, for some $\alpha \in \mathbb{R}$. That is,

$$F(pq) + F(p(1-q)) + F((1-p)q) + F((1-p)(1-q))$$
$$= \{p^\alpha + (1-p)^\alpha\}\{F(q) + F(1-q)\} \qquad (9.4.15)$$
$$+ \{q^\alpha + (1-q)^\alpha\}\{F(p) + F(1-p)\}$$

for all $p, q \in]0, 1[$. Before we present our main result, we prove a lemma which is analogous to Lemma 8.4.2. Also, recall the notation $A_k(p) = p^k + (1 - p)^k, p \in]0, 1[$.

Lemma 9.4.3. *Let $F :]0, 1[\rightarrow \mathbf{C}$ satisfy (9.4.15) for some $\alpha \in \mathbf{N} \cup \{0\}$, $p, q \in]0, 1[$, and let F be a polynomial of the form*

$$F(p) = \sum_{k=0}^{n} a_k \, p^k, \qquad p \in]0, 1[, \tag{9.4.16}$$

with $a_n \neq 0$ and $a_k \in \mathbf{C}, 0 \leq k \leq n$. Moreover let

$$h(p) := F(p) + F(1 - p). \tag{9.4.17}$$

Then the following equations hold:

$$a_n[1 + (-1)^n]A_n(q) = a_n[1 + (-1)^n]A_\alpha(q) + T_{\alpha,n}(q), \ n \geq 1; \tag{9.4.18}$$

where

$$T_{\alpha,n}(q) + \begin{cases} \binom{\alpha}{n}(-1)^n h(q) & \text{if } \alpha \geq n+1 \\ [1 + (-1)^\alpha]h(q) & \text{if } \alpha = n \\ 0 & \text{if } \alpha \leq n - 1 \, ; \end{cases} \tag{9.4.19}$$

$$\begin{aligned} a_{n-1}&[1 + (-1)^{n-1}]A_{n-1}(q) + (-1)^{n-1} n a_n A_n(q) \\ &= A_\alpha(q)\left[a_{n-1}[1 + (-1)^{n-1}] + (-1)^{n-1} n a_n\right] \\ &+ T_{\alpha,n-1}(q), n \geq 2; \end{aligned} \tag{9.4.20}$$

$$\begin{aligned} a_{n-2}&[1 + (-1)^{n-2}]A_{n-2}(q) \\ &+ (-1)^{n-2}\left[(n-1)a_{n-1}A_{n-1}(q) + a_n\binom{n}{2}A_n(q)\right] \\ &= A_\alpha(q)\Big(a_{n-2}[1 + (-1)^{n-2}] \\ &+ (-1)^{n-2}\left[(n-1)a_{n-1} + a_n\binom{n}{2}\right]\Big) \\ &+ T_{\alpha,n-2}(q), n \geq 3; \end{aligned} \tag{9.4.21}$$

and

$$a_{n-3}[1 + (-1)^{n-3}]A_{n-3}(q) + (-1)^{n-3}\Big[(n-2)a_{n-2}A_{n-2}(q)$$

$$+ a_{n-1}\binom{n-1}{2}A_{n-1}(q) + a_n\binom{n}{3}A_n(q)\Big]$$

$$= A_\alpha(q)\Big[[1 + (-1)^{n-3}]a_{n-3} + (-1)^{n-3}\Big((n-2)a_{n-2}$$

$$+ a_{n-1}\binom{n-1}{2} + a_n\binom{n}{3}\Big)\Big] + T_{\alpha,n-3}(q), n \geq 4.$$

(9.4.22)

If in addition n is an odd integer then we have

$$(n^2 - 4n)a_n = 4a_{n-1} \quad if \quad \alpha = n-1 \ (n \geq 3), \tag{9.4.23}$$

$$4(1-n)a_{n-1} = n^2 a_n \quad if \quad \alpha = n \quad (n \geq 3), \tag{9.4.24}$$

$$-4a_{n-1} = n^2 a_n \quad if \quad \alpha < n-1 (n \geq 3). \tag{9.4.25}$$

If n is an odd integer, then

$$\left.\begin{array}{l} n^2 - n(\alpha + 3) + 2\alpha = 0 \quad if \ \alpha \neq 1 \\ n^2 - 3n = 0 \quad if \ \alpha = 1 \end{array}\right\} \ (0 \leq \alpha \leq n-2, \ n \geq 3) \tag{9.4.26}$$

Proof: The proof is similar to the proof of Lemma 8.4.2. The left-hand side $L_F(p,q)$ of (9.4.15) is again given by

$$L_F(p,q) = \sum_{k=0}^{n} p^k \Big[a_k A_k(q) + (-1)^k \sum_{i=k}^{n} a_i \binom{i}{k} A_i(q)\Big] \tag{8.4.14}$$

(see Lemma 8.4.2), whereas the right-hand side $R_F(p,q)$ of (9.4.15) is a little bit more complicated than $R_f(p,q)$ in (8.4.15) and (8.4.16):

$$R_F(p,q) = A_\alpha(q) \sum_{k=0}^{n} a_k A_k(p) + A_\alpha(p)h(q)$$

$$= A_\alpha(q) \sum_{k=0}^{n} a_k \Big[p^k[1 + (-1)^k] + \sum_{l=0}^{k-1}\binom{k}{l}(-1)^l p^l\Big] \tag{9.4.27}$$

$$+ \Big[p^\alpha[1 + (-1)^\alpha] + \sum_{l=0}^{\alpha-1}\binom{\alpha}{l}(-1)^l p^l\Big]h(q)$$

(here we made use of (8.4.16)). Thus (9.4.27) has in comparison with $R_f(p,q)$ in Lemma 8.4.2 one additional summand, namely $A_\alpha(p)h(q)$, whereas the first summand of $R_F(p,q)$ is the same as that of $R_f(p,q)$ (when $A_\alpha(q)$ is replaced by $h(q)$). If we now compare the coefficients of p^n, p^{n-1} and p^{n-3} in the equation $L_F(p,q) = R_F(p,q)$ we get equations (8.4.24), (8.4.25) and (8.4.26) with $A_\alpha(q)$ instead of $h(q)$, and in each of these three equations we obtain the additional summand $T_{\alpha,n}(q)$, $T_{\alpha,n-1}(q)$ and $T_{\alpha,n-3}(q)$, respectively (which results from the term $A_\alpha(p)h(q)$; see (8.4.15)). Thus we arrive exactly at (9.4.18) (with (9.4.19)), (9.4.20) and (9.4.22). In completely the same manner we obtain by comparison of the coefficients of p^{n-2} in $L_F(p,q) = R_F(p,q)$

$$a_{n-2}A_{n-2}(q) + (-1)^{n-2}\left[a_{n-2}\binom{n-2}{n-2}A_{n-2}(q)\right.$$

$$+ a_{n-1}\binom{n-1}{n-2}A_{n-1}(q) + a_n\binom{n}{n-2}A_n(q)\bigg]$$

$$= A_\alpha(q)\left(a_{n-2}[1 + (-1)^{n-2}] + (-1)^{n-2}\left[a_{n-1}\binom{n-1}{n-2}\right.\right.$$

$$\left.\left.+ a_n\binom{n}{n-2}\right]\right) + T_{\alpha,n-2}(q),$$

where $n - 3 \geq 0$. But this equation is (9.4.21).

Now suppose n is odd. Then we get from (9.4.20) in case $\alpha = n - 1$ or $\alpha = n$

$$na_n A_n(q) = na_n A_{n-1}(q) + 2h(q), \qquad (9.4.28)$$

respectively

$$2a_{n-1}A_{n-1}(q) = 2a_{n-1}A_n(q) + nh(q). \qquad (9.4.29)$$

Now comparison of the coefficients of q^{n-1} in (9.4.28) and (9.4.29) yields

$$na_n\binom{n}{n-1} = 2na_n + 2\left[2a_{n-1} + a_n\binom{n}{n-1}\right],$$

respectively

$$4a_{n-1} = 2a_{n-1}\binom{n}{n-1} + 2na_{n-1} + na_n\binom{n}{n-1}.$$

But these two equations give (9.4.23) and (9.4.24).

Next, suppose $\alpha < n-1$ $(\alpha \in \mathbb{N} \cup \{0\}$, $n \geq 3)$ and n is odd. From (9.4.20) we obtain

$$2a_{n-1}A_{n-1}(q) + na_n A_n(q) = A_\alpha(q)[2a_{n-1} + na_n], \qquad (9.4.30)$$

$q \in]0,1[$. By comparison of the coefficients of q^{n-1} in (9.4.30) we have

$$a_{n-1} = -\frac{n^2}{4} a_n,$$

which is (9.4.25). This equation implies $a_n \neq 0$ and $a_{n-1} \neq 0$ (otherwise we have $a_n = a_{n-1} = 0$ which is impossible by supposition). Putting (9.4.25) into (9.4.30) we arrive at

$$na_n(-2A_n(q) + nA_{n-1}(q) - (n-2)A_\alpha(q)) = 0. \qquad (9.4.31)$$

Finally we divide (9.4.31) by na_n (since $na_n \neq 0$) and compare the coefficients of q in (9.4.31) to get (9.4.26). This finishes the proof.

Now we are ready to prove the main result in this section, giving the form of all measurable solutions of (9.4.15) for all $\alpha \in \mathbb{R}$. In Ebanks, Sahoo and Sander (1990a) this result was first established for all $\alpha \in \mathbb{N}$, and later extended by Ebanks (1992) to all $\alpha \in \mathbb{C}$. The proof presented below combines some new ideas, and contains many simplifications of the original proof.

Theorem 9.4.4. *A map $F :]0,1[\to \mathbb{C}$ is a measurable solution of equation (9.4.15) for some $\alpha \in \mathbb{C}$, if and only if*

$$F(p) = \phi_\alpha(p) + b\,p^\alpha \log p, \qquad p \in]0,1[, \qquad (9.4.32)$$

where ϕ_α is given by

$$\phi_\alpha(p) = \begin{cases} a\,(4p - 3), & \text{if } \alpha = 0, \\[2mm] a\,(4p^3 - 9p^2 + 5p) + c, & \text{if } \alpha = 1, \\[2mm] a\,(4p^3 - 3p^2 - p), & \text{if } \alpha = 2, \\[2mm] a\,(8p^3 - 9p^2 + p), & \text{if } \alpha = 3, \qquad (9.4.33) \\[2mm] a\,(12p^5 + 15p^4 - 40p^3 + 15p^2 - 2p), & \text{if } \alpha = 4, \\[2mm] a\,(-48p^5 + 75p^4 - 40p^3 + 15p^2 - 2p), & \text{if } \alpha = 5, \\[2mm] 0, & \text{otherwise,} \end{cases}$$

and where a, b, c are arbitrary complex constants.

If F is real-valued then the measurable solutions of (9.4.15) for $\alpha \in \mathbb{R}$ are again given by (9.4.32) and (9.4.33) where a, b, c are real constants.

Proof: An immediate but tedious calculation shows that F given by (9.4.32) and (9.4.33) satisfies (9.4.15). Since the proof of the converse is long we first give an outline of the main steps.

Step 1. Let $\alpha \in \mathbb{C}$, and suppose that $F :]0,1[\to \mathbb{C}$ is measurable on $]0,1[$ and satisfies (9.4.15). Then F has the form

$$F(p) = bp^{\alpha} \log p + \psi_{\alpha}(p), \quad p \in]0,1[, \tag{9.4.34}$$

where ψ_{α} is a polynomial and b is a constant.

Step 2. Since (9.4.15) is linear in F, and since the function $p \mapsto bp^{\alpha} \log p$ satisfies (9.4.15) for any $b \in \mathbb{C}$, we need to examine only the polynomial $\psi_{\alpha}(p)$ in (9.4.34). In the following let $F :]0,1[\to \mathbb{C}$ be that polynomial:

$$F(p) = \psi_{\alpha}(p) = \sum_{k=0}^{n} a_k\, p^k. \tag{9.4.35}$$

Then the continuous extension $\overline{F} : [0,1] \to \mathbb{C}$ of F satisfies

$$\overline{F}(0) = 0 = \overline{F}(1) \quad \text{if } \mathbf{Re}(\alpha) > 0 \text{ and } \alpha \neq 1. \tag{9.4.36}$$

Thus, if $\alpha \in \mathbb{N}$ we have

$$a_0 = 0, \ F(p) = \sum_{k=1}^{n} a_k\, p^k \text{ and } \sum_{k=1}^{n} a_k = 0 \ \text{ provided } \alpha \geq 2. \tag{9.4.37}$$

Moreover, without loss of generality we assume that n is odd in (9.4.37). (If F is of even degree, we rename that degree as $n - 1$ and add a new term $a_n\, p^n$ with $a_n = 0$.) Thus we assume that

$$a_n \neq 0 \quad \text{or} \quad a_{n-1} \neq 0. \tag{9.4.38}$$

Step 3. If F is a polynomial solution (9.4.35) of (9.4.15) for $\alpha \notin \mathbb{N} \cup \{0\}$ then $F \equiv 0$.

Step 4. If F is a polynomial solution (9.4.35) of (9.4.15) for $\alpha \in \{0, 1\}$ then

$$F(p) = \begin{cases} 4ap - 3a & \text{if } \alpha = 0 \\ 4ap^3 - 9ap^2 + 5ap + c & \text{if } \alpha = 1, \end{cases} \qquad (9.4.39)$$

where a and c are constants.

Step 5. Let F be a polynomial solution (9.4.35) with odd n, where $a_n \neq 0$ or $a_{n-1} \neq 0$. Moreover let $\alpha \in \mathbb{N} \cup \{0\}$, $\alpha \geq 2$. If α is even, then $n \leq \alpha + 1$. Thus F has the form

$$F(p) = \sum_{k=1}^{\alpha+1} a_k \, p^k, \quad \sum_{k=1}^{\alpha+1} a_k = 0. \qquad (9.4.40a)$$

If $3 \leq \alpha$ and α is odd, then $n \leq \alpha$ which implies

$$F(p) = \sum_{k=1}^{\alpha} a_k \, p^k, \quad \sum_{k=1}^{\alpha} a_k = 0. \qquad (9.4.40b)$$

Step 6. If $\alpha \geq 2$ is even so that the "degree" of F in (9.4.35) is $n = \alpha + 1$ then $\alpha = 2$ or $\alpha = 4$ (so that $n = 3$ or $n = 5$, respectively). If $\alpha \geq 3$ is odd so that $n = \alpha$ in (9.4.35), then $\alpha = 3$ or $\alpha = 5$ (so that again $n = 3$ or $n = 5$, respectively). This means that besides the two polynomials in (9.4.39), only four other nonzero polynomials can occur: Two polynomials of degree 3 (if $\alpha = 2$ or $\alpha = 3$) and two polynomials of degree 5 (if $\alpha = 4$ or $\alpha = 5$).

Step 7. We determine explicitly the coefficients of the polynomials of degree 3 and degree 5 mentioned in Step 6 and obtain together with (9.4.39) all polynomials in (9.4.33).

Verification of Step 1. If $\alpha \in \mathbb{C}$ and if $F :]0, 1[\to \mathbb{C}$ is a measurable solution of (9.4.15), then we can apply Theorem 8.2.1. Thus F has the form

$$F(p) = \sum_{j=1}^{M} \sum_{k=0}^{m_j-1} c_{jk} \, p^{\lambda_j} \log^k p, \quad p \in]0, 1[, \qquad (8.4.6)$$

where $M, m_1, \ldots, m_M \in \mathbb{N}$, $\lambda_1, \ldots, \lambda_M, c_{jk} \in \mathbb{C}$, $1 \leq j \leq M$, $0 \leq k \leq m_j - 1$. Moreover we have $\sum_{j=1}^{M} m_j \leq 20$ if $f(p) + f(1-p)$ and $p^\alpha + (1-p)^\alpha$ are linearly dependent and $\sum_{j=1}^{M} m_j \leq 35$ otherwise. Substituting (8.4.6) (with F instead of f) into (9.4.15) we get, using (8.4.5) and Lemma 8.4.1

$$\sum_{j=1}^{M} \sum_{l=0}^{m_j-1} d_{jl}(q) A_{jl}(p) = A_\alpha(p) h(q) + A_\alpha(q) \sum_{j=1}^{M} \sum_{l=0}^{m_j-1} c_{jl} A_{jl}(p), \qquad (9.4.41)$$

the second case both α and n are odd so that $\alpha < n$ gives $\alpha \leq n-2 < n-1$. Thus we can apply (9.4.26) in Lemma 9.4.3 and get that

$$n = \frac{1}{2}(\alpha + 3 \pm \sqrt{(\alpha-1)^2 + 8}.) \tag{9.4.52}$$

Since $n \in \mathbb{N} \cup \{0\}$ we must have

$$(\alpha - 1)^2 + 8 = x^2$$

for some positive integer x. This is only possible (for $\alpha \geq 2$) if $\alpha = 2$, which results in $n = 1$ or $n = 4$. But $n = 1$ contradicts our assumption $n > \alpha + 1$ or $n > \alpha$, and $n = 4$ is a contradiction since n is odd. Thus Step 5 together with (9.4.40) is verified.

Verification of Step 6. We consider the two cases identified in Step 5.

Case 1. Suppose $\alpha \geq 2$, α is even, and $n \leq \alpha + 1$. Without loss of generality we put $n = \alpha + 1$. We shall prove that there are only two possibilities which lead to nonzero F: $\alpha = 2$ or $\alpha = 4$. Suppose to the contrary that $\alpha \geq 6$ so that $n = \alpha + 1 \geq 7$. We distinguish now two sub-cases. If $a_n = 0$ we obtain from (9.4.20) together with $\alpha = n - 1$ (and n odd) again (9.4.28), that is $h = 0$. This implies $L_F(p, q) = 0$. As in Step 3 we get $F \equiv 0$. Now let $a_n \neq 0$. Then (9.4.20) implies (9.4.28) and thus

$$h(q) = \frac{1}{2} n a_n (A_n(q) - A_{n-1}(q)). \tag{9.4.53}$$

Using the fact that n is odd and $\alpha = n - 1$, substitution of (9.4.53) into (9.4.22) yields

$$2a_{n-3}A_{n-3}(q) + (n-2)a_{n-2}A_{n-2}(q)$$
$$+ a_{n-1}\binom{n-1}{2}A_{n-1}(q) + a_n\binom{n}{3}A_n(q) \tag{9.4.54}$$
$$= b_{n-1}A_{n-1}(q) + \frac{1}{2}\binom{n-1}{2}n a_n [A_n(q) - A_{n-1}(q)],$$

$q \in]0, 1[$, where

$$b_{n-1} := 2a_{n-3} + (n-2)a_{n-2} + a_{n-1}\binom{n-1}{2} + a_n\binom{n}{3}.$$

Thus (9.4.54) is an equation of the form

$$\alpha_n A_n(q) + \alpha_{n-1} A_{n-1}(q) + \alpha_{n-2} A_{n-2}(q) + \alpha_{n-3} A_{n-3}(q) = 0, \qquad (9.4.55)$$

$q \in\,]0, 1[$, where

$$\alpha_n = \left[\frac{1}{2}n\binom{n-1}{2} - \binom{n}{3}\right] a_n = \frac{1}{2} a_n \binom{n}{3}.$$

(The other three coefficients can be calculated but they will not be needed here.) Since n is odd and $n \geq 7$ we get from Theorem 8.2.5(a), in particular, $\alpha_n = 0$ which leads to the contradiction $a_n = 0$. Thus the first part of Step 6 is verified.

Case 2. Now we suppose $\alpha \geq 3$, α is odd, and $n \leq \alpha$; without loss of generality put $n = \alpha$. We want to show that only $\alpha = 3$ and $\alpha = 5$ can lead to nonzero polynomial F. Thus let us assume that $\alpha \geq 7$ which says that $n = \alpha \geq 7$. We proceed as in Case 1 and distinguish now the cases $a_{n-1} = 0$ and $a_{n-1} \neq 0$. Equation (9.4.20) with $\alpha = n$ implies (9.4.29). Thus if $a_{n-1} = 0$ we get again $h = 0$, $L_f(p, q) = 0$ and thus $F \equiv 0$. Now suppose $a_{n-1} \neq 0$. From (9.4.20) and (9.4.29) we obtain

$$h(q) = \frac{2}{n} a_{n-1}(A_{n-1}(q) - A_n(q)). \qquad (9.4.56)$$

Since n is odd and $\alpha = n$, substitution of (9.4.56) into (9.4.22) leads to

$$2a_{n-3} A_{n-3}(q) + (n-2) a_{n-2} A_{n-2}(q)$$
$$+ a_{n-1} \binom{n-1}{2} A_{n-1}(q) + a_n \binom{n}{3} A_n(q) \qquad (9.4.57)$$
$$= b_{n-1} A_n(q) + \frac{2}{n} \binom{n}{3} a_{n-1}(A_{n-1}(q) - A_n(q)).$$

This is again an equation of the form (9.4.55) where

$$\alpha_{n-1} = \left[\frac{2}{n}\binom{n}{3} - \binom{n-1}{2}\right] a_{n-1} = -\frac{1}{3}\binom{n-1}{2} a_{n-1}.$$

Theorem 8.2.5(a) implies, since n is odd and $n \geq 7$, $\alpha_{n-1} = 0$ so that we get the contradiction $a_{n-1} = 0$. Step 6 is now established.

Verification of Step 7. Because of Step 6 we now consider the remaining four cases in which possible nonzero polynomial solutions of (9.4.15) exist.

<u>Case 1.</u> Let F be a solution of the form (9.4.40a) where $\alpha = 2$ and $n = \alpha + 1 = 3$. Since $\alpha = n - 1$ we get $4a_2 = -3a_3$ from (9.4.23) with $n = 3$. Because of $a_1 + a_2 + a_3 = 0$ (and $a_0 = 0$) we get $a_1 = -\frac{1}{4}a_3$. Putting $a = \frac{a_3}{4}$ we get the third solution in (9.4.33).

<u>Case 2.</u> If F satisfies (9.4.40b) with $\alpha = 3$ and $n = \alpha$ then (9.4.24) yields $8a_2 = -9a_3$. Again $a_1 + a_2 + a_3 = 0$ leads to $a_1 = \frac{1}{8}a_3$. Putting $a = \frac{1}{8}a_3$ we get the fourth line of (9.4.33), namely

$$F(p) = 8ap^3 - 9ap^2 + ap.$$

<u>Cases 3 and 4.</u> By Case 3 we denote the case $\alpha = 4$ and $n = \alpha + 1 = 5$ and by Case 4 the case $\alpha = 5$ and $n = \alpha$. Then (9.4.53) in Case 4 and (9.4.56) in Case 5 are again valid. Using these expressions we get from a substitution of (9.4.40) into (9.4.15)

$$\sum_{k=0}^{n} a_k A_k(p) A_k(q) = h(p) A_\alpha(q) + h(q) A_\alpha(p)$$

$$= \begin{cases} \dfrac{1}{2} n a_n \big[A_n(p) A_{n-1}(q) & \text{if } n = \alpha + 1 = 5 \\ \quad + A_{n-1}(p)(A_n(q) - 2A_{n-1}(q)) \big], & \\ \dfrac{2}{n} a_{n-1} \big[A_n(q) A_{n-1}(p) & \text{if } \alpha = n = 5. \\ \quad + A_n(p)(A_{n-1}(q) - 2A_n(q)) \big], & \end{cases}$$

Thus we get (as in Step 7 of Theorem 8.4.3)

$$\sum_{k=0}^{5} \beta_k A_k(p) = 0, \quad p \in]0, 1[, \tag{9.4.58}$$

where

$$\beta_k = a_k A_k(q) \ (k = 0, 1, 2, 3), \quad \text{in Case 3 and Case 4,} \tag{9.4.59}$$

$$\left. \begin{aligned} \beta_4 &= -\frac{1}{5} a_4 (-5 A_4(q) + 2 A_5(q)) \\ &= -\frac{1}{5} a_4 (-4 A_3(q) + A_2(q)), \quad \text{in Case 4,} \end{aligned} \right\} \tag{9.4.60}$$

$$\left.\begin{aligned}
\beta_5 &= \frac{1}{2}a_5(2A_5(q) - 5A_4(q)) \\
&= \frac{1}{2}a_5(-4A_3(q) + A_2(q)), \qquad \text{in Case 3.}
\end{aligned}\right\} \qquad (9.4.61)$$

Here we made use of the identity following equation (8.4.68). Because of the representation (9.4.58) we get from Theorem 8.2.5(b)

$$a_3 A_3(q) = -4a_2 A_2(q) - 10(a_1 A_1(q) + 2a_0 A_0(q)). \qquad (9.4.62)$$

In Case 3 we also obtain from (8.2.45)

$$\frac{1}{2}a_5(-4A_3(q) + A_2(q)) = -2a_2 A_2(q) - 6(a_1 A_1(q) + 2a_0 A_0(q)), \quad (9.4.63)$$

whereas Case 4 leads to

$$-\frac{1}{5}a_4(-4A_3(q) + A_2(q)) = 5a_2 A_2(q) + 15(a_1 A_1(q) + 2a_0 A_0(q)). \quad (9.4.64)$$

These last three equations are of the form (8.2.46) and thus (8.2.47) implies (using $a_0 = 0$; see (9.4.37))

$$4a_2 = -30a_1, \ a_3 = 20a_1 \text{ in Cases 3 and 4,} \qquad (9.4.65)$$

$$\frac{1}{2}a_5 + 2a_2 = -18a_1, \ -2a_5 = 12a_1 \text{ in Case 3,} \qquad (9.4.66)$$

$$-\frac{1}{5}a_4 - 5a_2 = 45a_1, \ \frac{4}{5}a_4 = -30a_1 \text{ in Case 4.} \qquad (9.4.67)$$

Using $a_1 + a_2 + a_3 + a_4 + a_5 = 0$ and (9.4.65), (9.4.66) respectively (9.4.65), (9.4.67), we arrive in Case 3 at (with $a_1 = -2a$)

$$\left.\begin{aligned}
a_2 &= -\frac{30}{4}a_1 = 15a, \ a_3 = 20a_1 = -40a, \\
a_5 &= -6a_1 = 12a, \ a_4 = 15a,
\end{aligned}\right\} \qquad (9.4.68)$$

and in Case 4 (with $a_1 = -2a$)

$$\left.\begin{aligned}
a_2 &= -\frac{30}{4}a_1 = 15a, \ a_3 = 20a_1 = -40a \\
a_4 &= -\frac{150}{4}a_1 = 75a, \ a_5 = -48a.
\end{aligned}\right\} \qquad (9.4.69)$$

But (9.4.68) and (9.4.69) yield exactly the two remaining polynomials of degree 5 in (9.4.33). Because of (9.4.34), (9.4.35), Step 4 and Step 7, the first part of the theorem is completely proven.

The solutions (9.4.32) and (9.4.33) are real-valued if and only if $\alpha \in \mathbb{R}$. Thus we get as in Theorem 8.4.3 that $a, b, c \in \mathbb{R}$ for real-valued solutions of (9.4.15). This finishes the proof.

With Theorem 9.4.1 and Theorem 9.4.4, we can now characterize the information measures which have the measurable sum property and are (2,2)-additive of degree (α, β), for all pairs (α, β) of real numbers.

Theorem 9.4.5. *Let α, β be real numbers. The sequence of maps $I_n : \Gamma_n^o \to \mathbb{R}$ $(n = 2, 3, ...)$ is (2,2)-additive of degree (α, β) and has the sum form with a measurable generating function, if and only if $\{I_n\}$ is given by*

$$I_n = b\, H_n^{(\alpha,\beta)} + E_n^{(\alpha,\beta)},$$

where

$$
E_n^{(\alpha,\beta)} =
\begin{cases}
a[3H_n^{(0,1)} - 1], & \text{if } \alpha = \beta = 0, \\[2mm]
a[2H_n^{(3,1)} - 3H_n^{(2,1)}] + c[H_n^{(0,1)} + 1], & \text{if } \alpha = \beta = 1, \\[2mm]
a[2H_n^{(3,1)} - H_n^{(2,1)}], & \text{if } \alpha = \beta = 2, \\[2mm]
a[4H_n^{(3,1)} - 3H_n^{(2,1)}], & \text{if } \alpha = \beta = 3, \\[2mm]
a[6H_n^{(5,1)} + 7H_n^{(4,1)} - 16H_n^{(3,1)} + 4H_n^{(2,1)}], & \text{if } \alpha = \beta = 4, \\[2mm]
a[24H_n^{(5,1)} - 35H_n^{(4,1)} + 16H_n^{(3,1)} - 4H_n^{(2,1)}], & \text{if } \alpha = \beta = 5, \\[2mm]
0, & \text{otherwise,}
\end{cases}
$$

$(n = 2, 3, ...)$, where a, b, c are arbitrary constants.

Remark 9.4.6. Theorem 9.3.1 is an improvement of a result of Kannappan and Sander (1989). Many new ideas and refinements of known results were necessary to obtain Theorem 9.3.1 (for example the lemmas in section 6.5 and section 9.2) in its present form. If $M_1(p) = p^\alpha$ and $M_2(p) = p^\beta$ in Theorem 9.3.1, then the 1-dimensional version for $\ell \geq 3$ and $m \geq 3$ was presented by Kannappan and Sahoo (1986b, and also 1987a) and Losonczi (1986). (For the measurable solutions see Kannappan and Sahoo (1987b).) The 2-dimensional version of Theorem 9.3.1 for $\ell \geq 3$ and $m \geq 3$ was established by Kannappan and Sahoo (1988a). If $\ell = 3$, $m = 2$ then the 1-dimensional case of Theorem 9.3.1 was given by Kannappan and Sahoo (1989), whereas the k-dimensional case for $\ell \geq 3$, $m \geq 2$ was settled by Sahoo and Sander (1989). All results of Section 9.4 are due to the authors. In Ebanks, Sahoo and Sander (1990a), Theorem 9.4.4 was proved for $\alpha, \beta \in \mathbb{N}$. Recently, Ebanks (1992) established Theorem 9.4.4 for all $\alpha, \beta \in \mathbf{C}$.

9.5. Summary

In the first chapters of this book we came across the following two important functional equations (where $G : D_2 \to \mathbb{R}$, and where $\varphi, M : J \to \mathbb{R}$ with M multiplicative):

$$G(x,y) + G(x+y, z) = G(x, z) + G(x+z, y) \qquad (4.4.3)$$

for all $(x, y, z) \in D_3$, and

$$\varphi(x) + M(1-x)\varphi\left(\frac{y}{1-x}\right) = \varphi(y) + M(1-y)\varphi\left(\frac{x}{1-y}\right) \qquad (4.4.15)$$

for all $(x, y) \in D_2$. These equations are the characteristic functional equation for branching measures of information and the fundamental equation of information, respectively. Both equations were completely solved without any regularity assumptions. Thus the corresponding information measures, given by (2.1.1) and (3.1.1), respectively, could be characterized.

The situation is a little bit different in the case of sum form information measures which are (ℓ, m)-additive of type λ or of type (M_1, M_2). Here the essential equations are the fundamental sum form equations

$$\sum_{i=1}^{\ell} \sum_{j=1}^{m} f(p_i q_j) = \sum_{i=1}^{\ell} \sum_{j=1}^{m} f(p_i)f(q_j), \qquad (8.1.2)$$

respectively

$$\sum_{i=1}^{\ell} \sum_{j=1}^{m} F(p_i q_j) = \sum_{i=1}^{\ell} \sum_{j=1}^{m} [M_1(p_i)F(q_j) + M_2(q_j)F(p_i)], \qquad (5.4.7)$$

for $P \in \Gamma_\ell^o$, $Q \in \Gamma_m^o$, where $f, F, M_1, M_2 : J \to \mathbb{R}$, and where M_1 and M_2 are multiplicative. If $\ell = m = 2$ in (8.1.2) or (5.4.7) then the measurable solutions were determined in the 1-dimensional case (if M_1 or M_2 is not a projection in (5.4.7)). If M_1 and M_2 are projections in (5.4.7) then all measurable solutions were presented in the k-dimensional case for fixed $\ell \geq 2$, $m \geq 2$. But in these last two instances nearly nothing is known concerning the general solutions of (8.1.2) and (5.4.7), respectively. If $\ell = 2$ and $m \geq 3$, or $\ell \geq 3$ and $m = 2$ (and if in (5.4.7) M_1 or M_2 is not a projection) then in many cases the general solution was given.

So the problem remains to determine the general solutions of (8.1.2) and (5.4.7) in the just mentioned cases when one or both of ℓ and m equals 2. As already pointed out (cf. Remark 7.2.4, Lemma 8.3.3, Remark 8.3.6, Step 1 of Theorem 8.4.3, and Theorem 9.4.1) these open problems can be reduced to the problem of solving equation (6.3.40). Thus the remaining problems in this book can be solved by determining the general solution of the fundamental equation of sum form information measures

$$ f(pq) + f(p(1-q)) + f((1-p)q) + f((1-p)(1-q)) = 0, $$

for $p, q \in J$, where $f : J \to \mathbb{R}$.

REFERENCES

[1] S.H.S. Abou-Zaid (1984), *Functional Equations and Related Measurements.* M. Phil. Thesis, Dept. of Pure Math. University of Waterloo, Waterloo, Canada.

[2] J. Aczél (1965), *The general solution of two functional equations by reduction to functions additive in two variables and with the aid of Hamel bases.* Glasnik Mat.-Fiz. Astronom. 20, 65-72.

[3] J. Aczél (1966), *Lectures on Functional Equations and Their Applications.* Academic Press, New York.

[4] J. Aczél (1978), *Some recent results on characterizations of measures of information related to coding.* IEEE Trans. Inform. Theory, IT-24, 592-595.

[5] J. Aczél (1979), *Derivations and functional equations basic to characterizations of information measures.* C.R. Math. Rep. Acad. Sci. Canada 1, 165-168.

[6] J. Aczél (1980a), *A mixed theory of information - V. How to keep the (inset) expert honest.* J. Math. Anal. Appl. 75, 447-453.

[7] J. Aczél (1980b), *A mixed theory of information - VII. Inset information functions of all degrees.* C.R. Math. Rep. Acad. Sci. Canada 2, 125-129.

[8] J. Aczél (1980c), *Functions partially constants on rings of sets.* C.R. Math. Rep. Acad. Sci. Canada 2, 159-164.

[9] J. Aczél (1980d), *Information functions of degree* $(0, \beta)$. Utilitas Math. 18, 15-26.

[10] J. Aczél (1980e), *Information functions on open domain III.* C.R. Math. Rep. Acad. Sci. Canada 2, 281-285.

[11] J. Aczél (1981a), *Notes on generalized information functions.* Aequationes Math. 22, 97-107.

[12] J. Aczél (1981b), *Derivations and information functions (A tale of two surprises and a half).* In Contributions to Probability. Academic Press, New York-London-Sydney-San Francisco, 191-200.

[13] J. Aczél (1983), *A new theory of generalized information measures, recent results in the 'old' theory and some 'real life' interpretations of old and new information measures.* Jahrb. Ueberblicke Mathematik, 25-35.

[14] J. Aczél (1984a), *Measuring information beyond communication theory - Why some generalized information measures may be useful, others not.* Aequationes Math. 27, 1-19.

[15] J. Aczél (1984b), *Some unsolved problems in the theory of functional equations, II.* Aequationes Math. 26, 255-260.

[16] J. Aczél (1984c), *Some recent results on information measures, a new generalization and some 'real life' interpretations of 'old' and new measures.* J. Aczél (ed.) Functional Equations: History, Applications and Theory, D. Reidel Publishing Company, pp.175-189

[17] J. Aczél (1986), *Characterizing information measures: Approaching the end of an era.* In Lecture Notes in Computer Science, Vol. 286 (Uncertainty in Knowledge-Based Systems), 359-384, Springer.

[18] J. Aczél (1987), *A Short Course on Functional Equations.* D. Reidel Publishing Company, Dordrecht-Boston-Lancaster-Tokyo.

[19] J. Aczél, J. Baker, D. Z. Djokovic, Pl. Kannappan and F. Rado (1971), *Extensions of certain homomorphisms of semigroups to homomorphisms of groups.* Aequationes Math. 6, 263-271.

[20] J. Aczél and Z. Daróczy (1975), *On measures of information and their characterizations.* Academic Press, New York-San Francisco-London.

[21] J. Aczél and Z. Daróczy (1978), *A mixed theory of information-I. Symmetric, recursive and measurable entropies of randomized systems of events.* RAIRO Informat. Theor. 12, 149-155.

[22] J. Aczél and J. Dhombres (1989), *Functional Equations in Several Variables.* Cambridge University Press, Cambridge-New York-New Rochelle-Melbourne-Sydney.

[23] J. Aczél, B. Forte and C.T. Ng (1974), *Why the Shannon and Hartley entropies are 'natural'.* Adv. in Appl. Probab. 6, 131-146.

[24] J. Aczél and PL. Kannappan (1978), *A mixed theory of information-III. Inset entropies of degree β.* Inform. and Control 39, 315-322.

[25] J. Aczél and PL. Kannappan (1982), *General two-place information functions.* Resultate Math. 5, 99-106.

[26] J. Aczél and C.T. Ng (1981), *On general information functions.* Utilitas Math. 19, 157-170.

[27] J. Aczél and C.T. Ng (1983), *Determination of all semisymmetric recursive information measures of multiplicative type on n positive discrete probability distributions.* Linear Algebra Appl. 52/53, 1-30.

[28] R. Baer (1934), *Erweiterung von Gruppen und ihren Isomorphismen.* Math. Z. 38, 375-416.

[29] M. Behara (1990), *Additive and nonadditive measures of entropy.* John Wiley and Sons, New York, Chichester, Brisbane, Singapore, Toronto.

[30] M. Behara and P. Nath (1973), *Additive and non-additive entropies of finite measurable partitions.* In Probability and Information Theory, II (Lecture Notes in Math.) Vol. 296, Springer, Berlin, pp.102-138.

[31] M. Behara and P. Nath (1974), *Entropy and utility.* Information, Inference and Decision (Ed. G. Monges). Theory and Decision Library, 1, 145-154, D. Reidel, Dordrecht.

[32] T. Borges (1967), *Zur Herleitung der Shannonschen Information.* Math. Z. 96, 282-287.

[33] N. Bourbaki (1966), *Elements of mathematics, general topology.* Addison-Wesley, Reading, Mass.-Palo Alto-London, Don Mills, Ont.

[34] T.W. Chaundy and J.B. McLeod (1960), *On a functional equation.* Edinburgh Math. Notes 43, 7-8.

[35] I. Csiszár (1978), *Information measures: A critical survey.* In Trans. Seventh Prague Conf. Information Theory, Statistical Decision Functions, Random Processes. Prague: Academia, pp.73-86.

[36] Z. Daróczy (1967a), *Uber eine Characterisierung der Shannonschen Entropies.* Statistica (Bologna) 27, 199-205.

[37] Z. Daróczy (1967b), *Uber die Charakterisierungen der Shannonschen Entropies.* Proc. Colloq. om Information Theory (Debrecen, 1967), Vol I, János Bolyai Math. Soc. Budapest, 135-139

[38] Z. Daróczy (1969), *On the Shannon measure of information* (in Hungarian). Magyar Tud. Akad. Mat. Fiz. Oszt. Közl. 19, 9-24. In Selected Trans. in Math. Stat. and Prob. 10 (1971), 193-210.

[39] Z. Daróczy (1970), *Generalized information functions.* Inform. and Control 16, 36-51.

[40] Z. Daróczy (1971), *On the measurable solutions of a functional equation.* Acta. Math. Acad. Sci. Hungar. 22, 11-14.

[41] Z. Daróczy and A. Járai (1979), *On the measurable solution of a functional equation arising in information theory.* Acta. Math. Acad. Sci. Hungar. 34, 105-116.

[42] Z. Daróczy and I. Kátai (1970), *Additive zahlentheoretische Funktionen und das Mass der Information.* Ann. Univ. Budapest Eötvös Sect. Math. 13, 83-88.

[43] Z. Daróczy and Gy. Maksa (1979), *Nonnegative information functions.* Analytic Function Methods in Probability and Statistics, Collq. Math. Soc. János Bolyai 21, 65-76.

[44] K.R. Davidson and C.T. Ng (1981), *Information measures and cohomology.* Utilitas Math. 20, 27-34.

[45] T.M.K. Davison and B.R. Ebanks (1995), *Cocycles on cancellative semigroups.* Publicationes Math. Debrecen 46, 137-147.

[46] N.G. DeBruijn (1951), *Functions whose differences belong to a given class.* Nieuw Arch. Wiskunde 23, 194-218.

[47] P.A. Devijver (1977), *Entropies of degree β and lower bounds for the average error rate.* Inform. and Control 34, 222-226.

[48] G. Diderrich (1975), *The role of boundedness in characterizing Shannon entropy.* Inform. and Control 29, 149-161.

[49] G. Diderrich (1978), *Local boundedness and the Shannon entropy.* Inform. and Control 36, 292-308.

[50] G. Diderrich (1979), *Continued fractions and the fundamental equation of information.* Aequationes Math. 19, 93-103.

[51] G. Diderrich (1986), *Boundedness on a set of positive measure and the fundamental equation of information.* Publ. Math. Debrecen 33, 1-7.

[52] J. Dieudonné (1960), *Foundations of Modern Analysis*, Vol. 1, Academic Press, New York-San Francisco, London.

[53] B.R. Ebanks (1976), *On a generalized recursivity and entropies of degree α.* Inform. and Control 32, 33-42.

[54] B.R. Ebanks (1978), *The branching property in generalized information theory.* Adv. Appl. Probab. 10, 788-802.

[55] B.R. Ebanks (1979), *Branching measures of information on strings.* Canadian Math. Bull. 22, 433-448.

[56] B.R. Ebanks (1980), *Branching and generalized-recursive inset entropies.* Proc. Amer. Math. Soc. 79, 260-267.

[57] B.R. Ebanks (1981), *A characterization of separable utility functions.* Kybernetika 17, 244-255.

[58] B.R. Ebanks (1982a), *The general symmetric solution of a functional equation arising in the mixed theory of information.* C.R. Math. Rep. Acad. Sci. Canada 4, 195-200.

[59] B.R. Ebanks (1982b), *Kurepa's functional equation on semigroups.* Stochastica 6, 39-55.

[60] B.R. Ebanks (1983a), *Generalized-recursive entropies*. Utilitas Math. 23, 5-21.

[61] B.R. Ebanks (1983b), *Measures of vector information with the branching property*. Kybernetika 19, 263-269.

[62] B.R. Ebanks (1984a), *Measures of inset information on open domain-III. Weakly regular, symmetric, β-recursive entropies*. C.R. Math. Rep. Acad. Sci. Canada 6, 159-164.

[63] B.R. Ebanks (1984b), *Kurepa's functional equation on Gaussian semigroups*, in: J. Aczél (ed.), Functional Equations: History, Applications and Theory. D. Reidel Publ. Co., 167-173.

[64] B.R. Ebanks (1984c), *Polynomially-additive entropies*. J. Applied Probab. 21, 179-185.

[65] B.R. Ebanks (1985), *Measurable solutions of functional equations connected with information measures on open domains*. Utilitas Math. 27, 217-223.

[66] B.R. Ebanks (1986a), *Measures of inset information on open domains-II: Additive inset entropies with measurable sum property*. Probab. Th. Rel. Fields 73, 517-528.

[67] B.R. Ebanks (1986b), *On the equation $F(X) + M(X)G(X^{-1}) = 0$ for additive F and multiplicative M on the positive cone of $\mathbb{R}^n$*. C.R. Math. Rep. Acad. Sci. Canada 8, 247-252.

[68] B.R. Ebanks (1989a), *On the equation $F(X) + M(X)G(X^{-1}) = 0$ on K^n*. Linear Algebra Appl. 125, 1-17.

[69] B.R. Ebanks (1989b), *Generalized characteristic equation of branching information measures*. Aequationes Math. 37, 162-178.

[70] B.R. Ebanks (1990), *Branching inset entropies on open domains*. Aequationes Math. 39, 100-113.

[71] B.R. Ebanks (1992), *Determination of measurable sum form information measures satisfying (2,2)-additivity of degree (α, β)-II: The whole story*. Radovi Mat. 8, 159-169.

[72] B.R. Ebanks (1994), *Additive k-dimensional inset entropies with measurable sum property*. J. Math. Anal. Appl. 187, 952-960.

[73] B.R. Ebanks (1996), *Fundamental equation of information revisited*. Aequationes Math. 51, 86-99.

[74] B.R. Ebanks and L. Losonczi (1992), *On the linear independence of some functions*. Publ. Math. Debrecen 41, 135-146.

[75] B.R. Ebanks and Gy. Maksa (1986), *Measures of inset information on the open domain -I. Inset entropies and information functions of all degrees.* Aequationes Math. 30, 187-201.

[76] B.R. Ebanks and W. Sander (1986), *A mixed theory of information-IX. Inset information functions of degree* $(0, \beta)$. Utilitas Math. 30, 63-78.

[77] B.R. Ebanks, P.L. Kannappan, C.T. Ng (1987), *Generalized fundamental equation of information of multiplicative type.* Aequationes Math. 32, 5-17.

[78] B.R. Ebanks, P.L. Kannappan, C.T. Ng (1988), *Recursive inset entropies of multiplicative type on open domains.* Aequationes Math. 36, 268-293.

[79] B.R. Ebanks, P.K. Sahoo and W. Sander (1990a), *Determination of measurable sum form information measures satisfying (2,2)-additivity of degree* (α, β). Radovi Matematicki 6, 77-96.

[80] B.R. Ebanks, P.K. Sahoo and W. Sander (1990b), *General solution of two functional equations concerning measures of information.* Results in Math. 18, 10-17.

[81] B.R. Ebanks. PL. Kannappan, P.K. Sahoo and W. Sander (1997), *Characterizations of sum form information measures on open domains*, Aequationes Math, 54, 1-30.

[82] J. Erdös (1959), *A remark on the paper "On some functional equations" by S. Kurepa.* Glasnik Mat. Fiz. Astronom Ser.II, 14, 3-5.

[83] D.K. Faddeev (1956), *On the concept of entropy of a finite probabilistic scheme* (Russian). Uspehi Mat. Nauk (N.S.) 11, No. 1, 67, 227-231.

[84] B. Forte (1977), *Subadditive entropies for a random variable.* Boll. Un. Mat. Ital. B (5) 14, 118-133.

[85] B. Forte and C.A. Bortone (1977), *Non-symmetric entropies with the branching property*, Utilitas Math. 12, 3-23.

[86] B. Forte and Z. Daróczy (1968), *A characterization of Shannon's entropy.* Boll. Un. Mat. Ital. B (4) 1, 631-635.

[87] B. Forte and R. Gupta (1985), *Additive and subadditive entropies for discrete n-dimensional random vectors.* Aequationes Math. 28, 269-287.

[88] B. Forte and C.T. Ng (1973), *On a characterization of entropies of degree* β. Utilitas Math. 4, 193-205.

[89] B. Forte and C.T. Ng (1974), *Entropies with the branching property.* Ann. Mat. Pura ed Appl. 101, 355-373.

[90] B. Forte and C.T. Ng (1975), *Derivation of a class of entropies including those of degree β*. Inform. and Control 28, 335-351.

[91] S. Guiasu (1977), *Information theory with applications*. McGraw Hill, New York-London.

[92] K-G. Grosse-Erdmann (1989), *Regularity properties of functional equations and inequalities*. Aequationes Math. 37, 233-251.

[93] J. Havrda and F. Charvat (1967), *Quantification method of classification processes. Concept of structural a-entropy*. Kybernetika (Prague) 3, 30-35.

[94] E. Hewitt and K. Stromberg (1965), *Real and abstract analysis*. Springer, Berlin-Heidelberg-New York.

[95] Ja. Hincin (1953), *The concept of entropy in the theory of probability* (Russian). Uspehi Mat. Nauk 8, No. 3 (55), 3-20.

[96] M. Hosszú (1963), *On a functional equation treated by S. Kurepa*. Glasnik Mat.-Fiz.i Astr. 18, 59-60.

[97] M. Hosszú (1971), *On the functional equation $F(x + y, z) + F(x, y) = F(x, y + z) + F(y, z)$*. Periodica Math. Hungar. 1, 213-216.

[98] N. Jacobson (1975), *Lectures in Abstract Algebra, vol. III*. Springer-Verlag, see pp. 81-83.

[99] A. Járai (1979), *On measurable solutions of functional equations*. Publ. Math. Debrecen 26, 17-35.

[100] A. Járai (1982), *Regularity properties of functional equations*. Aequationes Math. 25, 52-66.

[101] A. Járai (1986), *On regular solutions of functional equations*. Aequationes Math. 30, 21-54.

[102] A. Járai (1995), *A Steinhaus type theorem*. Publ. Math. Debrecen 47, 1-13.

[103] A. Járai and W. Sander (1997), *A regularity theorem in information theory*. Publ. Math. Debrecen. 50, 339-357.

[104] B. Jessen, J. Karpf and A. Thorup (1968), *Some functional equations in groups and rings*. Math. Scand. 22, 257-265.

[105] A. Kaminski and PL. Kannappan (1977), *A note on some theorems in information theory*. Bull. Acad. Polon. Sci. Ser. Sci. Math. 25, 925-928.

[106] A. Kaminski and PL. Kannappan (1980), *Distributional solutions in information theory II*. Ann. Polon. Math. 38, 289-294.

[107] A. Kaminski, PL. Kannappan and J. Mikusinski (1979), *Distributional solutions in information theory I.* Ann. Polon. Math. 36, 101-110.

[108] A. Kaminski and J. Mikusinski (1974), *On the entropy equation.* Bull. Acad. Polon. Sci. Ser. Sci. Math. 22, 319-323.

[109] PL. Kannappan (1975), *On various characterizations of generalized directed divergence.* Indian J. Pure Appl. Math. 6, 655-667.

[110] PL. Kannappan (1978), *Note on generalized information function.* Tohoku Math. J. 30, 251-255.

[111] PL. Kannappan (1980a), *On some generalized functional equations in information theory.* Aequationes Math. 20, 149-158.

[112] PL. Kannappan (1980b), *A mixed theory of information - IV. Inset-inaccuracy and directed divergence.* Metrika 27, 91-98.

[113] PL. Kannappan (1980c), *One some functional equations from additive and nonadditive measures-I.* Proc. Edinburgh Math. Soc. 23, 145-150.

[114] PL. Kannappan (1981), *One some functional equations from additive and nonadditive measures-IV.* Kybernetika (Prague) 17, 394-400.

[115] PL. Kannappan (1982), *One some functional equations from additive and nonadditive measures-V.* Utilitas Math. 22, 141-147.

[116] PL. Kannappan (1983), *On a multiplace functional equation related to information measures and functions.* Kybernetika (Prague) 19, 110-120.

[117] PL. Kannappan and C.T. Ng (1973), *Measurable solutions of functional equations related to information theory.* Proc. Amer. Math. Soc. 38, 303-310.

[118] PL. Kannappan and C.T. Ng (1974), *A functional equation and its application to information theory.* Ann. Polon. Math. 30, 105-112.

[119] PL. Kannappan and C.T. Ng (1979), *Representations of measures of information.* Trans. of the Eighth Prague Conference, Academic Publ. House of the Czec. Acad. Sci. Prague, C, 203-207.

[120] PL. Kannappan and C.T. Ng (1980), *On functional equations and measures of information II.* J. Appl. Prob. 17, 271-277.

[121] PL. Kannappan and C.T. Ng (1983), *On a generalized fundamental equation of information.* Can. J. Math. 35, 862-872.

[122] PL. Kannappan and C.T. Ng (1985), *On functional equations and measures of information I.* Publ. Math. Debrecen 32, 243-249.

[123] PL. Kannappan and P.N. Rathie (1972a), *An application of a functional equation to information theory.* Ann. Polon. Math. 26, 95-101.

[124] PL. Kannappan and P.N. Rathie (1972b), *On a characterization of directed divergence.* Inform and Control 22, 163-171.

[125] PL. Kannappan and P.N. Rathie (1973a), *An axiomatic characterization of generalized directed-divergence.* Kybernetika (Prague) 9, 330-337.

[126] PL. Kannappan and P.N. Rathie (1973b), *On various characterizations of directed divergence.* In Translations of the Sixth Prague Conference on Information Theory, Statistical Decision Functions and Random Processes (Prague 1971), Academia, Prague, pp. 331-339.

[127] PL. Kannappan and P.N. Rathie (1974), *On a generalized directed-divergence function.* Czechoslovak Math. J. 24 (99), 5-14.

[128] PL. Kannappan and P.N. Rathie (1975a), *On the solution of a functional equation connected with inaccuracy.* J. Indian Math. Soc. 39, 131-147.

[129] PL. Kannappan and P.N. Rathie (1975b), *On generalized information function.* Tohoku Math. J. 27, 207-212.

[130] PL. Kannappan and P.N. Rathie (1978), *On a generalized directed divergence and related measures.* In Transactions of the seventh Prague Conference on Information Theory, Statistical Decision Functions, Random Processes and of the Eighth European Meeting of Statisticians (Tech. Univ. Prague 1974), Vol. B, Academia, Prague, pp. 255-265.

[131] PL. Kannappan and P.K. Sahoo (1985a), *On a functional equation connected to sum form nonadditive information measures on an open domain.* C.R. Math. Rep. Acad. Sci. Canada 7, 45-50.

[132] PL. Kannappan and P.K. Sahoo (1985b), *On a functional equation in two variables connected to sum form information measures on an open domain.* Indian J. Math. 27, 33-40.

[133] PL. Kannappan and P.K. Sahoo (1985c), *On a functional equation connected to sum form nonadditive information measures on open domains-III.* Stochastica, 9, 111-124.

[134] PL. Kannappan and P.K. Sahoo (1986a), *On a functional equation connected to sum form nonadditive information measures on an open domain-I.* Kybernetika (Prague) 22, 268-275.

[135] PL. Kannappan and P.K. Sahoo (1986b), *On the general solution of a functional equation connected to the sum form information measure-I.* Publ. De L'Institut Math. 40(54), 57-62.

[136] PL. Kannappan and P.K. Sahoo (1986c), *On the general solution of a functional equation connected to the sum form information measure-III.* Internat. J. Math. & Math. Sci., 9, 545-550.

[137] PL. Kannappan and P.K. Sahoo (1986d), *On the general solution of a functional equation connected to the sum form information measure-IV.* Utilitas Math. 30, 191-197.

[138] PL. Kannappan and P.K. Sahoo (1987a), *On the general solution of a functional equation connected to the sum form information measure-II.* Mathematica (Cluj), 29(52), 131-137.

[139] PL. Kannappan and P.K. Sahoo (1987b), *On a functional equation connected to sum form nonadditive information measures on an open domain-II.* Glas. Mat., 22(42),343-351.

[140] PL. Kannappan and P.K. Sahoo (1988a), *On the general solution of a functional equation connected to the sum form information measure-V.* Acta Math. Univ. Commen (Czechoslovakia), 54/55, 89-102.

[141] PL. Kannappan and P.K. Sahoo (1988b), *Weighted entropy of degree β on open domain.* Proc. of the Ramanujan Centennial Inter. Conference, RMS Publication No 1, 119-125.

[142] PL. Kannappan and P.K. Sahoo (1989), *Representation of sum form information measures with additivity of type (α, β) on open domain.* In Computing and Information, R. Janicki and W.W. Kockodaj (Editors), Elsevier Science Publishers B.V. (North-Holland), 243-253.

[143] PL. Kannappan and P.K. Sahoo (1990a), *Representation of sum form information measures with weighted additivity of type (α, β) on open domain.* Jour. Math. Phy. Sci., 24, 89-99.

[144] PL. Kannappan and P.K. Sahoo (1990b), *Parametrically additive sum form weighted information measures.* Advances in Computing and Information, S.G. Akl, F. Fiala and W. Koczkodaj (eds), Canadian Scholars' Press Inc., Toronto, 26-31.

[145] PL. Kannappan and P.K. Sahoo (1991), *Parametrically additive sum form information measures.* In Constantine Caratheodory: An International Tribute, T.M. Rassias (ed). 574-580.

[146] PL. Kannappan and P.K. Sahoo (1993), *Sum form equation of multiplicative type.* Acta. Math. Hungar. 61, 203-217.

[147] PL. Kannappan and W. Sander (1982), *A mixed theory of information - VIII. Inset measures depending upon several distributions.* Aequationes Math. 25, 177-193.

[148] PL. Kannappan and W. Sander (1989), *On entropies with the sum property on open domain.* Analysis 9, 253-267.

[149] D.G. Kendall (1963), *Functional equations in information theory*. Z. Wahrsch Verw. Gebiete 2, 225-229.

[150] D.F. Kerridge (1961), *Inaccuracy and inference*. J. Roy. Statist. Soc. Ser. B 23, 184-194.

[151] M. Kuczma (1972), *Note on additive functions of several variables*. Uniw. Ślaski w Katowicach Prace Nauk. Prace Mat. 2, 49-51.

[152] M. Kuczma (1985), *An introduction to the theory of functional equations and inequalities*. PWN, Uniw Ślaski, Warszawa-Krakow-Katowice.

[153] S. Kullback (1959), *Information theory and statistics*. John Wiley and Sons, New York.

[154] S. Kurepa (1956), *On some functional equations*. Glasnik Mat.-Fiz. Astronom Ser. II, 11, 3-5.

[155] A. G. Kurosh (1956), *The theory of groups*. Chelsea, New York.

[156] M. Laczkovich (1980), *Functions with measurable differences*. Acta Math. Acad. Sci. Hungar. 35, 217-235.

[157] T. Y. Lam (1973), *The algebraic theory of quadratic forms*. Benjamin Cummings. Reading, Mass., London, Amsterdam, Don Mills, Ontario, Sydney, Tokyo.

[158] J. Lawrence (1981), *The Shannon kernel of a non-negative information function*. Aequationes Math. 23, 233-235.

[159] J. Lawrence, G. Mess and F. Zorzitto (1979), *Near derivations and information functions*. Proc. Amer. Math. Soc. 76, 117-122.

[160] P.M. Lee (1964), *On the axioms of information theory*. Ann. Math. Statistics 35, 415-418.

[161] L. Losonczi (1981), *A characterization of entropies of degree α*. Metrika 28, 237-244.

[162] L. Losonczi (1985a), *Functional equations of sum form*. Publ. Math. Debrecen 32, 57-71.

[163] L. Losonczi (1985b), *Sum form equations on an open domain I*. C.R. math. Rep. Acad. Sci. Canada 7, 85-90.

[164] L. Losonczi (1986), *Sum form equations on an open domain II*. Utilitas Math. 29, 125-132.

[165] L. Losonczi (1991), *Remark*. Proceedings of the 28th International Symposium on Functional Equations, Aequationes Math. 41, 302-303.

[166] L. Losonczi (1993a), *Measurable solutions of a functional equation related to (2,2)-additive entropies of degree α*. Publ. Math. Debrecen 42, 109-137.

[167] L. Losonczi (1993b), *Measurable solutions of functional equations of sum form.* Acta Math. Hung. 61, 165-182.

[168] L. Losonczi (1997), *A structure theorem for sum form functional equations.* Aequationes Math. 53, 141-154.

[169] L. Losonczi and Gy. Maksa (1982), *On some functional equations of the information theory.* Acta. Math. Acad. Sci. Hungar. 39, 73-82.

[170] Gy. Maksa (1980), *Bounded symmetric information functions.* C.R. Math. Rep. Acad. Sci. Canada 2, 247-252.

[171] Gy. Maksa (1981a), *The general solution of a functional equation related to the mixed theory of information.* Aequationes Math. 22, 90-96.

[172] Gy. Maksa (1981b), *On near derivations.* Proc. Amer. Math. Soc. 81, 406-408.

[173] Gy. Maksa (1981c), *On the bounded solutions of a functional equation.* Acta. Math. Acad. Sci. Hungar. 37, 445-450.

[174] Gy. Maksa (1982), *Solution on the open triangle of the generalized fundamental equation of information with four unknown functions.* Utilitas Math. 21C, 267-282.

[175] Gy. Maksa (1987), *The general solution of a functional equation arising in information theory.* Acta. Math. Acad. Sci. Hungar. 49, 213-217.

[176] Gy. Maksa and C.T. Ng (1986), *The fundamental equation of information on open domain.* Publ. Math. Debrecen 33, 9-11.

[177] A.M. Mathai and P.N. Rathie (1976), *Recent contributions to axiomatic definitions of information and statistical measures through functional equations.* Essays in Probability and Statistics, pp. 607-633, Shinko Tsusho, Tokyo.

[178] P. Nath (1976), *On some functional equations and their applications.* Publ. De L'Institut Math. 20 (34), 191-201.

[179] C.T. Ng (1974), *Representation of measures of information with the branching property.* Inform. and Control 25, 45-56.

[180] C.T. Ng (1979), *Measures of information with the branching property over a graph and their representation.* Inform. & Control 41, 214-231.

[181] C.T. Ng (1980a), *Information functions on open domains.* C.R. Math. Rep. Acad. Sci. Canada 2, 119-123.

[182] C.T. Ng (1980b), *Information functions on open domains II.* C.R. Math. Rep. Acad. Sci. Canada 2, 155-158.

[183] C.T. Ng (1980c), *On the functional equation* $f(x) + f(y) = \alpha(1 - x)f(\frac{y}{1-x}) + \alpha(1 - y)f(\frac{x}{1-y})$. In Proceedings of the Eighteenth International Symposium on Functional Equations (Waterloo-Scarborough, Ont 1980), Centre for Information Theory, Faculty of Mathematics, University of Waterloo, Canada, 1980, p.27.

[184] C.T. Ng (1987), *The equation* $F(X) + M(X)G(X^{-1}) = 0$ *and homogeneous biadditive forms.* Linear Algebra Appl. 93, 255-279.

[185] A. Ostrowski (1929), *Mathematische Miszellen. XIV. Uber die Funktionalgleichung der Exponentialfunktion und verwandte Funktionalgleichungen.* Jahresber. Deutsch. Math.-Verein. 38, 54-62.

[186] C.F. Picard (1978), *Mesures d'information avec preference ne possedent pas la propriete de branchement.* In Theorie de l'information. Colloques Internationaux du C.N.R.S. 276 C.N.R.S., Paris, pp. 125-139.

[187] P.N. Rathie and PL. Kannappan (1971), *On a functional equation connected with Shannon's entropy.* Funkcial. Ekvac. 14, 153-159.

[188] P.N. Rathie and PL. Kannappan (1972), *A directed-divergence function of type β.* Inform. and Control 20, 38-45.

[189] P.N. Rathie and PL. Kannappan (1973a), *On a new characterization of directed divergence in information theory.* In Translations of the Sixth Prague Conference on Information Theory, Statistical Decision Functions and Random Processes (Prague 1971), Academia, Prague, pp. 733-745.

[190] P.N. Rathie and PL. Kannappan (1973b), *An inaccuracy function of type β.* Ann. Inst. Statist. Math. 25, 205-214.

[191] L. Reich (1992), *On linear relations among analytic functions arising in the theory of sum form equations and in the theory of linear differential equations.* Grazer Math. 316, 169-180.

[192] D. Roy (1980), *Axiomatic characterization of second order information improvement.* J. Combin. Inform. Syst. Sci. 5, 107-111.

[193] W. Rudin, *Principles of Mathematical Analysis.* Second Edition. McGraw-Hill, New York, 1964.

[194] P.K. Sahoo (1983), *On some functional equations connected to sum form information measures on open domains.* Utilitas Math. 23, 161-175.

[195] P.K. Sahoo (1986), *Theory and Applications of Some Measures of Uncertainty.* Ph.D. Thesis, Dept. of Applied Math., University of Waterloo, Waterloo, Canada.

[196] P.K. Sahoo (1995a), *Determination of all additive sum form information measures of k positive discrete probability distributions.* J. Math. Anal. Appl. 194, 235-249.

[197] P.K. Sahoo (1995b), *Three open problems in functional equations.* Amer. Math. Monthly 102, 741-742.

[198] P.K. Sahoo and W. Sander (1989), *Sum form information measures on open domain.* Radovi Mat. 5, 261-270.

[199] P.K. Sahoo, S. Soltani, A.K.C. Wong and Y.C. Chen (1988), *A survey of thresholding techniques.* Computer Vision, Graphics & Image Processing, 41, 233-260.

[200] P.K. Sahoo, C. Wilkins and J. Yeager (1997), *Threshold selection using Renyi's entropy.* Pattern Recognition, 30, 71-84.

[201] W. Sander (1977), *Verallgemeinerungen eines Satz von H. Steinhaus.* Manuscripta Math. 16, 11-25.

[202] W. Sander (1978), *Verallgemeinerte Cauchy-Funktionalgleichungen.* Aequationes Math. 18, 357-369.

[203] W. Sander (1985a), *Remark 16.* In The Twenty-second International Symposium on Functional Equations in Oberwolfach, Germany 1984. Aequationes Math. 29, 94-95.

[204] W. Sander (1985b), *Remark on a fuzzy mixed theory of information.* Proceedings of the Twenty-third International Symposium on Functional Equations (Gargnano, Italy 1985). Centre for Information Theory, Faculty of Mathematics, University of Waterloo, Canada, pp.50-51.

[205] W. Sander (1987a), *A mixed theory of information-X. Information functions and information measures.* J. Math. Anal. Appl. 126, 529-546.

[206] W. Sander (1987b), *The fundamental equation of information and its generalizations.* Aequationes Math. 33, 150-182.

[207] W. Sander (1988), *Information measures on the open domain.* Analysis 8, 207-224.

[208] W. Sander (1990), *Weighted additive information measures.* Internat. J. Math. and Math. Sci. 13, 417-424.

[209] W. Sander (1994), *On a generalized fundamental equation of information.* Results in Math. 26, 372-381.

[210] D. Schreier (1926a), *Über die Erweiterung von Gruppen I.* Monatsh. Math. Phys. 34, 165-180.

[211] D. Schreier (1926b), *Über die Erweiterung von Gruppen II*. Abh. Math. Sem. Univ. Hamburg 4, 321-346.

[212] J.-P. Serre (1959), *Groupes Algébriques et corps de classes*. Hermann. Paris.

[213] C.E. Shannon (1948a), *A mathematical theory of communication*. Bell System Tech. J. 27, 379-423.

[214] C.E. Shannon (1948b), *A mathematical theory of communication*. Bell System Tech. J. 27, 623-656.

[215] C.E. Shannon (1956), *The bandwagon*. IRE Trans. Information Theory, IT-2, 3.

[216] B.D. Sharma and I.J. Taneja (1975), *Entropy of type (α, β) and other generalized measures in information theory*. Metrika 22, 205-215.

[217] I.J. Taneja (1977), *On the branching property of entropy*. Ann. Pol. Math. 35, 67-75.

[218] I.J. Taneja (1979), *Some contributions to information theory -I (a Survey: On measures of information)*. J. Combin. Inform. System Sci. 4, 253-274.

[219] H. Theil (1967), *Economics and information theory*. North-Holland, Amsterdam Rand McNally, Chicago.

[220] H. Tverberg (1958), *A new derivation of the information function*. Math. Scand. 6, 297-298.

[221] I. Vajda (1968) *Axioms for a-entropy of a generalized probability scheme* (Czech). Kybernetika 4 (Prague), 105-112.

[222] I. Vajda (1989), *Theory of Statistical Inference and Information*. Kluwer Academic Publishers, Dordrecht-Boston-London.

[223] O. Zariski and P. Samuel (1963), *Commutative Algebra* Vol.I. Van Nostrand, Princeton.

Index

Additive extension, 9, 10, 56, 156
Additivity
    of degree α, 112
    property, 107
Algebraic, 67
Associativity, 34, 35, 112

Baire property, 26
Bernoulli-l'Hospital theorem, 4
Binomial theorem, 187, 203
Borel set, 23
Branching property, 29-31, 41, 43, 49, 53, 103

Cancellativity, 34, 39, 40
Cauchy difference, 41, 224
Characteristic polynomial, 184
Compact subsets, 24
Cone, 2
Continuous extension, 3, 213, 251

Derivation, 77, 88, 157
Deviation, 5
Differential equation, 118, 176, 184
Directed divergence
    of Shannon type, 5
    of degree α, 5, 53
Distributive law, 62
Divergence, 5
Domain
    closed, 3, 6, 33
    open, 3, 6, 29
    restricted, 7
    starlike, 41
Double difference property, 98, 99